# More is Different

## PRINCETON SERIES IN PHYSICS

***Edited by Paul J. Steinhardt (published since 1976)***

Studies in Mathematical Physics: Essays in Honor of Valentine Bargmann *edited by Elliott H. Lieb, B. Simon, and A. S. Wightman*

Convexity in the Theory of Lattice Gases *by Robert B. Israel*

Works on the Foundations of Statistical Physics *by N. S. Krylov*

Surprises in Theoretical Physics *by Rudolf Peierls*

The Large-Scale Structure of the Universe *by P. J. E. Peebles*

Statistical Physics and the Atomic Theory of Matter, From boyle and Newton to Landau and Onsager *by Stephen G. Brush*

Quantum Theory and Measurement *edited by John Archibald Wheeler and Wojciech Hubert Zurek*

Current Algebra and Anomalies *by Sam B. Treiman, Roman Jackiw, Bruno Zumino, and Edward Witten*

Quantum Fluctuations *by E. Nelson*

Spin Glasses and Other Frustrated Systems *by Debashish Chowdhury* (*Spin Glasses and Other Frustrated Systems* is published in co-operation with World Scientific Publishing Co. Pte. Ltd., Singapore.)

Weak Interactions in Nuclei *by Barry R. Holstein*

Large-Scale Motions in the Universe: A Vatican Study Week *edited by*

*Vera C. Rubin and George V. Coyne, S.J.*

Instabilities and Fronts in Extended Systems *by Pierre Collet and Jean-Pierre Eckmann*

More Surprises in Theoretical Physics *by Rudolf Peierls*

From Perturbative to Constructive Renormalization *by Vincent Rivasseau*

Supersymmetry and Supergravity (2nd ed.) *by Julius Wess and Jonathan Bagger*

Maxwell's Demon: Entropy, Information, Computing *edited by Harvey S. Leff and Andrew F. Rex*

Introduction to Algebraic and Constructive Quantum Field Theory *by John C. Baez, Irving E. Segal, and Zhengfang Zhou*

Principles of Physical Cosmology *by P. J. E. Peebles*

Scattering in Quantum Field theories: The Axiomatic and Constructive Approaches *by Daniel Iagolnitzer*

QED and the Men Who Made It: Dyson, Feynman, Schwinger, and Tomonga *by Silvan S. Schweber*

The Interpretation of Quantum Mechanics *by Roland Omnès*

Gravitation and Inertia *by Ignazio Ciufolini and John Archibald Wheeler*

The Dawning of Gauge Theory *by Lochlainn O'Raifeartaigh*

More is Different *edited by N.Phuan Ong and Ravin N. Bhatt*

# MORE IS DIFFERENT

## *Fifty Years of Condensed Matter Physics*

Edited by
N. Phuan Ong and Ravin N. Bhatt

*Princeton Series in Physics*

PRINCETON UNIVERSITY PRESS • PRINCETON AND OXFORD

Published by Princeton University Press, 41 William Street, Princeton, New Jersey 08540
In the United Kingdom: Princeton University Press, 3 Market Place,
Woodstock, Oxfordshire OX20 1SY



Library of Congress Cataloging-in-Publication Data

More is different : fifty years of condensed matter physics / edited by
N. Phuan Ong and Ravin N. Bhatt.
p. cm. — (Princeton series in physics)
Includes bibliographical references.
ISBN 0-691-08865-9 (cloth : alk. paper) — ISBN 0-691-08866-7 (pbk. : alk. paper)
1. Condensed matter. I. Ong, N. Phuan, 1948– II. Bhatt, Ravin N., 1952– III. Series.

QC173.454 .M67 2001
530.4'1—dc21 2001027847

British Library Cataloging-in-Publication Data is available

The publisher would like to acknowledge the editors of
this volume for providing the camera-ready copy
from which this book was printed

Printed on acid-free paper. ∞

www.pup.princeton.edu

Printed in the United States of America

1 3 5 7 9 10 8 6 4 2

1 3 5 7 9 10 8 6 4 2
(pbk.)

CONTENTS

# PREFACE

The start of the new millenium seems to be an appropriate time to celebrate the achievements of Nobel Laureate Philip W. Anderson. Anderson's numerous fundamental contributions in Condensed Matter Physics and related fields of science have had a deep and lasting impact. Clearly, the task of adequately recognizing and surveying this broad influence spanning five decades deserves more than a 5-day workshop. Nonetheless, in recent years, a large segment of the community has felt strongly that such an event should be organized, if for nothing else, to express our collective gratitude for his insights. Complicating the task was Phil's aversion to formal (and scientifically barren) celebrations and his reluctance to participate in their organization. A recent dictum of Phil's (adapted from a source that needs no formal citation) – that Condensed Matter Physics is a field in perpetual revolution – summarizes succinctly his belief that many opportunities remain untapped. The fat lady has not sung, as Phil is fond of quoting. Hence, he has repeatedly discouraged all suggestions, discreet or overt, for a conference in his honor.

Fortunately, the promise of safe refuge at Aspen from millenial madness and the promise of excellent skiing conditions were sufficient to convince Phil and Joyce – and a sizeable segment of the community – that the time was right for such a workshop. This volume is a collection of articles based on talks given at the workshop "Fifty years of Condensed Matter Physics" held at the Institute of Physics at Aspen during January 9th to 15th, 2000. The title reflects both the span of Phil's professional career to date (his first paper, based on his Ph.D. thesis, appeared in 1949), and the period during which Condensed Matter Physics has undergone explosive growth.

Some measure of Anderson's pervasive influence on the field is provided by the list of physical phenomena and ideas associated with his name: the Anderson Model, the Anderson-Higgs mechanism, Anderson's theorem, Anderson localization, the Edwards-Anderson order parameter, the ABM State, Anderson's pseudspin model $\cdots$. An equally impressive set of phenomena, though not usually associated with a name, are instantly and universally recognized by physicists as Andersonian in maturation, if not at birth (orthogonality catastrophe, superexchange, dirty superconductors, rigidity and broken symmetry, spinons $\cdots$). In a 5-day conference, it is impossible to do justice to any subset of these topics, much less provide a reasonably broad survey of these ideas. In selecting the speakers, we have opted for a varied list of topics to reflect his broad interests. We were guided by advice from an informal panel of leading researchers in the field and from Anderson himself. The talks lean heavily towards surveying the current status of actively developing fields. The selection was influenced by Anderson's preference for engaging problems at the forefront of current research, rather than celebrating past accomplishments. We hope that, with such a slant, the proceedings will be of greater interest and benefit to future researchers.

In the first article, Anderson recalls the circumstances surrounding the drafting of his well-known 1972 article "More is Different", and takes a look back on the large and unanticipated impact it has had (particularly on how scientists today view fields outside their own narrow purview). The influence of "More is Different" has been described by Schweber [1]: "In 1972 Philip Anderson, ··· challenged the radical theory reductionist view ··· Anderson believes in emergent laws ··· each level has its own "fundamental" laws and its own ontology. Translated into the language of particle physicists, Anderson would say each level has its Lagrangian and its set of quasistable particles. ··· Developments in quantum field theory and in the use of renormalization-group methods have given strong support to Anderson's views. ··· In fact, renormalization-group methods have changed Anderson's remark ··· from a folk theorem into an almost rigorously proved assertion."

The idea of electron localization in a disordered metal had its beginnings in Anderson's 1958 article "On the absence of diffusion ···." Ironically, progress in this field, though steady, was slow in the following two decades, and it was *after* the Nobel Prize, in the late 70's that the confluence of new scaling arguments and an improved understanding of electron interaction effects, together with carefully crafted experiments in two-dimensional systems (thin films and semiconductor structures), led to rapid development of the subject in the period '78-'82. T. V. Ramakrishnan reviews the history and key ideas in localization, while a description of recent experimental puzzles in both two and three dimensional systems, showing that the subject remains alive and vibrant, is provided by Myriam Sarachik.

In 1987, the layered cuprates were found to exhibit superconductivity at high temperatures. At the outset, Anderson recognized the problem of high-temperature superconductivity as a problem spawned by 'doping' into the 2D Mott insulating state, that involves holes hopping in a spin-1/2 background with strong antiferromagnetic correlations. The resonating-valence-bond theory was introduced in his 1987 Science paper (possibly the most influential theoretical paper on cuprate superconductivity). Different theoretical approaches derived to varying degrees from this paper are described in articles by Patrick Lee, Ganapathy Baskaran, and T. Senthil and Matthew Fisher. Research on the cuprate problem has benefited enormously from the technique of angle-resolved photoemission spectroscopy

[1] Silvan S. Schweber, *Physics, Community and the Crisis in Physical Theory*, Physics Today **46**, 34 (1993).

which is reviewed here by Juan Carlos Campuzano. Bernhard Keimer presents new neutron diffraction results with emphasis on the magnetic 'resonance peak' which seems to be a key defining signature of the cuprates.

Electron correlation provides another mechanism for converting metals into insulators, distinct from disorder. These insulators, known as Mott insulators, occur in many transition metal oxides and sulfides. The successes of the dynamical mean-field approach in strong-correlation problems are reviewed by Gabriel Kotliar. The increasingly persuasive case for unconventional $p$-wave pairing in the superconducting states of strontium ruthenate is summarized in Yoshi Maeno's article. New evidence supporting $p$-wave pairing in the organic superconductors (Bechgaard salts) are described by Stewart Brown *et al.*, who also discuss the intriguing magneto-transport properties in the field-induced spin-density-wave state. Much of what we know about $p$-wave pairing derives from the superfluid state of $^3$He. New features of the phase diagram observed when $^3$He is confined in the pores of silica aerogel are described by Doug Osheroff *et al.* Non Fermi-liquid signatures, a theme propounded most emphatically by Anderson in the cuprates, appear to be increasingly apparent in the heavy-electron systems based on the rare earths and actinides. The current experimental situation is surveyed in the articles by Hans Rudi Ott and by Frank Steglich *et al.*

Anderson has also made seminal contributions to problems in which the effects of disorder and frustration (arising from competing interactions) are important. In particular, the problem of the glass transition and slow relaxation has long been a favorite subject of Anderson. Marc Mézard surveys recent theoretical progress in understanding glassy behavior. The role of frustration in determining the low-temperature thermodynamic behavior of the pyrochlores – the spin analogs of ordinary ice – is described by Arthur Ramirez.

Anderson has been a pioneer in applying the rich diversity of ideas in Condensed Matter Physics to disciplines ranging from Astrophysics and Biophysics to Computer Science and Economics. John Hopfield's article explains why Nature has adopted a large-$N$ approach to the intriguing problems of olfaction and color discrimination. An interesting application of charge correlation arguments to a currently topical problem (the overscreening of DNA and other macromolecules by multivalent ions in solution) is discussed by Boris Shklovskii *et al.* The application of scaling and renormalization ideas to phenomena such as the spread of forest fires and measles and the distribution of galaxies effects is discussed by Per Bak and Kan Chen. Luciano Pietronero *et al.* describe the application of fractal distributions to the analysis of galaxy distribution. In 1986, Anderson and Yaotian Fu pioneered the application of the methods of statistical mechanics to the investigation of numerically intractable problems such as the travelling salesman problem and other 'NP'-complete problems. The article by Scott Kirkpatrick and Bart Selman reviews recent evidence on the relationship between phase transitions and NP-complete problems.

The conference included a session of personal reminiscences by some of

Phil's colleagues of long-standing, Elihu Abrahams, David Thouless, David Pines and Doug Osheroff. Doug disclosed what many in the audience had long suspected. While a good theorist adopts the mean-field approximation when he or she can get away with it, Phil has applied it to the art of brewing coffee with reckless abandon. At the end of each of the numerous formal dinners hosted by Joyce and him, he would carefully poll the guests for decaf or regular. Holding up the fingers of both hands, he then walks into the kitchen and proceeds to brew one large pot using the correct proportion of decaf and regular beans.

We reproduce in the following page Elihu's memorable poem, which recognized Phil not just as a scientist par-excellence, but as a wonderful colleague and friend.

We wish to thank all the speakers and participants for their role in making the workshop such an enjoyable and memorable affair, both scientifically and socially. In particular, we express our deep gratitude to Jane Kelly and Deborah Pease of the Aspen Center for Physics, who handled all administrative aspects at the conference site with admirable efficiency, and to Connie Brown of the Electrical Engineering Department at Princeton University for handling the large traffic of applications for the conference. Finally, we are grateful to Lucent Bell Laboratories, the Princeton Department of Physics and the Princeton Materials Institute for the generous support which made the event possible.

N. Phuan Ong
Ravin N. Bhatt

*Aspen Center for Physics*
*50 Years of Condensed Matter Physics*
*In honor of Philip W. Anderson*
*January, 2000*

## 1950 to Y2K

PWA, PWA
1950 to Y2K

Each winter after holiday,
In Aspen we meet to convey
Some new ideas to friends who may
Have many helpful things to say.
Discussions carry on all day
On ski trails and in gondolay.
Bad weather does not us dismay;
We know fun physics will hold sway.

For fifty years he's had a say
(1950 to Y2K)
About condens'd matter per se.
This week in Aspen every day
We are honoring his dossier.
So with red and white and rosé
Happy good cheer and a rondelay:
Here's a toast to PWA.

PWA, PWA
1950 to Y2K

Notice the great variety:
One, superconductivity
Types one and two, clean and dirty,
And, natch, physics of high Tc.

Two, ferroelectricity.
Glassy relaxation is three.
Four, defects and topology.
Five, waves of charge/spin density.

With deep originality,
Phil managed with authority
To assign some reality
To the elusive property
OFTHEGROUNDSTATEOFTHETHREE-
DIMENSIONALQUANTUMANTIFERROMAGNET.
Only with great ability
Could one find with such probity
PWA's catastrophe
Of the orthogonality.

PWA, PWA
1950 to Y2K

So much we owe to Philip A.
That often we become blasé.
With superexchange, we've made hay
In sev'ral fields - recall Monet.
Double exchange has a heyday,
Anderson-Higgs is here to stay,
As well as helium three A.
(But neutron stars are far away).

Poor man's renormalization,
Anderson localization,
Heavy fermionization,
And valence optimization.
Importance of quantization
For solid state observation.
The charge and spin separation
And local moment formation.

PWA, PWA
1950 to Y2K

Anderson's science on display
Teaches physics as cabaret.
And try as theoreticians may

Experiment has final say.
Beyond the science of day to day,
In politics - a role to play.
Now here's a brief communiqué:
Humility wont often pay!

PWA, PWA
1950 to Y2K

We note that Phil the douser may
Discover just where waters lay,
Or find for us some Courvoisier,
In absence of the sommelier.
Another thing I'd like to say:
Joyce A. does often save the day
Before some Phil thing goes astray
And looks like it might ricochet.

This doggerel might never end
But composing more would send
Me definitely round the bend.
So I have nothing to append.

PWA, PWA
1950 to Y2K

We gather'd here in fine array
To Phil our respects to pay.
Since new physics is his forte,
He's given us great games to play.
We wish him well for Y2K.
Much more good stuff from him we pray,
To keep us going on our way.
So to Phil Anderson - olé!

-Elihu Abrahams, 1/12/00

## Chapter 1

# More is Different – One More Time

Philip W. Anderson
Joseph Henry Laboratories of Physics
Princeton University, Princeton, N.J. 08544-0708

In the Spring of 1967, just before my first term as Cambridge's first–and last–part-time professor, I spent a pleasant month at La Jolla with, among others, some old friends from Bell who had been recruited there by Walter Kohn to give a jump-start to the infant physics department. I had an appointment as Regents' lecturer, the only requirement being to give one more or less public lecture. I think this was the first time I had ever had such an assignment; I was nervous and worked hard on it. Afterwards I heard tell that the lecture was quite incomprehensible, though 30 years later I learned by chance that at least one listener, specifically Christiane Caroli (visiting from Paris), had taken the message seriously. Nearly five years later I wrote up an edited version and, somewhat to my surprise, Science accepted it: this was "More is Different" [1].

This was the late sixties when even established economic verities were being seriously questioned. Such books as Schumacher's "Small is Beautiful" were around, and in England one of the environmental movement's slogans was "more is worse"–to which the reply of the establishment was of course "more is better" (foreshadowing the Reagan era which opponents–and even some supporters–characterized with the slogan "greed is good".) So it was natural to suggest that more was merely different.

Sociologists of science posit that there is a personal or emotional subtext behind much scientific work, and that its integrity is therefore necessarily compromised. I agree with the first but reject the second. I think "More is Different" embodies these truths. The article was unquestionably the result of a buildup of resentment and discontent on my part and among the condensed matter physicists I normally spoke with. 1967 was a temporary maximum of arrogance among the particle physics establishment, riding high in government advisory circles (this was the heyday of JASON and of the RAND corporation), and in possession of funding at a level which made international travel a commonplace and afforded overheads which made their employment profitable for any university. There was for instance difficulty in getting condensed matter colleagues recognised by the NAS, and many physics departments in major universities such as Yale, Columbia, and Princeton had only token representation of the field of condensed matter.

Viki Weisskopf was by no means the most narrowminded of the nuclear and particle types. "More is Different" is a reply to his article dividing science into "intensive" (for practical purposes, particle physics) vs "extensive" (the rest). I had always considered him a friend (and do now)–he sat on my thesis committee, and much of the early work in my thesis field was his–and this made it particularly hard to take. Looking back, I was right to be disturbed–in avoiding pejorative terms like "fundamental" he was attempting to camouflage a message which was identical to those said less delicately by others: that particle physics was the only truly intellectually challenging specialty, the others, especially the solid state physics (as it was then actually called) , dear to my heart, were "mere chemistry".

My article accepted Viki's point that science is largely hierarchical in that the subject matter of one science is the "substrate" on which or out of which another builds the objects of its interest. For the essentially emotional reasons described above, my concern was to show that there is nonetheless intellectual autonomy: one cannot assume that the laws of the one science are trivial consequences of those of the other. It was by way of example that I used my newly generalized idea of broken symmetry to illustrate the processes by which this kind of autonomy arises. I wanted to explain in detail the way in which, in at least one sort of situation, truly novel properties and concepts could emerge from a simpler substrate. Broken gauge symmetry, which is exhibited by superconductors, was perfect for my purposes because many of the greatest theoretical physicists had tried to understand the problem of superconductivity before it was broken by the "phenomenologist" John Bardeen. Bloch and Wigner had even supposed that they had "proofs" that no solution of the sort existed. Bloch's proof was simply based on a misunderstanding of the experimental facts, but Wigner's error is crucial and generic, and is worth discussing. He based the "superselection rules" on the idea that it would be meaningless to have coherence between states with different numbers of particles, i e that ground states are always discrete eigenstates of the symmetries of the problem. This misses the possibility of "quasidegeneracy", that macroscopic systems can have Goldstone "zero modes" which restore the broken symmetry. Thus the "ground state" is often not best described as such a symmetry eigenstate. This concept was an idea I had encountered in my theory of antiferromagnetism in 1952, and which came into its own in understanding superconductivity.

"Emergence" was a term from evolutionary biology with which I, like most physicists, was unfamiliar at that time. For over a century biologists had speculated that life "emerged" from non-living matter without any divine (or other) intervention, "by accident" as Dawkins puts it. In fact, they saw the whole evolutionary process as emergence of the more complex from the simpler. In a sense, MID is just putting the concept of emergence in a physical context and generalizing it. Many biological examples of emergence are examples of its kind of scale change. But if one listens to the great synthesizer of evolutionary theory, Ernst Mayr, talking about emergence even today, it is clear that he is discussing a concept which is a little different, broader, and vaguer. He considers the "emergence" of the functionality of a hammer from the combination of a stick and a stone – what evolutionists call an exaptation – as a valid instance of emergence, while I was focussing on the effect of scale change. Only in the case of scale change, I felt, is there a clearly sufficient argument for qualitatively new concepts to appear.

The reader of MID today should realise that it was written in 1967, before much of the technical apparatus with which we understand the macroscopic, $N \rightarrow \infty$ limit had been created. This was before the renormalisation group had been applied in statistical physics (and more or less simultaneously with the elecroweak theory's publication, and long before its acceptance as the triumph of broken symmetry in particle physics). Kadanoff had only begun to state the idea of universality formally, though many of us, myself included, felt in our bones that something of the sort existed. Universality is the most primitive way of showing that the same macroscopic results can follow from very different microscopic causes.

Even farther in the future (1975) was to come the formal topological theory of order parameter defects such as vortices, flux lines, domain boundaries and the like, which allows the macroscopic system wide and unexpected flexibility in its behavior. Nonetheless, there was quite enough to make it clear that the emergence of new concepts and properties is almost inevitable when one makes up a macroscopically condensed phase. The process by which this happens is straightforwardly clear and has a precise description in terms of the theory of broken symmetry. All of the properties which characterise such phases – crystal structure, metallicity, macroscopic quantum coherence, elasticity, and so on – have no meaning in a world of individual atoms, and logically arise only when one puts together many atoms. One goes first, conceptually, to the limit of a very large system, and then backs off to the finite one that one actually has. Playing this game, I pointed out, is particularly fascinating in nuclear physics, where nuclei containing of order only 100 nucleons exhibit unmistakeable evidence of such properties as shape and superfluidity, which in principle are only definable on the macroscopic level.

As I have tried to explain, my main goal was to demonstrate the intellectual autonomy of the higher level phenomena from the tyranny of the fundamental equations which constitute the "theory of everything" (this is a phrase which was to be invented thirty years in the future.) Though the modern theory of chaos in deterministic dynamical systems– the idea of sensitive dependence on initial conditions– had yet to cross the horizon of a conventionally trained physicist like myself, quasidegeneracy of the quantum state of a macroscopic system has the equivalent effect of divorcing the future from the past as far as rote, mechanical computation using the laws of atomic physics is concerned, in spite of the fact that these are exactly known. LaPlace's perfect computer is not a physically realizable machine, and determinism via the laws of physics is a nonsense.

Nonetheless, a perverse reader could postulate a sufficiently brilliant genius–a super-Einstein–who might see at least the outlines of the phenomena at the new scale; but the fact is that neither Einstein nor Feynman succeeded in solving. superconductivity. Imagine how much more difficult it would have been to predict the phenomenon of superconductivity a priori without the actual thing in front of you, than it was for Bardeen, Cooper and Schrieffer to follow experimental hints to guide them to a theory.

It was only in the last few paragraphs of the article that I spelled out a moral for the general structure of scientific knowledge: that scale changes in a wide variety of instances can lead to qualitative change in the nature of phenomena. It is a general rule that emergence of concepts and entities which are novel in some fundamental way occurs at the point where scales change materially. Life takes advantage of the fact that aperiodic macromolecules can carry information in the ordering of their constituents, a kind of broken symmetry going beyond that of an ordinary condensed phase. Again, organisms composed of many cells, with their differentiation of function and of cell types, constitute a leap of imagination relative to protists. The further generalisation to social and economic organisation seems obvious. Money and markets, for instance, are unnecessary when society is organised on the tribal level only, but seem usually to appear spontaneously when larger units or wider regions become organized.

In the years after MID my own work often involved the theme that complex generic behavior can arise from simplicity. Localisation was in the back of my mind in the original article, and in the '70's I picked it up again, as well as the theme of generating magnetic spins from a non-magnetic substrate; and I came to see these as generic examples of the concept of emergence rather than as isolated ideas. With the spin glass another was added, one which in fact led my group away from physical systems into the "emerging" (in another sense) science of complexity. We became involved in theories of complex optimisation and evolutionary landscapes. John Hopfield even carried the spin glass idea into his theories of brain function.

In the meantime MID was percing along and acquiring a small following. The first I knew of this was when I received a phone call out of the blue, in summer 1977, asking me to speak at a neurosciences meeting at Keystone CO that winter. When I answered the phone, Gene Yates was relieved to find that I really (and still) existed. He had no idea of what else I had done and was flabbergasted by the prize announcement and pleased that I came to his meeting nonetheless. He put me on a program with an interesting variety of characters– Ralph Abrahams from the Chaos collective at Santa Cruz, an early General Systems theorist named Art Iberall, and a somewhat hyper neurophysiologist named Arnie Mandell (who later won a MacArthur award for his studies of altered consciousness). It turned out that I thoroughly enjoyed the kind of broad-gauge, openminded discussions I had with these and others of Gene's friends, and when he asked me to a meeting in Dubrovnik on self-organization [2] which he organized in 1980 I was glad to go and meet a cross-section of the fundamental thinkers on biology–Leslie Orgel, Gunther Stent, Harold Morowitz, Steve Gould, Brian Goodwin, and many others, and to continue on my learning curve.

Thus, starting from MID and the spin glass, and an interesting if abbreviated sabbatical in 1981 helping with John Hopfield's course at Caltech on the physics of information, I gradually became socialised into the community of scientists who were thinking about the general themes of complexity, self-organisation, and emergent properties in general. Thus it was very natural for me to attend the first two organising workshops in 1984 and 1985 [3] of what became the Santa Fe Institute, then taking shape in the minds of George Cowan, Peter Carruthers, Murray Gell-Mann, David Pines and other senior scientists associated with Los Alamos. An early organising meeting at Aspen and the decision to help run the first workshops on economics [4] in 1986-7 (using a little expertise picked up in courses my wife and I attended in Cambridge) left me permanently attached to SFI.

By this time MID was really humming along. The title became almost a mantra for the work of the Santa Fe Institute and for the science of complexity in general (whatever that means). SFI started out as an interdisciplinary institute focusing on the subjects which grow out of making connections between existing fields. We found ourselves becoming interested, in a number of instances, in studying the emergence of the more complex level from the simpler one: how, in a number of cases, more complex behaviors resulted from the interactions of a number of simpler "agents". In economics, ecology, immunology, archaeology, neurophysiology we became to an extent captivated by "agent-based modeling"– the use of the computer to demonstrate such phenomena, in particular. This kind of work is by no means the only activity of SFI but it has become something of a trademark.

In a recent article I applied a similar line of reasoning to MID in the opposite direction, towards the origins of our physical universe. The Big Bang involved at least two thermodynamic phase transitions, one of which has been the subject of fairly intensive speculation: the phase transition to broken electroweak symmetry, which has been conjectured to be the cause of an "inflationary" era in cosmology. From the first the question of whether there should not be visible traces in the form of topological defects in the putative Higgs field (monopoles, cosmic strings) has been discussed. But there are even deeper and more subtle questions to be answered. Perhaps the most striking of all instances of emergence is the emergence of the classical world of identifiable, distinguishable objects in space, obeying causality with a definite sense of time, out of the microscopic quantum description of the universe as a collection of quantum fields describing absolutely identical quanta moving in isotropic, homogeneous space-time. I have conjectured [5] that space-time itself might be an emergent property, born perhaps at the time when gravitational instabilities of the cosmos which eventually became clusters of galaxies began to form. Some cosmologists such as Lee Smolin have gone even further, but so far I haven't joined them.

Be that as it may, another thought is that the apparent difficulties and contradictions of quantum measurement theory are the result of attempting to apply, to systems at one scale, the concepts and properties that are appropriate to an entirely different scale: causality, rigidity, and so on. As I put it in another article [6], to an electron, the properties of the apparatus – Stern-Gerlach magnets, slits, and the like – are much more mysterious than the properties of the electron are to us. These objects have the strange property that they can act merely as boundary conditions for the electron, that is they can act on it without changing their quantum state. The complexity of the quantum description of such objects makes it in principle impossible to follow an atomic-scale object once it has interacted with a macroscopic object and hence their wave functions have become entangled.

The original article may have been too concise to express my full meaning. It is not a prescrip;tion for ignoring reductionist ideas, and indulging in what is called "holistic" thinking, in which one ignores the physical or biological substrate upon which a given science feeds. Just a few days ago I received a copy of a correspondence arguing about reductionism vs this kind of holism in which I was quoted as supporting both sides, and this was not a unique case.

I think the original article was clear in advocating reductionism in the sense of the assertion that the basic laws of physics, chemistry and biology hold under all known circumstances– magic doesn't happen. That being said, what was a little new is that this does not imply what I called "constructionism", (more recently, others have called it "strong" or "strict-sense" reductionism): that the consequences of those laws can be worked out in detail or that they seriously restrict the endless possibilities of nature or even our free will–the former being demonstrable, the latter a conjectural corollary.

Then the question arises, whether there is any point in the reductionist program– if you can't work the consequences out in detail, why bother with the underlying laws at all? I gave my own personal answer to this question in the article, that understanding on that kind of level is infinitely satisfying; but there is a practical answer as well. As science becomes more complex and unavailable to the general public, the primitive, Baconian model of science which is taught – or was when I grew up – in high-school textbooks is no longer adequate. Again and again, groups of scientists working in isolation have succeeded in convincing themselves that black is white by the most reliable-seeming "studies", or even by simply repeating some set of seemingly rational propositions to each other often enough. Scientists are not immune to self-interest or egregious error.

In reality, academic scientists no longer rely solely on the direct experimental method of one hypothesis, one test to decide whether a given proposition is correct. One way of making sure that science is correct has been emphasised by such writers as Merton and Ziman [7] namely the social structure of science and its character as "organised skepticism". I would propose that as science matures an equal or greater role is played by tying results in to the exponentially growing web of consistent knowledge, and one of the best ways one can do that is by showing that phenomena in one science can be explained from the basic laws of that science's substrate subject. For instance, genetics became enormously more powerful and believable as we explored the mechanism via structural chemistry and then molecular biology. And when claims of cold fusion by an isolated coterie of specialists hit the headlines, the importance of cross-checking against fundamental knowledge in related fields became apparent – as well as the inefficacy, for correcting error, of "direct", "benchtop" measurements, by interested parties.

The message for which I was groping in 1967 has become much clearer to me with the passage of time. It was born of the realisation that science is no longer a collection of isolated communities, each applying the Baconian "scientific method" of empiricism and Popper's paradigm of "falsifiability" within its own bailiwick. The Newtonian mode, unification, has taken over from Bacon, and science is becoming an interconnected whole. But in the process of unification we were in danger of being victimized by those who appear to own the most universal, most microscopic laws: those who strive to achieve the "theory of everything" and discover the fundamental particles of which the universe is made. If they owned the fundamentals, they claimed, they could deduce all the rest. I fired the first salvo in rebuttal: that I saw the "theory of everything" as the theory of almost nothing. The actual universe is the consequence of layer upon layer of emergence, and the concepts and laws necessary to understand it are as complicated, subtle and, in some cases, as universal as anything the particle folks are likely to come up with. This also makes it possible to believe that the structure of science is not the simple hierarchical tree that the reductionists envision, but a multiply connected web, each strand supporting the others. Science, apparently, like everything else, has become qualitatively different as it has grown.

I rest my case.

## BIBLIOGRAPHY

[1] P. W. Anderson, Science **177**, 393 (1972).

[2] F. E Yates, ed., *Self-Organizing Systems*, (Plenum Press, NY, 1987). The amplified proceedings of the 1980 Dubrovnik conference.

[3] D. Pines, ed., *Emerging Syntheses in Science,* Santa Fe Institute, Santa Fe, 1985. Reissued by Addison-Wesley, 1988. Some papers from the founding workshop.

[4] P. W. Anderson, K. Arrow and D. Pines eds., *The Global Economy as an Evolving Complex System* (Addison Wesley, Reading MA, 1988).

[5] P. W. Anderson, *Measurement in Quantum Theory and Complex Systems.*, in *The Lessons of Quantum Theory.*, de Boer, Dal and Ulfbeck, eds. (Elsevier, 1986).

[6] P. W. Anderson, *Is measurement itself an emergent property?*, Complexity **3**, 14 (1997).

[7] J. M. Ziman, *Reliable Knowledge*, (Cambridge Univ. Press, Cambridge, 1978).

# CHAPTER 2

# LOCALIZATION YESTERDAY, TODAY AND TOMORROW

T. V. RAMAKRISHNAN

Centre for Condensed Matter Theory, Department of Physics
Indian Institute of Science, Bangalore 560 012, India

ABSTRACT

Anderson showed nearly forty years ago that excitations in a sufficiently random medium are spatially localized. In this talk, I describe the idea in its original experimental context. Early applications are then outlined. The second phase of this field is characterized by a theory for the onset and development of localization as reflected in measurable quantities. Finally, I discuss a number of poorly understood phenomena where localization-related physics may be important, e.g. the glass transition, superconductor insulator transition, the metal-insulator transition in two dimensions, decoherence near $T = 0$, and the unusual electrical behavior of manganites.

## 2.1 ABSENCE OF DIFFUSION IN RANDOM LATTICES

### 2.1.1 EXPERIMENTAL BACKGROUND

The idea of localization [6], like many of the major insights due to Anderson, arose from a surprising experimental result. The technique of ENDOR (electron nuclear double resonance) developed by Feher [2] enabled him to monitor simultaneously the spin state of an electron and the nuclear spin coupled to it via the hyperfine interaction. In lightly doped Si:P, the shallow bound donor electron is coupled to $^{29}$Si nuclear spins (5% isotopic abundance in Si). The actual magnetic field experienced by the electron depends on the number and location of the $^{29}$Si spins within its wave function spread. Out of this distribution of magnetic fields, a narrow region, i.e., a particular set of electrons, is picked out by ESR. By monitoring (via NMR) the relaxation of nuclear spins coupled to these electrons, one can follow the change in the electronic spin state. Feher found [2] that the electron spin relaxation time was of order a few seconds. A simple perturbative estimate made at the time by E. Abrahams (unpublished) using the known exchange mediated coupling between donor electron spins predicts a spin flip lifetime about $10^6$ times smaller. This million-fold discrepancy suggested to Anderson the possibility that the *spread* of exchange couplings was crucial, not their average magnitude. If the spread or randomness is large enough, the electron spin does not diffuse, but stays unflipped in a locally favored orientation.

### 2.1.2 THE LOCALIZATION IDEA

Anderson transformed the problem of spin localization to a more general one of a particle on a lattice, with a site energy $\epsilon_i$ (at site $i$) that varies randomly and independently for each site over a range $-W \leq \epsilon_i \leq W$. The particle hops from a site to one of its $z$ nearest neighbors with amplitude $t$. Anderson computed the probability that a particle, starting from a particular site and hopping randomly on the lattice, returns to the starting site. He found this probability to be nonzero for $(zt/W) \leq \lambda_c$ where $\lambda_c$ is a number of order unity. Its precise value depends on the energy of the particle, the lattice type and dimensionality. Thus the idea that localization is generic to a strongly disordered extended system was born.

Most of the technical effort in Anderson's paper was connected with the problem of summing the random-walk series consisting of various paths returning to the origin. However, the crucial new idea was that in a disordered system, the *distribution* of physical quantities matters, *not* their average. For example, the quantity of interest in site-localization is the probability $P(x)$ that $[Im\ \{\Sigma(is)\}/s]$, a random function, has a value $x$ for large values of $x \simeq s^{-1}$. Here, $\Sigma$ is the on-site self-energy for 'energy' $is$. It turns out that $P(1/s) \sim s^{3/2}$ i.e. vanishingly small. This supports localization, provided the power series (in $t$) for $\Sigma$ converges. On the other hand, the *average* value of $[Im\{\Sigma\,(i\,s)\}/s]$ diverges as $1/s$ for small $s$, suggesting delocalization always. This notion, that the distribution of physical quantities in random systems determines physical behavior, is now commonplace in the field of disordered systems.

The concept of localization in a random system is the basis for understanding the electronic nature of all systems that are nonperiodic and insulating. Prior to this, it was assumed that a system could be insulating (at $T = 0$) only if the band of occupied states is separated from that of unoccupied states by an energy gap. Such band gaps are possible only in periodic systems. In nonperiodic systems there is no gap in the density of states, but there could be a change in their nature (localized-to-extended) as a function of (electron) energy. A mobility gap without a gap in the density-of-states ('a gap without a gap') is a consequence of localization. Further, while in periodic, strongly-correlated systems, a Mott insulating phase can occur for an odd number of electrons per unit cell (which ought to have partially filled bands in one-electron theory), the Mott phase has a real charge excitation gap; there are no states in this gap. Both band and Mott insulators are possible only for commensurate electron number (even and odd per unit cell, respectively). The Anderson localization mechanism is the only generic one for an insulating ground state with arbitrary (incommensurate) electron filling.

In addition to these qualitative consequences for disordered electronic systems, the localization idea plays an essential role in determining the physical nature of all disordered systems.

### 2.1.3 RELATED DEVELOPMENTS

At around the time that Anderson developed the concept of localization (the late fifties), several related ideas were being spawned, many of which would come together later and enrich our understanding of localization-related physics. I. M. Lifshitz [3] realized that exponentially rare fluctuations in a random medium, e.g. large regions from which the repulsive random potential is absent, can make room for large localized states of low energy. One thus has the picture shown in Fig. 2.1 for localized states in a repulsive random potential. For energies near zero, the lowest possible value, states are Lifshitz localized by rare fluctuations. As energy increases, the length scale of disorder and the de Broglie wavelength become comparable, and one has Anderson localization by typical fluctuations.

In 1957, Landauer [4] established a connection between the conductance of a one-dimensional, disordered system in units of $e^2/\hbar$ and the transmission coefficient of an electron wave through it. This is a remarkable, fully-quantum approach to conductance (with a universal scale), hitherto regarded as the inverse of classical dissipative viscous resistance to electron flow. Further, on compounding transmission $T$ from small bits that together make up a one-dimensional system, one finds non-Ohmic, localized behavior, i.e. $T(L) \sim exp\,(-L/\xi)$.

Kohn [5] connected the sensitivity of the ground state energy $E$ of a many-body system to electron phase changes (achieved via a gauge transformation $k$) with the imaginary part of the electrical conductivity $\sigma''\,(\omega)$, i.e. $(\partial^2 E/\partial k^2) \simeq [\omega\,\sigma''\,(\omega)]_{\omega \to 0}$. This gives some insight into the nature of the insulating state. Similar ideas were later developed and applied extensively to disordered, one-electron systems by Thouless [6].

In the late fifties, quantum many-body perturbation theory for electron transport was developed, starting with either the Kubo formula or the density matrix [7]. The approach was largely applied to systems with weak disorder. About twenty years later, a localizing process was identified, starting from the metallic end, and using many-body theory for electron transport [8].

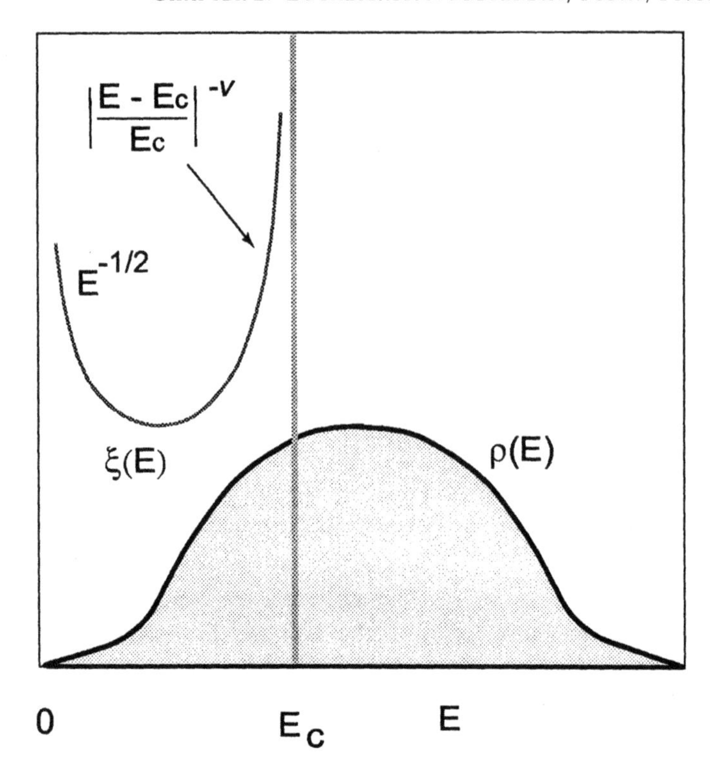


Figure 2.1: Schematic picture of localization length $\xi_{loc}$ and density of states $\rho(E)$ (along the $y$-axis) for a three-dimensional, noninteracting electronic system with a purely repulsive random potential, as a function of electron energy $E$ (along the $x$-axis). Lifshitz localization ($\xi \to \infty$ as $E \to 0$) and corresponding density of states for energy $E$ close to zero, and Anderson localization characterized by a mobility edge $E_c$, are shown.

### 2.1.4 Consequences of Localization

While the localization idea was born in 1958, its implications were not actively or immediately pursued because of various other developments in solid-state physics (e.g. the BCS theory of superconductivity, magnetism in metals) and the feeling that electron interactions would strongly affect localized states, possibly delocalizing them. Starting in the early sixties, the consequences of localization for disordered electronic systems became apparent, largely through the sustained efforts of Mott and his coworkers [9]. Mott recognized clearly that the very existence of glassy/amorphous (nonperiodic) *insulators* was due to the fact that electronic states in a whole range of energies can be localized. In periodic solids, a true gap in the electronic energy spectrum is possible, while this cannot happen if the system is nonperiodic. Thus the transparency and insulating nature of glass had been a longstanding fundamental puzzle. Because of disorder, states in a certain energy band (within mobility edges) are localized. If the Fermi level lies in that band, the system is insulating. This is the basic guiding principle for the non-conducting electronic behavior of disordered systems.

The conductivity of such Anderson insulators was shown by Mott to depend on temperature $T$ as $exp\{-(T_0/T)^{1/(d+1)}\}$, where $d$ is the spatial dimensionality. The underlying process of phonon or interaction induced variable-range hopping involves transitions from occupied to optimal, unoccupied states, *both* localized. This novel process is possible only because there is a continuous band of localized states around the Fermi energy, distributed appropriately in space.

In general, to zeroth order (as discussed by Mott), interaction can cause the electrons to localize when the Coulomb correlations are short-ranged, disorder is absent, and the density is commensurate [10]. In the disordered Anderson insulator, the long-range nature of Coulomb interactions gives rise to characteristic effects on the density of states and the temperature dependence of conductivity, as pointed out by Efros and Shklovskii [11].

Mott also revived and developed the idea of Ioffe and Regel that the description of electrons as waves scattered off random obstacles does not make sense if the wavelength is less than the mean-free-path $\ell$. Thus the minimum mean-free-path for extended electronic states is $\ell \simeq \lambda \simeq 2\pi k_F^{-1} \simeq a$, an atomic length. This leads to a (Mott) minimum metallic conductivity $\sigma_{min}$ equal to $(3\pi^2)^{-1}(e^2/\hbar a)$ in three dimensions, and to the universal value $e^2/h$ in two dimensions.

A large body of evidence, strongly suggesting a change in electron transport regime for this value of $\sigma_{min}$ in three and two dimensions, was put together by Mott and was looked for (and found) by many experimentalists, who were directly stimulated by his work [12].

Finally, there is the possibility of a transition at $T = 0$, from a metallic to an insulating phase with increasing disorder. According to the idea of a minimum metallic conductivity, the conductivity jumps from $\sigma_{min}$ to zero at this Anderson transition. The dielectric constant is argued to diverge as a power-law on approaching critical disorder from the insulating side.

### 2.1.5 LOCALIZATION AS A GENERAL FEATURE OF DISORDER

In the late sixties and seventies, Anderson and collaborators pioneered the application of localization ideas to a variety of disordered, many-body systems. For example, Hertz, Fleishman and Anderson [13] argued that in a disordered interacting system of bosons, macroscopic occupation of the lowest energy localized state, or of a narrow range of localized states is opposed by repulsive interaction. Interaction in a disordered Bose system therefore plays a crucial role in limiting the occupation of localized states, and thus in effectively promoting condensation into the lowest *extended* state (i.e. superfluidity). For spin glasses, the spin freezing transition was connected with the occupation of localized eigenstates of the random exchange matrix $J_{ij}$ [14].

The effect of a deformable lattice and disorder acting together to localize electronic states was discussed first by Anderson [15]. The important idea of negative$-U$ centres led to a description of the trapping of electron pairs in glassy semiconductors, rationalizing a number of unusual electrical and magnetic properties of these systems [16].

The period was also marked by many serious efforts [17] to understand more clearly and sharpen the delocalization condition originally obtained by Anderson. This led to a better understanding of how localization is inevitable, but also to its dependence on the dimensionality, on lattice connectivity, electron energy, and type of randomness.

## 2.2 SCALING AND WEAK LOCALIZATION

### 2.2.1 THOULESS CONDUCTANCE

Starting in the early 1970's Thouless and collaborators [6, 18, 19] developed and applied the idea that the conductance of a disordered one-electron system can be related to the sensitivity of the energy levels of a finite sample of the system (e.g. a hypercube of linear dimension $L$ and 'volume' $L^d$) to changes in the boundary condition. Suppose the average energy level separation in a narrow energy range is $\Delta W$. Suppose further that the system is perturbed at the boundary, e.g. the boundary condition is changed from periodic to antiperiodic. This shifts the individual energy levels in the region of interest by, say $\Delta E(L)$. The dimensionless ratio $\Delta E(L)/\Delta W(L)$ was identified by Thouless with conductance at scale $L$, i.e.

$$\frac{\Delta E(L)}{\Delta W(L)} = g_{Th}(L) = \frac{G_{Th}(L)}{e^2/\hbar}. \tag{2.1}$$

This is a very quantum-mechanical description of conductance. Actually, $G_{Th}(L)$ is directly related with the imaginary part $\sigma''(\omega)$ of the conductivity. However, for reasonable distributions of $\sigma''(\omega)$, $G_{Th}(L)$ can be related to the real part $\sigma'(\omega)$ on length scale $L$.

This graphic and suggestive representation of conductance is also natural for numerical exploration. For a two-dimensional system, Thouless and coworkers [19] found a transition from a nearly $L$-independent (metallic) conductance to an exponential $L$-dependence for a given disorder at a definite electron energy (the mobility edge); the conductance near the mobility edge is $(25\ \text{k}\Omega)^{-1} \simeq (e^2/h)$. However, Licciardello and Thouless noticed a weak but persistent scale dependence for the conductance in the metallic regime, in contrast to the expectation of scale independence. In the light of later developments, weak scale dependence is the precursor of localization on an exponentially large length scale, and the numerically inferred mobility edge in two dimensions is a regime of rapid crossover to a localization length smaller than the system size.

The Thouless picture suggests a scale-dependent renormalization of conductance. A change of the scale size $L$ by $\delta L$ could be thought of as adding more of the system at the boundaries; this changes $\Delta E(L)$. One can argue that the relative change in $\Delta E(L)$, or in $g(L)$ is proportional $(\delta L/L)$, and depends on $g(L)$ at scale $L$;

$$\left(\frac{\delta g(L)/g(L)}{\delta L/L}\right) = \beta(g(L)). \tag{2.2}$$

An argument of this kind was made by Thouless in his 1978 Les Houches Lectures [19]. Wegner had suggested in 1976 a real-space finite-size scaling of the Anderson model [20], and argued heuristically that the conductivity ought to go to zero near critical disorder as a power of the distance from it.

### 2.2.2 SCALING THEORY

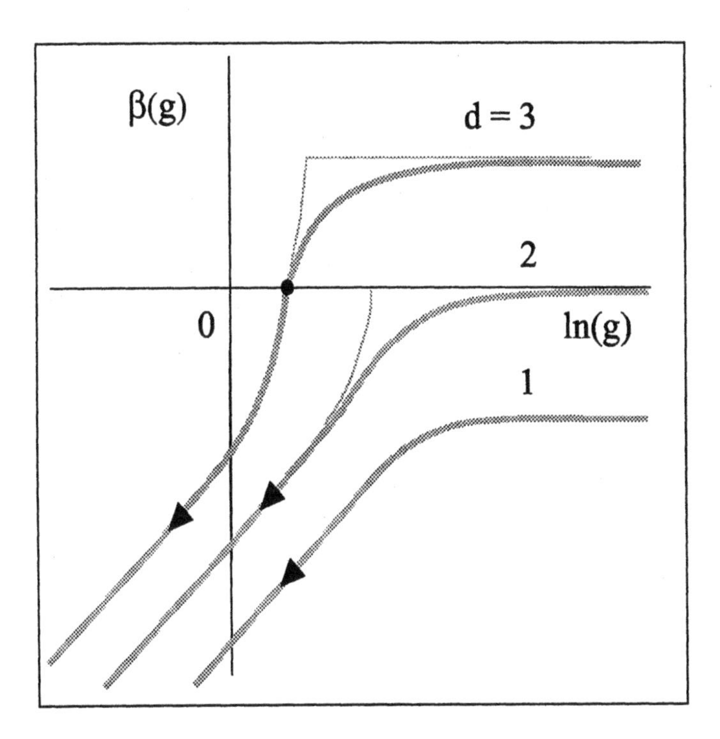


Figure 2.2: The scaling of conductance $g$ in three, two and one dimensions. The scaling function $\beta(g)$ is plotted against the $\ln(g/g_c)$ where $g_c$ is the quantum scale $(e^2/\hbar\pi^2)$, for dimensions $d = 3, 2$ and 1. The dotted line is the minimum metallic conductivity prediction.

In late 1978, Anderson, who had just come back from Les Houches, had the scaling arguments of Thouless [19] very much on his mind, and said 'why not do it?'. Assuming that there is a scaling curve, one knows the limiting forms for large $g$ (Ohm's-law metal) and small $g$ (an Anderson insulator in which, crudely speaking, $g(L) \simeq g_c exp(-L/\xi)$). Connecting these with a smooth, monotonic, one-parameter family of curves, one obtains the scaling curves $\beta(g)$ shown in Fig. 2.2 [8]. We note that the $x$-axis is $\ln(g)$ namely logarithm of the conductance in natural quantum units e.g. $g_c = (e^2/\hbar\pi^2)$.

The scaling curve has several implications. In three dimensions, there is a critical disorder $g_c^{3d}$, and a quantum critical point (continuous $T = 0$ transition). If the microscopic length scale ($\ell$) conductance $g_0$ is larger than $g_c^{3d}$, the system scales at large $L$ to a metal, ie $g(L \to \infty) \sim \sigma L$ with $\sigma \sim (e^2/\hbar\xi)$ and $\xi \sim [(g_0 - g_c^{3d})/g_c^{3d}]^{-\nu}$. The correlation length exponent $\nu = (d-2)^{-1} = 1$. The system is an insulator for $g_0 < g_c^{3d}$, and localization length $\xi_{loc}$ diverges with the same exponent. In two dimensions, there is no metallic state (at $T = 0$); the system always scales to an insulator, no matter how weak the disorder.

### 2.2.3 WEAK LOCALIZATION

So far, there had been no serious progress in understanding the onset of localization starting from the side of the disordered metal. All theoretical work was concerned with delocalization, namely in assessing the stability of localized states. Indeed, Thouless wrote in his 1978 Les Houches lectures [19] that "For various reasons, no satisfactory theory of localization has been developed by means of a study of the stability of delocalized states". In my opinion this situation changed in 1979. The scaling hypothesis implies that the small deviation from ohmic behavior for large $g$ is of the form

$$\beta(g) = (d-2) - (a/g), \tag{2.3}$$

where the first term on the right is the Ohm's law result, and the second term is the leading correction. It turns out that this term can actually be calculated using diagrammatic many-body perturbation theory. For example, in two dimensions one finds that

$$G(L) = \sigma_0 - \frac{e^2}{\hbar\pi^2} \ln\left(\frac{L}{\ell}\right), \tag{2.4}$$

where $\ell$ is the mean-free-path. This implies a $\beta(g)$ in Eq. 2.3 with $a = 1$, if $g$ is in units of $e^2/\hbar\pi^2$). The new term in Eq. 2.4 arises from a singular back-scattering process first considered by Abrahams *et al.* [8, 21, 22, 23, 24].

The process in indicated in Fig. 2.3a as a correction to conductivity, and in Fig. 2.3b in terms of the trajectories of an electron propagating from one point $\vec{r}$ to another point $\vec{r}'$. We notice that the reduction in conductivity arises from an interference process in which an electron is multiply scattered by the random medium from a state $\vec{k}$ to a state $\vec{k}'$ and another electron with momentum close

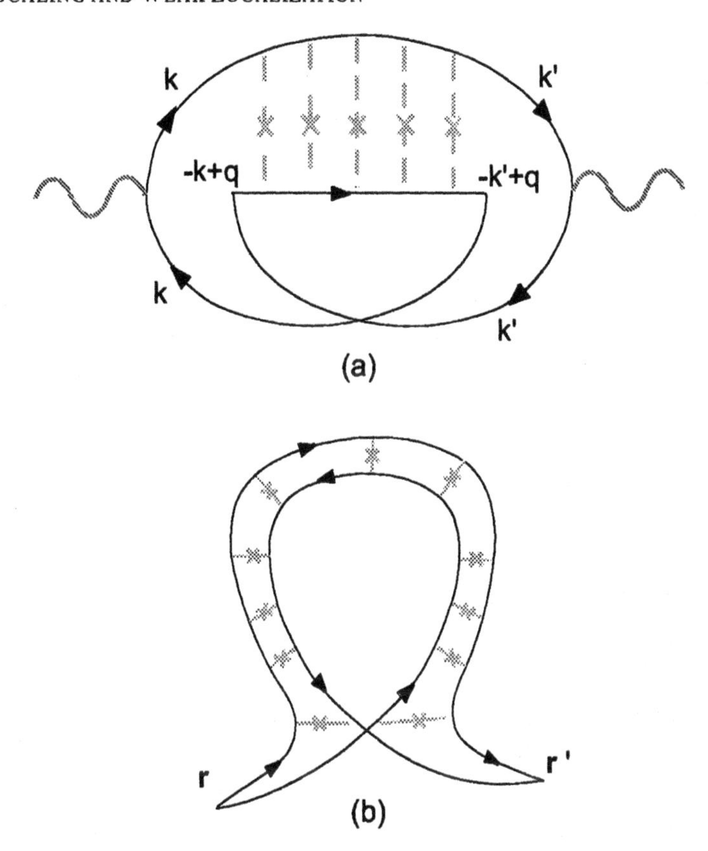


Figure 2.3: The singular backscattering process leading to localization. Figure (a) shows its contribution to conductivity. The diagram describes the interference between an electron and another of nearly opposite momentum being scattered by the random potential. Figure (b) shows the same process in real space; constructive interference between a self-intersecting electron path, and another nearly identical path (through the same random medium) oppositely traversed, is implied.

to $-\vec{k}$ undergoes exactly the same sequence of scatterings as the first one, ending up with a momentum close to $-\vec{k}\,'$ (see Fig. 2.3a). This singular backscattering process leads to a strong correlation between current fluctuations $\vec{k}$ and $-\vec{k}$, so that it reduces the conductivity. The nature of this process becomes clearer when it is described in real space (Fig. 2.3b). Consider an electron propagating from $\vec{r}$ to $\vec{r}\,'$. The action is a sum of contributions from many paths. Two of these are shown in Fig. 2.3b. In one, the electron traverses a (nearly) self-intersecting path in the random medium. In another, it traverses the same (or nearby) self-intersecting path in the opposite sense. On impurity configuration averaging, the two equal contributions $A$ add, so that the amplitude for return to origin is

$2A$. The probability is $4|A|^2$, which is twice the probability $|A|^2 + |A|^2$ in the absence of constructive interference [24]. Thus, return to origin probability in a disordered system is enhanced by quantum interference effects. This is the beginning of localization. Its size, and effects on physical properties can be precisely calculated.

A number of consequences for anomalous transport properties of disordered two- and three-dimensional systems follow, e.g. a logarithmic *increase* of the resistance of a thin film with decreasing temperature; a negative magnetoresistance varying as $\ln H$ in large fields $H$ in two dimensions, and as $\sqrt{H}$ in three dimensions. In two dimensions, the localization length depends exponentially on the microscopic disorder or short length-scale conductivity as inferred from Eq. 2.4. One has

$$\xi_{loc}^{2d} = \ell\, exp[\sigma_o/g_c] = \ell\, exp(\pi k_F \ell/2).$$

This relation has been experimentally verified recently [25], thus reinforcing the idea that 'weak localization' continuously goes over to describe the localized regime as well.

At about the time that the scaling and weak localization ideas were being developed, Altshuler and Aronov showed [26] that interaction between diffusing electrons (diffusive density fluctuations) leads to novel retardation effects. These have consequences for the density of electronic states (e.g. a $\sqrt{|E - E_F|}$ dip in three dimensions) and other thermodynamic quantities, and for transport as well.

There is a rich variety of characteristic transport and other anomalies in disordered interacting electronic systems [27, 30] that can be understood in terms of interaction/localization effects.

As disorder and effective interaction increase, the electronic system undergoes a transition to an insulating phase. This quantum transition of a disordered conducting electron liquid to insulating glass has been actively explored by McMillan [31], Castellani, di Castro, Lee and others [32].

### 2.2.4 THEORETICAL DEVELOPMENTS

The eighties saw, in addition to experimental activity, a diversity of theoretical approaches to localization and interaction effects in disordered systems. They greatly reinforce and extend our understanding of phenomena in this part of physics. A sample is given below.

(i) **Field Theory:**

Wegner and coworkers [33] mapped the quantum particle (electron) diffusion problem into an unusual kind of field theory. The Green's function $G_{\mathbf{r}\mathbf{r}'}(E+i\eta)$ for a particle of energy $E+i\eta$ propagating from $\mathbf{r}$ to $\mathbf{r}'$ can be written as a functional integral that weights different paths appropriately. Quenched randomness is described by $n$ field replicas with $n \to 0$. The quantity $\eta$ is like a symmetry breaking field (e.g. $H_{ext}$), and the discontinuity $\frac{i}{2\pi}\left[G^{+}_{rr'}(E+i\eta) - G^{-}_{rr'}(E-i\eta)\right] = N(E)$ namely the density of states is like the magnetization. Since $N(E)$ is nonzero and noncritical, we have a problem of nonzero magnetization in both phases. The quantity of interest, electron diffusion $D$, is like the stiffness for transverse fluctuations of fixed length spins. The $n$ spin fields effectively interact with each other *via* the disorder.

The mapping itself is nonperturbative, and the localized phase is one in which the 'spin wave' stiffness vanishes at large distances. Thus, a renormalization group or scaling approach is natural for analyzing localization and its onset as a function of short length scale disorder or 'bare' stiffness $D_o$. This has been implemented perturbatively by Wegner and coworkers [33], and by Hikami [34] who also developed a diagrammatic analogy. They find that to order $\epsilon^3$ where $\epsilon = d-2$, $\beta(g)$ has the form Eq. 2.3. A nonperturbative, self consistent theory for the $\beta$ function has been derived by Hikami following an earlier diagrammatic theory due to Vollhardt and Wölfle [35]. The Vollhardt-Wölfle scaling function is exact in both small and large-$g$ limits, and bridges them through an approximate, self-consistent Hartree-like approach.

The field theory enables one to explore the scaling of more complicated field averages. For example, the generalized inverse participation ratio

$$P_k(E) = \overline{\sum_i |\psi_i(r)|^{2k}\, \delta(E-\epsilon_i)/\rho(E)}$$

for localized states near the mobility edge $E_c$ was first calculated by Wegner [36], and shown to be $P^k(E) \sim (E_c - E)^{\pi_k}$ where $\pi_k = (k-1)\tilde{D}_k = (k-1)(2\epsilon^{-1}+1-k)$ to leading order in $\epsilon = (d-2)$. The above behavior implies for the critical wavefunction an anomalous scaling with respect to system size $L$, namely $P_k(L)|_{(E=E_c)} = L^{-(k-1)(\tilde{D}_k/\nu)}$. Such a scaling, characterized by an infinite number of exponents $D_k = (\tilde{D}_k/\nu)$, implies a multi-fractal wave function. This is graphically confirmed by detailed, beautiful numerical work (see below). The theoretical estimate of $D_2$ for $d=3$ is $D_2 = 1$. Other quantities, such as spectral compressibility and local fluctuations in density of states, can also be calculated.

It was realized by Finkelshtein [37] that a similar bosonic field theory can be developed for the disordered interacting electron gas. The scaling of effective spin and charge interactions and of the diffusion leads to several dynamic universality classes for the metal-insulator transition explored actively by Finkelshtein [37] as well as Lee, Castellani, di Castro and Kotliar [32] and by Kirkpatrick and Belitz [38]. The results have been applied to experimental systems. The approach is fundamentally one of Landau Fermi-liquid theory with strongly energy-dependent parameters, some of which, however, scale to strong coupling and acquire a singular energy dependence.

(ii) **Numerical Approach**

In the last twenty years, the phenomenal development of computational approaches, coupled with ideas such as the scaling theory of localization, finite-size scaling, and Thouless and Landauer conductances, has led to increasingly accurate and extensive studies of one-electron localization in disordered systems [39]. Indeed, our present quantitative, reliable knowledge of the Anderson transition in three dimensions [39, 40] or localization in a strong magnetic field largely derives from numerical work. For example, in three dimensions, the transition is known to be universal, with the same exponent $\nu = 1.47 \pm 0.02$ [40] (for the divergence of the localization length) for different distributions of disorder. [The $\epsilon = (2-d)$ expansion estimate is $\nu = 1$; there are some unresolved difficulties with the $\epsilon$ expansion, e.g. at order $\epsilon^4$]. The distribution of conductances becomes very broad, the critical wave function is highly ramified (multi-fractal), and diffusion is strongly non-classical. Quantities characterizing the critical disorder regime have been obtained numerically. The exponent $D_2$ related to the inverse participation ratio ($k = 2$) is found to be $D_2 = 1.3 \pm 0.2$ [33, 36], while the theoretical estimate is $D_2 = 1$ [33]. The on-site autocorrelation function $C(t)$ decays as $t^{-0.43}$ for critical disorder [40], while for diffusing particles the exponent (in $3d$) ought to be -3/2, and for critically slowed down diffusion ($D(L) \sim L^{-1}$) it ought to be -1.

The *logarithm* of the transmission probability distribution has been obtained and appears to be a universal function, albeit different in the localized and extended regimes [39]. The critical behavior of spectral compressibility has been investigated by a number of authors [54].

(ii) **Diversification**

We briefly mention here a number of developments that have deepened our understanding of localization. A semiclassical approach to weak localization was developed by Chakravarty and Schmid [41] who showed how self-intersecting path contributions arise as quantum corrections. The energy levels of a disordered systems are random, characterized in the Thouless picture by an average level spacing and level stiffness or sensitivity to perturbations. There is a well-developed theory of random Hamiltonian matrices, originally due to attempts at characterizing energy level spectra of nuclei. The connection between these two has been explored [42], and has been specially fruitful for understanding the localized regime, fluctuations in density of states, conductance distribution, and level spectra in mesoscopic systems. Quenched randomness has been formally dealt with in a number of ways; the replica field theory ($n \to 0$) approach has been mentioned already. Other approaches are dynamical field theory [43], and the supersymmetric field theory developed by Efetov [44]. The latter has been specially effective in answering several questions related to the behavior of mesoscopic systems [45].

In one dimension, the effect of randomness can be explored in great depth, both numerically and analytically [39]. In the present context (as well as in other areas of condensed matter physics), the variety of ideas and systems that can be realized theoretically in one-dimensional disordered systems makes them a fertile source of new insights. The broad distribution of conductances was first discussed by Anderson *et al.* [46] using the Landauer transmission idea [3]. The probability distribution of the scattering matrix has been investigated numerically [39], and a scaling equation for it has been derived for it by Mello, Peyrera and Kumar [47]. Unusual phenomena such as amplification by an active random medium [48], universal conductance fluctuations, and resonant transmission can be investigated in great detail [49].

### 2.2.5 MESOSCOPIC SYSTEMS AND PHENOMENA

This is a large and lively field [49, 50] which I leave out altogether. It is fed by the coming together of our ability to engineer and probe atomic or meso-scale structures, and the localization or disorder related ideas mentioned above. In these systems, disorder affects and often determines the stiffness of the underlying electronic energy level structure.

### 2.2.6 OTHER EXCITATIONS

As mentioned earlier, randomness in a medium can lead to localization of excitations in that medium. The paradigm is: Consider a quantity that is conserved. Its nonuniform density necessarily diffuses in a random medium. With increasing disorder, this diffusion slows down and finally ceases. For example, in an elastically deformable medium, the excitations are phonons. If they do not interact (harmonic phonons), the energy carried by a particular mode can be distributed only among other phonon modes with the same energy. With increasing disorder, diffusion of this energy can cease, and phonons with that energy are localized. Similarly, electromagnetic waves in a medium with random, real, positive dielectric constant can be localized if the randomness is large enough [51]. A precursor of such localization is enhanced back-scattering [52].

Photon localization in a random, non-absorbing dielectric is specially interesting since photon-photon interaction effects are negligibly small; ideal Anderson localization may be explored. Recently, localization of visible light passing through a fine micron-sized powder of GaAs has been experimentally realized [53]. It is quite likely that details of the Anderson transition and of wave functions near this transition will be probed experimentally using such systems.

### 2.2.7 SUMMARY

We now have a richly detailed picture of the Anderson localization transition in three dimensions, mainly through extensive numerical work that exhibits real space finite-size scaling. The transition is continuous, marked by a single divergent length scale, and also by a very broad (but universal) distribution of conductances, spectral fluctuations, and a characteristic multi-fractal wavefunction. Analytically, the perturbative renormalization group ($\epsilon$-expansion) approach predicts a continuous transition, one-parameter scaling, multi-fractal wavefunction near critical disorder, and exponents for various physical quantities. The exponent values are not accurate for $\epsilon = 1$, as is often the case with an $\epsilon$-expansion, especially around the lower critical dimension, and for random systems. The perturbative theory implies that other (higher-order gradient) couplings become relevant [54]. This is related to the broadening of the distribution of conductances. The significance of a one-parameter scaling theory for conductance (an average over a wide distribution) as the single relevant scaling variable is thus not clear. At present there is no (non-perturbative) analytical realization of the field theory mapping for the critical regime, say in three dimensions, that describes how the distribution of (the logarithm of) the conductance scales, and the nature of the wave function. Experimentally, while there are many disordered systems that have a metal-insulator transition, the observed effects are due not solely to disorder, but to electron interaction and disorder acting together in ways not fully unraveled. Perhaps, photon localization is the only example of the Anderson transition without interaction effects.

Incipient localization, and interaction in a random medium, cause a number of phenomena in metals (large and small pieces of them) that have been predicted and observed. There is every reason to believe that the same localizing process eventually leads to localized states, for strong disorder or at large enough length scales or at low enough temperatures.

In two dimensions, the ground state of a disordered non-interacting electronic system has only localized states. There is considerable experimental and numerical evidence for the details of how this happens, and for how the localization length depends on disorder. In one dimension, our knowledge is most complete, analytically and numerically.

## 2.3 TOMORROW

While there is considerable continuing activity in the above areas, especially in relation to mesoscopic phenomena, there are a number of fields where disorder and localization related effects seem to be involved, perhaps in new ways that we do not fully understand. Some of these, mentioned briefly below, are the glass transition and the breakdown of ergodicity in random systems, the superconductor insulator transition, the metal insulator transition in two dimensions, low temperature decoherence in disordered conductors, and the electrical behavior of magnetoresistance oxides. Except for the first, all the areas mentioned are directly related to puzzling experimental results.

### 2.3.1 GLASS AND SPIN GLASS

Early ideas on spin glasses, e.g. the discussion by Anderson [55], attempted to make contact with localization. The exchange matrix $J_{ij}$ is random for a spin glass, and can have localized eigenstates. One can imagine that as temperature is lowered, localized eigenstates are overwhelmingly favoured, leading to spin glass behavior. About twenty years ago, before the two decades of progress in localization theory and in the theory of spin glasses, Anderson [55] suggested that "The solution of the spin glass problem and that of the localization problem are bound up together. If we make any appreciable progress on one, it must be applicable to the other". This has not happened, though there has been a lot of progress in both. The progress has been largely along nonintersecting paths. There have been a few attempts, e.g. Ref. [56], but they have not led to a deeper understanding of spin glasses or of glasses.

In these systems, the state-counting notion of temperature, derived from the fixed total energy or microcanonical ensemble breaks down. This ergodicity breakdown [57] is due to a kind of localization, where because of disorder, local fluctuations $\delta E(\mathbf{r})$ in energy do not diffuse.

### 2.3.2 SUPERCONDUCTOR-INSULATOR TRANSITION

It has been known for nearly thirty years [58] that a thin film of a superconducting material, e.g. Al or Pb, if sufficiently disordered does not go superconducting. The ground state changes from superconductor to insulator with increasing disorder. The transition has been studied extensively both experimentally and theoretically. Early theoretical approaches [59] focussed on effects of disorder and interaction in either effectively increasing the coulomb interaction [60], or reducing density of states [61], or reducing the order parameter phase stiffness due to localization [62], all electronic effects.

Fisher and co-workers [63] developed a bosonic approach that regards the insulator as a superfluid of vortices with localized bosons (Cooper pairs) and the

Bose superfluid as a phase with localized vortices. The transition is a quantum phase transition in which with increasing interaction or disorder, phase stiffness decreases so that vortices proliferate, delocalize and condense. Basically, disorder plays two roles here. One, it is like interaction or temperature in a homogeneous system, promoting fluctuations and reducing the superfluid stiffness. Second, it localizes vortices in one phase and bosons in the other. This bosonic approach, with clear cut predictions for the scaling behavior of physical quantities, near $T = 0$ and critical disorder, has been very influential in organizing our thinking, and experimental results. For example, the resistance per square $R_\Box$ has the universal form $R_\Box = R^* f[(B - B_c)/T^{1/\nu_B z_B}]$ where $B_c$ is the critical field (at $T = 0$). A number of experimental results support this prediction [58, 64, 65]. However, it appears [66] that at low temperatures ($T \tilde{<} 150$ mK) this scaling breaks down, as if $T \to T + T_0$ where $T_0$ is a nonzero 'dissipation' temperature. Similar results have been obtained for Josephson junction arrays and in quantum Hall - insulator transitions. There are, in addition, persistent concerns having to do with the nonuniversality of the observed critical sheet resistance [58, 67] in contrast to the prediction of a universal value, the possible existence of a metallic $T = 0$ phase [67] between the superconductor and the insulator, the very small dynamic range of the critical regime (generally much less than a decade, often a few percent in $R_\Box$), the smallness of the Hall resistance (it is much smaller than the predicted value of $R^* \sim \hbar/e^2$), the difference between 'thickness' scaling exponents and field scaling exponents( [58],see N Markovic *et al.*) and finally the fact that there is a finite density of zero-energy electronic excitations as seen in tunnelling experiments [68].

There is considerable re-thinking going on in the field. There is the possibility of dissipative effects, e.g. due to diffusion of quantum vortices without their Bose condensation [69]. The idea of a Bose metal, i.e. a non-superfluid, non insulating quantum phase has been proposed [70]. There are persistent demands to bring back the electrons, simply because they seem to be there, i.e. zero-energy unpaired quasiparticle excitations seem to exist [67, 68] and may be central for dissipative processes or Cooper pair phase decoherence. In all the current theoretical bosonic scenarios, disorder or electron localization is either absent or plays a secondary role. However, in view of the fact that the transition does occur at a dissipative sheet resistance $R^*_\Box \sim (\hbar/e^2)$, localization effects ought to be significant.

### 2.3.3 METAL-INSULATOR TRANSITION IN 2D

There *is* a metal-to-insulator transition in clean, high mobility 2d inversion layer systems, e.g. Si, GaAs, Si-Ge, $\cdots$ [71]. The existence of such a transition came as a surprise to a generation that had grown up assuming that there is no metallic ground state in two dimensions even for weak disorder. However, that is in the absence of interactions, which are present and strong in the low-density inversion-layer systems. At the low electron densities at which the transition occurs, the $r_s$ values range from 7 to 27, so that the bare coulomb interaction energy is an order of magnitude larger than the kinetic energy. (The Wigner crystal $r_s$ for $2d$ is estimated to be about 35). Most theories treat disorder and interaction perturbatively, and thus do not directly address the coulomb interaction dominated regime.

There is by now, a fairly large amount of experimental evidence from transport measurements for a continuous $T = 0$ metal to insulator transition (some of this is reviewed by Sarachik in another talk at this conference). The resistance per square has been cast in a scaling form $R_{\square} = R^* f(T/T_0)$ where $T_0 \to 0$ as $n \to n_c$. While reasonable fits are obtained to a scaling form suggesting a quantum critical point, the data are far from the critical region, the range in $R_{\square}$ is small, typically less than a factor of two, and the measurements do not extend to very low temperatures, with $(T/T_F^0)$ generally larger than 0.1. There is a wide spread in the 'critical' exponents $z\nu$. Two other features of the data are the large drop in resistivity on the metallic side over a broad temperature range $\sim 1$ K (which is of order $T_F^0$), and the general absence of logarithmic terms. However, the most recent measurements, carried out somewhat away from $n_c$, but at very low temperatures, do show logarithmic terms of the expected size [72].

Perhaps, the most basic theoretical question is whether a non-Fermi liquid metallic state occurs in this disordered quantum coulomb liquid, and if it does, the nature of its transition to a Wigner glass or a pinned Wigner liquid. A number of non Fermi-liquid scenarios have been proposed [73]; there is also the likelihood that the transition is masked by various 'mundane' effects, e.g. temperature-dependent trap densities [74] or screening of charged impurities [74]. These have to be factored out. Whether the transition is a novel quantum critical point, or whether one has a 'high-temperature' metallic phase crossing over to an insulating phase at temperatures that rise steeply from exponentially low values to large values as $n \to n_c$ is a question that needs to be settled early, with a mix of experiments and insightful phenomenology.

### 2.3.4 DECOHERENCE IN DISORDERED CONDUCTORS

In disordered low dimensional conductors, localization effects are measurably cut off by electron phase randomizing inelastic collisions [75]. A number of mechanisms e.g. electron-phonon coupling, fluctuating electric fields due to electron electron interaction etc. have been proposed and discussed [75, 76]. All these have rates that vanish as some power of the temperature. One thus expects that as $T \to 0$, the decoherence time diverges. It was therefore a surprise when, in 1997, Mohanty, Jariwala and Webb presented data from quasi-$1d$ Au films implying the saturation of decoherence rates at low temperatures [77].

The phenomenon seems general [77, 78], and has been observed in a variety

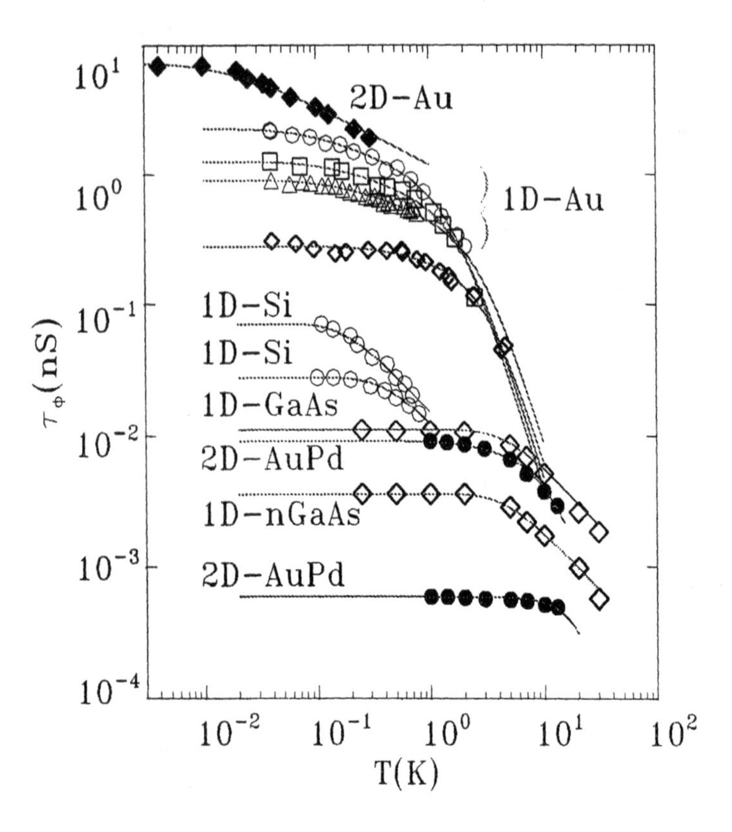


Figure 2.4: Inelastic or phase decoherence time $\tau$ in nanoseconds ($y$ axis) plotted for different quasi $1d$ and $2d$ systems as indicated. The data points and a theoretical fit (curves) are shown. (From Ref. 78).

of quasi-$1d$ and $2d$ systems, as well as in quantum dots. The saturation begins at $T \sim 1$ to 0.2 K. (Fig. 2.4).

Zero-temperature decoherence has, irrespective of its origin, several obvious consequences. The localization transition at $T = 0$ is strongly modified, depending on the relative size of $\xi_{loc}$ and $L_{in}$ [79]. The notion of a unique (non-degenerate) quantum many-body ground state with a sharp excitation spectrum becomes a fundamentally inaccessible idealization, with real consequences e.g. for persistent currents in mesoscopic rings [78].

For these and other reasons, the observation of Mohanty *et al.* [77, 78] has excited a great deal of interest and controversy in the field. Much of the theoretical effort has focussed on a search for effective decoherence mechanisms in interacting disordered systems. The idea that zero-point fluctuations limit $L_{in}$ appears not tenable on general grounds. Experiments on quantum dots [80] in which known amounts of 'noise' are fed in clearly suggest that their main effect is to raise the electron temperature and not to cause athermal decoherence. Other proposals involve $(1/f)$ noise, due presumably to slowly relaxing defects. Such noise extrapolated to low frequencies will lead to hysteretic effects that are not observed. At present, there is no fully satisfactory explanation for the phenomenon.

### 2.3.5 LOCALIZATION IN MANGANITES

This is a major field [81] in which orbital degeneracy, Jahn-Teller effect, Hund's rule determined spin state correlation and consequent double exchange, correlation and lattice dynamics play important roles, not to mention disorder, both dynamic and static. The richness of phases and phenomena is not well understood. I shall discuss here only one peculiar property, namely the apparent strong violation of the Mott $\sigma_{min.}$ condition, and its implications.

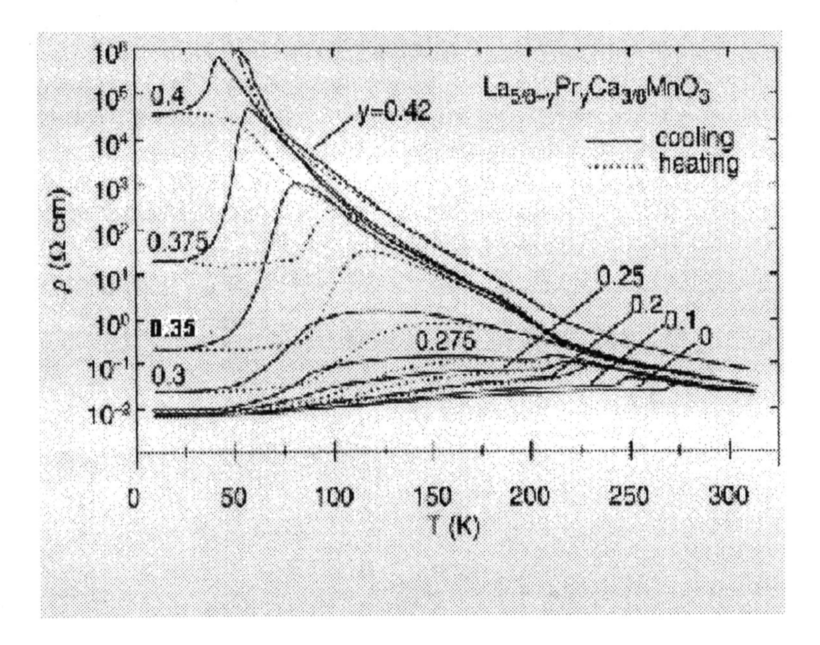


Figure 2.5: Electrical resistivity $\rho$ in $\Omega\,cm$ of $La_{(5/8-y)}Pr_yCa_{3/8}MnO_3$ ($y$ axis), as a function of temperature $T$ (in $K$), for compositions $y$ indicated (from Ref. 82).

In a number of doped manganites e.g. $Re_{0.7}Ca_{0.3}MnO_3$ (where $Re$ is a rare earth such as Pr, Nd $\cdots$), the low temperature phase is a ferromagnetic metal, as it is for $La_{0.7}(Ca,Sr)_{0.3}MnO_3$, with this difference: the low temperature resistivities can be a factor $10^3$ to $10^5$ larger than the Mott maximum value, expected to be 1 to 10 m$\Omega$ cm. In spite of this high resistivity value, the system does not go insulating down to low temperatures; the resistivity flattens out [81, 82]. A systematic study of this effect has been made by Cheong and coworkers [82]. The compound $La_{(\frac{5}{8}-y)}Pr_yCa_{\frac{3}{8}}MnO_3$ is investigated for different Pr content $y$ (Fig. 2.5). One observes an activated behavior for resistivity at high temperatures, with a transition to a ferromagnetic metallic state and a temperature independent resistivity at low temperatures. The low temperature resistivity value can be much larger than $\rho_{Mott}$, and depends strongly on the Pr concentration.

Cheong *et al.* propose an appealing two-coexisting-phase explanation of this strong apparent violation of Mott maximum metallic resistivity. Their argument, supported by direct electron diffraction images, is that the system consists of two phases, namely a ferromagnetic metallic phase, and an insulating charge (orbitally ?) ordered phase, with a characteristic superlattice peak. The relative proportions vary with $y$, and there is a percolative transition to an insulator for $y_c \tilde{>} 0.41$. The high resistivities are due to proximity of $y_c$. Since the grain sizes are large (microns), and there is evidence against compositional or charge inhomogeneities on this scale, the interesting question is: what is the origin of the insulating phase for arbitrary, incommensurate $y$? Is it a kind of Anderson insulator? If so, what is the disorder? How is it that at the same composition $y$, disorder is present in some parts of the system and absent in other parts, on a large enough scale so that both Anderson insulating and metallic phases coexist?

From its quiet beginnings more than four decades ago, localization, and more generally, the effect of randomness in many body systems, has emerged as a major theme of condensed matter physics, with many surprises still ahead. Throughout this long journey of ideas and phenomena, Anderson's creative spirit has shown the way.

## BIBLIOGRAPHY


[1] P W Anderson, Phys. Rev. **109** 1492 (1958).

[2] G Feher and E A Gere, Phys. Rev. **114** 1245 (1959).

[3] I M Lifshitz, Sov. Phys. Usp. **1** 549 (1965).

[4] R Landauer, IBM J. Res. Dev. **1** 223 (1958).

[5] W Kohn, Phys. Rev. A **133** 171 (1964).

[6] J T Edwards and D J Thouless, J Phys. **C5** 807 (1972).

[7] See for example the reviews of D. ter Haar, Rep. Progr. Phys. **24** 304 (1961); G V Chester, Rep. Progr. Phys. **26** 411 (1963).

[8] E Abrahams, P W Anderson, D C Licciardello and T V Ramakrishnan, Phys. Rev. Lett. **42** 673 (1979).

[9] N F Mott and E A Davis, *Electronic Processes in Non Crystalline Materials* (2nd ed., Clarendon, Oxford, 1979), is one summary of this body of work.

[10] N F Mott, Proc. Roy. Soc. London, Ser A. **167** 384 (1949); N F Mott, *Metal Insulator Transitions* (Taylor and Francis, London, 1990); For a recent review of the Mott transition, see A Georges *et al.*, Rev. Mod. Phys. **68** 13 (1996).

[11] A L Efros and B I Shklovskii, J Phys. C **8** L 49 (1975); B I Shklovskii and A L Efros, *Electronic Properties of Doped Semiconductors*, (Springer, Berlin, 1983).

[12] N F Mott, M Pepper, S Pollitt, R H Wallis and C H Adkins, Proc. Roy. Soc., London Series A **345** 169 (1975).

[13] J A Hertz, L Fleishman and P W Anderson, Phys. Rev. Lett. **43** 942 (1979).

[14] P W Anderson, in *Ill-Condensed Matter* (Eds. R Balian, R Maynard and G Toulouse, North Holland, Amsterdam, 1979), p 159-260.

[15] P W Anderson, Nature (London) **235** 163 (1972).

[16] P W Anderson, Phys. Rev. Lett. **34** 953 (1975).

[17] A Abou-Chacra, P W Anderson and D J Thouless, J Phys. C **6** 1734 (1973).

[18] See for example, D J Thouless, Phys. Rept. **13** 93 (1974).

[19] D J Thouless, in Ref.14 summarizes this work.

[20] F J Wegner, Z. Phys. B **25** 327 (1976).

[21] P W Anderson, E Abrahams, and T V Ramakrishnan, Phys. Rev. Lett. **43** 718 (1979).

[22] E Abrahams and T V Ramakrishnan, J Non-Cryst. Solids **35** 15 (1980).

[23] L P Gor'kov, A I Larkin and D E Khmelnitskii, JETP Lett. **30** 248 (1979).

[24] D E Khmelnitskii, Physica **126 B+C** 235 (1984).

[25] F W van Keuls, H Mathur, H W Jiang, and A J Dahm, Phys. Rev. B **56** 13263 (1997).

[26] B L Altshuler and A G Aronov, Solid State Commun. **39** 115 (1979).

[27] P A Lee and T V Ramakrishnan, Rev. Mod. Phys. **57** 287 (1985).

[28] T V Ramakrishnan in *Chance and Matter* (eds. J Souletie, J Vannimenus and R Stora, Elsevier, Amsterdam, 1987) p. 213-304.

[29] B L Altshuler and A G Aronov in *Electron Electron Interactions in Disordered Systems* (ed. M Pollak and A L Efros, North Holland, Amsterdam, 1984).

[30] H Fukuyama, in Ref. 41, p. 155.

[31] W L McMillan, Phys. Rev. B **24** 2739 (1981).

[32] C Castellani, C di Castro, P A Lee and M Ma, Phys. Rev. B **30** 527 (1984); C Castellani, C di Castro, P A Lee, M Ma, S Sorella and E Tabet, Phys. Rev. B **30** 1596 (1984); C Castellani, G Kotliar and P A Lee, Phys. Rev. Lett. **59** 323 (1987).

[33] F J Wegner, Z Phys. **35** 207 (1979); L Schäfer and F J Wegner, Z Phys. B **38** 113 (1980); A J McKane and M Stone, Ann. Phys. **131** 36 (1981).

[34] S Hikami, Phys. Rev. B **24** 2671 (1981).

[35] D Vollhardt and P Wölfle, Phys. Rev. Lett. **45** 482 (1980); Phys. Rev. B **22** 4666 (1980); Phys. Rev. Lett. **48** 699 (1982); see review in *Electronic Phase Transitions* (eds. W Hanke and Y V Kopaev, Elsevier, Amsterdam, 1992).

[36] F J Wegner, Z. Phys. B **36** 209 (1980).

[37] A M Finkelshtein, Sov. Phys. – JETP **57** 97 (1983); Z Phys. **56** 189 (1984); See also A M Finkelshtein, in *Physics Reviews 14* (Ed. I M Khalatnikov; Harwood, London, (1990)) p. 1-100.

[38] D Belitz and T R Kirkpatrick, Rev. Mod. Phys. **66** 261 (1994); T R Kirkpatrick and D Belitz, cond-mat/9707001, to appear in *Electron Correlations in the Solid State* (ed. N H March, Imperial College Press, London, 2000).

[39] See for example the review of B Kramer and A Mackinnon, Rep. Prog. Phys. **56** 1469 (1993).

[40] T Ohtsuki, K Slevin and T Kawarabayashi, Ann. Phys. (Leipzig) **8** 655 (1999) describes recent progress.

[41] S Chakravarty and A Schmid, Phys. Rep. **140** 193 (1986).

[42] K A Muttalib, J-L Pichard, and A D Stone, Phys. Rev. Lett. **59** 2475 (1987); K A Muttalib, Phys. Rev. Lett. **65** 745 (1990).

[43] M Horbach and G Schön, Physica A **167** 93 (1990).

[44] K B Efetov, Adv. Phys. **32** 53 (1983).

[45] A D Mirlin, cond-mat 9907126, Phys. Rep. (to appear).

[46] P W Anderson, D J Thouless, E Abrahams, and D S Fisher, Phys. Rev. B **22** 3519 (1980).

[47] P A Mello, P Peyrera and N Kumar, Ann. Phys. **181** 290 (1988).

[48] P Pradhan and N Kumar, Phys. Rev. B **50** 9644 (1994).

[49] C W J Beenakker, Rev. Mod. Phys. **69** 731 (1997).

[50] Y Imry, *Introduction to Mesoscopic Physics*, Oxford University Press (1997).

[51] See the review by S John, Physics Today **32** May 1991.

[52] P W Anderson, Phil Mag. B **52** 505 (1985).

[53] D W Wiersma, P Bartolini, A Lagendijk, R Righini, Nature **390** 671 (1997).

[54] V E Kravtsov and I V Lerner, Solid State Commun. **52** 593 (1984). N Kumar and A M Jayannavar, J Phys. C Solid State Phys. **19** L85 (1986); I V Lerner, Physica A **167** 1 (1990).

[55] P W Anderson, in Ref. 14, p. 214-219.

[56] See for example, V S Dotsenko, M V Feigelman and L B Ioffe, *Physics Reviews 15*, part 1 (Ed. I M Khalatnikov, Harwood, London, 1991).

[57] R G Palmer, Adv. Phys. **31** 669 (1982).

[58] Some reviews are: A F Hebard, in *Strongly Correlated Electronic Materials* (eds. K S Bedell *et al.*, Addison Wesley, Reading, Mass, 1993) p.251; Y Liu and A M Goldman, Mod. Phys. Lett. B **8** 277 (1994); A M Goldman and N Markovic, Physics Today **51** No. 11, p. 39 (1998); N Markovic *et al.*, cond-mat/9904168.

[59] See for example, references 27, 28, 30, 38 and the following review:- A M Finkelshtein, Physica B **197** 636 (1994).

[60] P W Anderson, K A Muttalib and T V Ramakrishnan, Phys. Rev. B **28** 117 (1983).

[61] S Maekawa and H Fukuyama, J Phys. Soc. Jpn. **51** 1380 (1982); D Belitz, Phys. Rev. B **40** 111 (1989); Y Oreg and A M Finkelshtein, Phys. Rev. Lett. **83** 191 (1999).

[62] T V Ramakrishnan, Physica Scripta, T **27** 24 (1989).

[63] M P A Fisher, Phys. Rev. Lett. **65** 923 (1990); M P A Fisher, G Grinstein and S M Girvin, Phys. Rev. Lett. **64** 587 (1990).

[64] A F Hebard and M A Paalanen, Phys, Rev. Lett. **65** 927 (1990).

[65] M A Paalanen, A F Hebard and R R Ruel, Phys. Rev. Lett. **69** 1604 (1992); A Yazdani and A Kapitulnik, Phys. Rev. Lett. **74** 3037 (1995).

[66] N Mason and A Kapitulnik, Phys. Rev. Lett. **82** 5341 (1999).

[67] P B Haviland *et al.*, Phys. Rev. Lett. **62** 2180 (1989); J M Valles, R C Dynes and J P Garno, Phys. Rev. Lett. **69** 3567 (1992).

[68] S Y Hsu, J A Chervenak and J M Valles, Jr., Phys. Rev. Lett. **75** 132 (1995) and references there.

[69] E Shimshoni, A Auerbach and A Kapitulnik, Phys. Rev. Lett. **80** 3352 (1998).

[70] P Das and S Doniach, Phys. Rev. B **60** 1261 (1999).

[71] S V Kravchenko *et al.*, Phys. Rev. Lett. **77** 4938 (1996); S V Kravchenko *et al.*, Phys. Rev. B **50** 8039 (1994); S V Kravchenko *et al.*, Phys. Rev. B **51** 7038 (1995).

[72] M Y Simmons, A R Hamilton, M Pepper, E H Linfield, P D Rose and D A Ritchie, cond-mat/9910368.

[73] S Chakravarty, S Kivelson, C Nayak and K Völker, cond-mat 9805383; Qimiao Si and C M Varma, Phys. Rev. Lett. **81** 4951 (1998).

[74] B L Altshuler and D L Maslov, Phys. Rev. Lett. **82** 145 (1999); B L Altshuler, D L Maslov and V M Pudalov, cond-mat/9909353; S Das Sarma and E H Hwang, cond-mat/ 0001057.

[75] B L Altshuler, A G Aronov and D E Khmelnitskii, J Phys. C **15** 7367 (1982); H Fukuyama and E Abrahams, Phys. Rev. B **27** 5976 (1983); A Stern, Y Aharonov and Y Imry, Phys. Rev. A **40** 3437 (1990).

[76] E Abrahams, P W Anderson, P A Lee and T V Ramakrishnan, Phys. Rev. B **24** 6783 (1981).

[77] P Mohanty, E M Q Jariwala and R A Webb, Phys. Rev. Lett. **78** 3366 (1997).

[78] P Mohanty, Ann. Phys. (Leipzig) **8** 549 (1999); P Mohanty, Physica B (to be published).

[79] M E Gershenson, Ann. Phys. (Leipzig) **8** 559 (1999) and references therein.

[80] A G Huibers *et al.*, Phys. Rev. Lett. **83** 5090 (1999).

[81] See for example *Colossal Magnetoresistance, Charge Ordering and Related Properties of Manganites* (eds. C N R Rao and B Raveau, World Scientific, Singapore, 1998); *Colossal Magnetoresistance Oxides* (ed. Y Tokura, Gordon and Beach, London, 1999); JMD Coey, M Viret and S von Molnar, Adv. Phys. **48** 167 (1999).

[82] M Uehara, S Mori, C H Chen and S -W Cheong, Nature **399** 560 (1990).

# CHAPTER 3

# METAL-INSULATOR TRANSITIONS IN DISORDERED SYSTEMS

MYRIAM P. SARACHIK

Department of Physics, City College of the City University of New York
New York, New York 10031

ABSTRACT

Metal-insulator transitions that occur in the limit of zero temperature in a variety of electronically disordered solids as a function of composition, dopant concentration, magnetic field, stress, or some other tuning parameter, have been the focus of study for many decades, and continue to be a central problem in condensed matter physics. A full overview of so broad a field would require considerably more space than I have been allotted. I will thus limit my presentation to two specific areas that have been of particular interest to me. In the first part I will give a brief overview of the recent history and current status of the so-called "critical exponent puzzle" in three-dimensional doped semiconductors and amorphous metal-semiconductor mixtures, including some interesting recent developments. The second part will be a brief summary of new findings in dilute 2D systems, such as silicon MOSFETs and GaAs/AlGaAs heterostructures, where the resistivity exhibits a metallic temperature dependence in some ranges of electron (hole) densities, raising the possibility of an unexpected metal-insulator transition in two dimensions. An equally intriguing property of these strongly interacting, low density 2D systems is their enormous magnetoresistance: for magnetic fields applied parallel to the electron plane, the resistivity increases dramatically by several orders of magnitude in response to relatively modest fields on the order of a few Teslas, saturating to a constant value at higher fields.

## 3.1 CRITICAL EXPONENT PUZZLE IN 3D

Starting from the Ioffe-Regel criterion, which asserts that the mean free path of an electron in a metallic system cannot be shorter than its wavelength, Mott [1] proposed in 1972 that there exists a minimum value of the conductivity, $\sigma_{min} = Ce^2/ha$, below which a system cannot be in the metallic phase. Here, $e$ is the electron charge, $h$ is Planck's constant, $a$ is the average distance between electrons at the critical concentration $n_c$, and $C$ was estimated by Mott to be a number between 0.15 and 0.3. The conductivity is thus expected to drop discontinuously to zero at a Mott transition driven by interactions. Some years later, the scaling theory of Abrahams, Anderson, Licciardello and Ramakrishnan [2] for disordered systems of non-interacting electrons implied that the transition is instead a continuous one, so that arbitrarily small values of conductivity are possible in the metallic phase. Theory soon followed which showed that the electrons (or holes) generally localize even more strongly when weak interactions are included [3]. Experiments in amorphous metal-semiconductor mixtures such as AuGe [4] and NbSi [5], as well as in some doped semiconductors, most notably the elegant stress-tuning measurements by Paalanen *et al.* in Si:P [6], provided strong evidence that the transition is indeed continuous.

Following these early seminal contributions, however, the field has been plagued for nearly two decades by controversy and conflicting results. On the one hand, a consensus emerged rather quickly regarding the behavior of the amorphous metal-semiconductor mixtures: tunneling experiments [4, 5] documented the appearance of a square-root singularity in the density of states as the transition is approached from the metallic side, consistent with theoretical predictions for interacting electrons [3]; the critical exponent $\mu$ that characterizes the (continuous) approach of the conductivity to the transition, $\sigma \approx (n - n_c)^\mu$, was found to be close to 1 in essentially all the amorphous systems examined. In contrast, conflicting results have been reported in doped semiconductors, and these materials continue to be the subject of debate and uncertainty as discussed in more detail below. It should be noted, however, that a number of investigators, most actively A. Moebius [7], continue to question the generally accepted view; based on a scaling analysis of the conductivity of amorphous CrSe, this school maintains that the transition is discontinuous with a minimum metallic conductivity, as originally postulated by Mott.

Electron interactions are known to play a role in the case of doped semiconductors. This was demonstrated by tunneling experiments in Si:B [8] which yielded results very similar to those found for GeAu and NbSi: a square-root singularity develops as the transition is approached from the metallic side, and a "Coulomb gap" appears in the insulator as the dopant concentration is further decreased. However, a great deal of disagreement persists concerning the critical behavior of the conductivity, with various laboratories reporting different results on different, and sometimes even the same, semiconductor systems.

Typical curves are shown in Fig. 3.1, where the conductivity of Si:B is plotted as a function of temperature $T$ in frame (a), and as a function of $T^{1/2}$ in frame

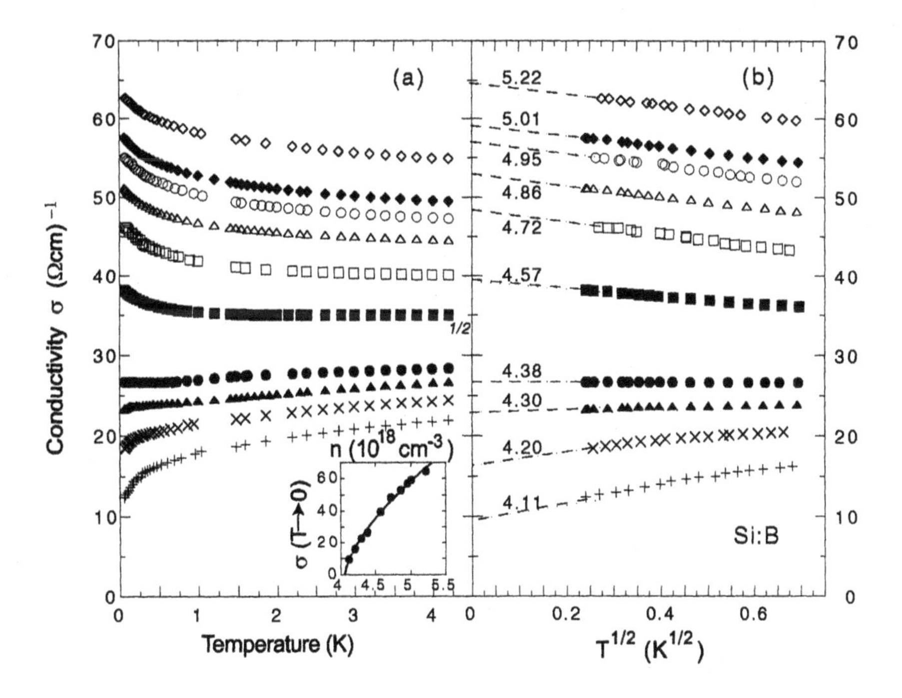


Figure 3.1: Conductivity of Si:B plotted as a function of: (a) temperature $T$; and (b) $T^{1/2}$. Each curve corresponds to a different dopant concentration, labelled in units of $10^{18}$ $\mathrm{cm}^{-3}$. The inset shows zero-temperature extrapolations plotted as a function of dopant concentration.

(b). The conductivity is given by [3]:

$$\sigma(T) = \sigma(0) + \Delta\sigma_{int} + \Delta\sigma_{loc} = \sigma(0) + mT^{1/2} + BT^{p/2}$$

The second term on the right-hand side is due to electron-electron interactions and the last term is the correction to the zero-temperature conductivity due to localization. The exponent $p$ reflects the temperature dependence of the scattering rate, $\tau_\phi^{-1} \propto T^p$ of the dominant phase-breaking mechanism responsible for delocalization, such as electron-phonon scattering or spin-orbit scattering. The last term is assumed small at very low temperatures, and the conductivity is generally plotted as a function of $T^{1/2}$, as in Fig. 3.1(b). There is little theoretical justification for using this expression within the critical range, where it is applied "by default" in a region where the behavior of the conductivity is not known. We note that the slope $m$ that determines the conductivity at low temperatures changes sign from negative to positive as the metal-insulator transition is approached. The significance of this change of sign has been a matter of debate, and lies at the heart of the controversy regarding the critical behavior.

The critical conductivity exponent $\mu$ is generally determined by the following procedure. For each sample with a given concentration (or for each value of stress, magnetic field, or other tuning parameter), a single, zero-temperature extrapolated value of the conductivity is deduced from data obtained at finite temperatures. The zero-temperature extrapolations are then plotted as a function of concentration (or stress, or field) to obtain the critical behavior, as indicated in the inset to Fig. 3.1(a). This experimental protocol has yielded conflicting values for $\mu$. Important examples are shown in the next two figures.

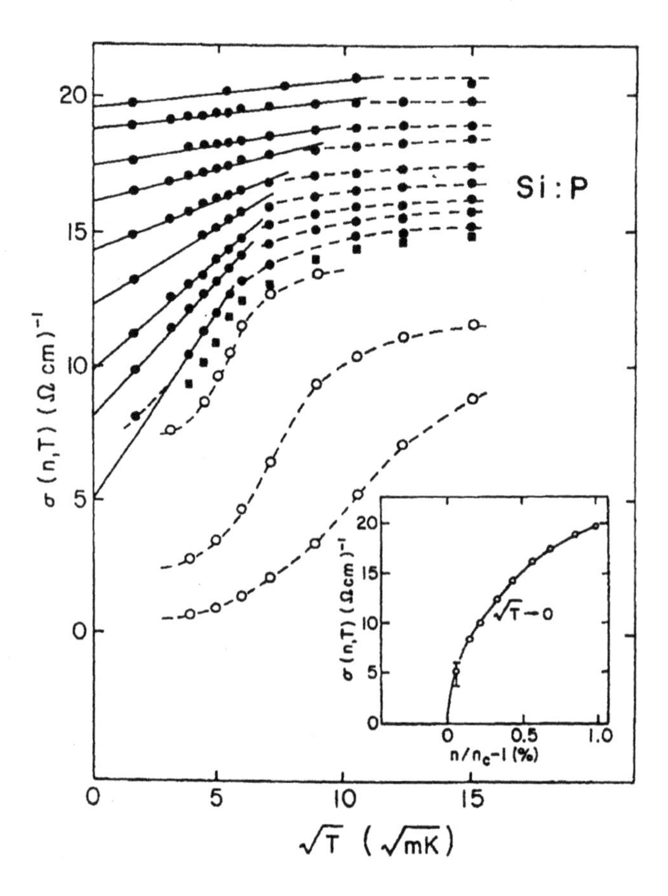


Figure 3.2: Data of Paalanen *et al.* (ref. 6) for the conductivity as a function of temperature of uniaxially stressed Si:P very near the metal-insulator transition; the inset shows zero-temperature extrapolations plotted as a function of stress, yielding an exponent $\mu = 1/2$.

Figure 3.2 shows early classic measurements of the conductivity of uniaxially-stressed Si:P taken down to unusually low temperatures by the Bell group [6]; fitted to the square-root temperature dependence of Eq. (1), the zero-temperature extrapolations yielded a critical exponent $\mu = 1/2$ (see the inset). If the correlation length exponent $\nu$ is equal to the conductivity exponent $\mu$, as expected within Wegner scaling [9], this violates a lower bound $\nu > 2/3$ in three dimensions calculated by Chayes *et al.* [10]. A possible solution to the exponent puzzle was subsequently proposed by the Karlsruhe group of H. v. Lohneysen [11] based on measurements in unstressed Si:P shown in Fig. 3.3.

Stupp *et al.* suggested that only those samples for which the low temperature slopes of the conductivity are positive are in the critical region and should be used to deduce the critical behavior. As shown in Fig. 3.3(b), the Karlsruhe experiments yielded a much larger exponent, $\mu \approx 1.3$, based on a restricted range of dopant concentrations very near a critical concentration that is assumed to be substantially smaller than the generally accepted value. The critical exponent near $1/2$ obtained by the Bell group was attributed by these authors to the improper inclusion of samples outside the critical region. The Bell group strongly disputed this claim, and attributed the large Karlsruhe exponent to inhomogeneities that cause "rounding" near the transition [12, 13]. Indeed, the unknown breadth of the critical region had been a source of some concern, and the Karlsruhe *ansatz* offered a relatively simple and attractive solution to a vexing problem. At the same time, however, carefully executed investigations by Itoh *et al.* [14] of the conductivity of neutron transmutation doped Ge:Ga on both sides of the metal-insulator transition provided strong evidence for the smaller exponent around $1/2$. To complicate matters further, Castner [15] contended that some of the positive-slope curves that were classified as metallic actually obey Mott variable-range hopping, placing these on the *insulating* side of the transition. The few who were directly involved in the issue divided rather sharply into opposing camps, while the rest of the community began to lose interest in a problem that was making little apparent headway.

It has been difficult to obtain reliable determinations of the critical behavior for a number of reasons. One important issue is whether the distribution of dopant atoms is statistically random. This problem can be minimized by doping through neutron transmutation, as has been done in Ge:Ga [14]. Another difficulty is that the approach to the transition is most often controlled experimentally by varying the dopant concentration, $n$, near its critical value, $n_c$, a method that entails the use of a discrete set of carefully characterized samples. This makes it difficult to do systematic, controlled studies on closely spaced samples very near the transition within the critical regime. This problem has been circumvented in a few studies where individual samples have been driven through the transition using a different tuning parameter such as uniaxial stress or magnetic field. The central problem, however, is that zero-temperature conductivities deduced from extrapolations from finite temperature measurements are uncertain and unreliable, particularly in the absence of any theory known to be valid in the critical region.

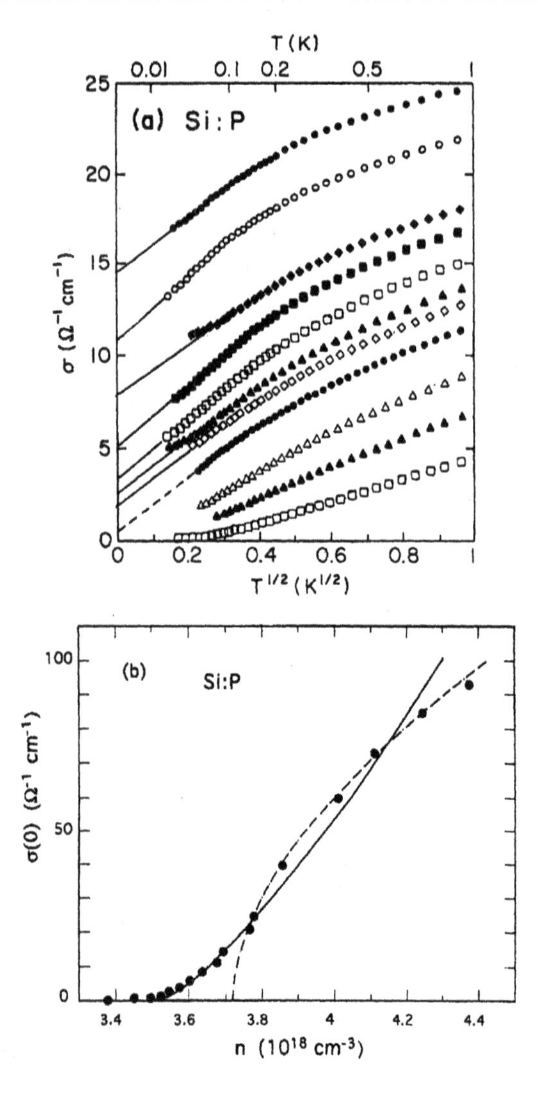


Figure 3.3: (a) Data of Stupp *et al.* (ref. 11) for the conductivity as a function of temperature of a series Si:P samples with closely spaced dopant concentrations very near the metal-insulator transition. (b) The conductivity extrapolated to zero-temperature as a function of dopant concentration. Stupp et al. (1993) claim that restricting the analysis to samples whose resistivities have positive slopes at low temperatures yields a true critical conductivity exponent $\mu \approx 1.3$ (solid line), while including points further from the transition (claimed to be outside the critical region) results in an apparent exponent $\mu \approx 1/2$ (dashed line).

A full scaling analysis that uses data obtained at all temperatures obviates the need for extrapolations to zero [7, 16]. Attempts to apply finite-temperature scaling to the conductivity of crystalline doped semiconductors had largely failed until quite recently (except in the presence of an externally applied magnetic field). Application by Belitz and Kirkpatrick [17] of finite-temperature scaling to data for Si:P gave satisfactory results only over a severely restricted range of temperature, and yielded critical conductivity exponents $\mu = 0.29$ for the Bell data, and $\mu = 1$ for the Karlsruhe data [13]. No data were available from either group for the insulating side of the transition; as shown below, the availability of data on both sides of the transition imposes important constraints.

Potentially important progress on this question was recently achieved in experiments on stress-tuned Si:B [18]. As shown in Fig. 3.4, full scaling of the conductivity with temperature and stress of the form [16]:

$$\sigma(n, T) = \sigma(n_c, T) F(\Delta n / T^{1/z\nu})$$

was demonstrated on both sides of the transition. Here $\sigma(n_c) \propto T^{\mu/z\nu}$, $n_c$ is the critical concentration, $\Delta n = (n - n_c)$, $\nu$ is the critical exponent that characterizes the divergence of the length scale, and $z$ and $\mu$ are the dynamical exponent and critical conductivity exponent, respectively. On a log-log scale, the inset to Fig. 3.4 shows the (unscaled) conductivity for various values of stress. The critical curve is denoted by the straight line corresponding to a power law; the temperature dependence at the critical point is found to be $\sigma \propto T^{1/2}$ in stressed Si:B. It is worth emphasizing again that the power of this method lies with the fact that *all* the data obtained at all temperatures are used in the scaled curves of Fig. 3.4, rather than a single zero-temperature extrapolation for each stress deduced from a full curve of the conductivity as a function of temperature. This imposes more stringent constraints, particularly when data are available on both sides of the transition, and yields critical exponents that are far more reliable and robust.

It is puzzling that the experiment on stress-tuned Si:B yielded a critical conductivity exponent $\mu = 1.6$ considerably larger than any previous determination. Instead of answering old questions, this raises new ones. For example, does the use of stress as a tuning parameter yield the same physics as varying the concentration, as had always been assumed? Although it has not resolved the controversy regarding the value of $\mu$, the Bogdanovich *et al.* experiment demonstrated that finite temperature-scaling can be applied. This has triggered a number of new attempts to use the same method in other cases. Among these are reports from Karlsruhe [19] of scaling for stressed Si:P (yielding $\mu = 1$), and by Itoh's group in Ge:Ga [20].

I will close this section by showing some surprising results of a re-analysis of old data taken at City College on unstressed samples of Si:B. Encouraged by the full scaling form that was successfully applied to stressed Si:B, an equivalent analysis was attempted for data obtained earlier by Peihua Dai where dopant

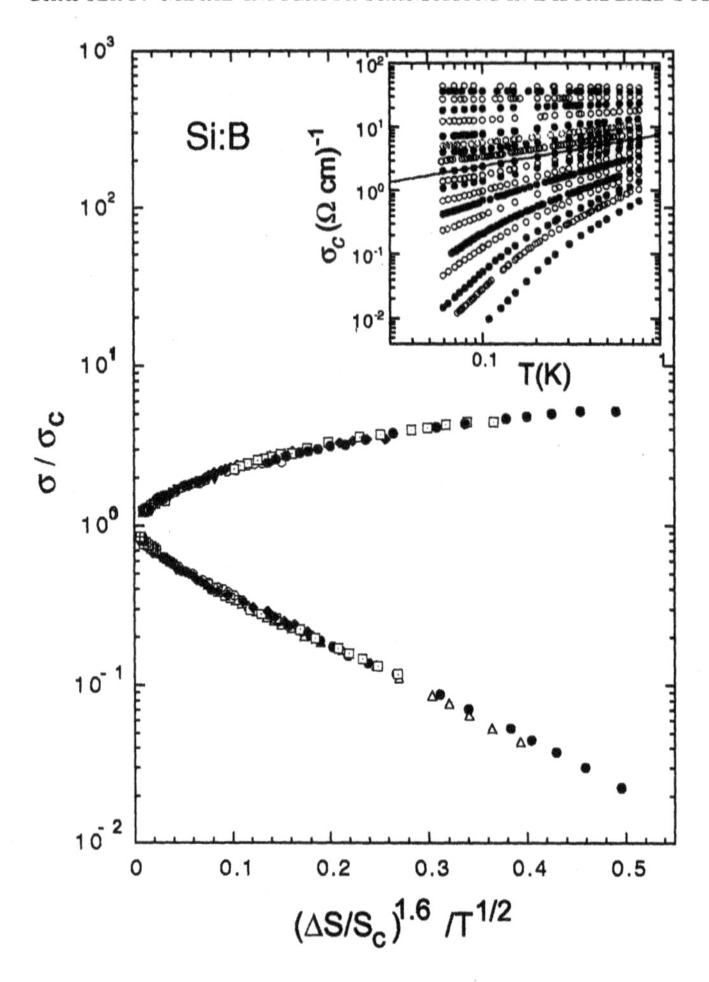


Figure 3.4: For different values of the tuning parameter, $S$, the normalized conductivity, $\sigma/\sigma_c$, of uniaxially stressed samples of Si:B is shown on a log-log scale as a function of the scaling variable $[(\Delta S)/S_c]/T^{1/z\nu}$, with $z\nu = 3.2$. Here $\Delta S = (S - S_c)$, where $S_c$ is the critical stress; the critical temperature dependence at the transition is $S_c \propto T^{1/2}$. The inset shows the unscaled conductivity as a function of temperature on a log-log scale for different values of stress.

concentration rather than stress was used to tune through the transition. As mentioned earlier, in the case of stressed Si:B the critical curve exhibits a power-law dependence on temperature, $\sigma \propto T^{1/2}$; more generally, the critical temperature dependence of the conductivity has been reported in various different semiconductor systems as either $T^{1/2}$ or $T^{1/3}$.

Although the data can be manipulated to yield scaled curves for either the metallic or the insulating branches, neither of these choices for the critical T-

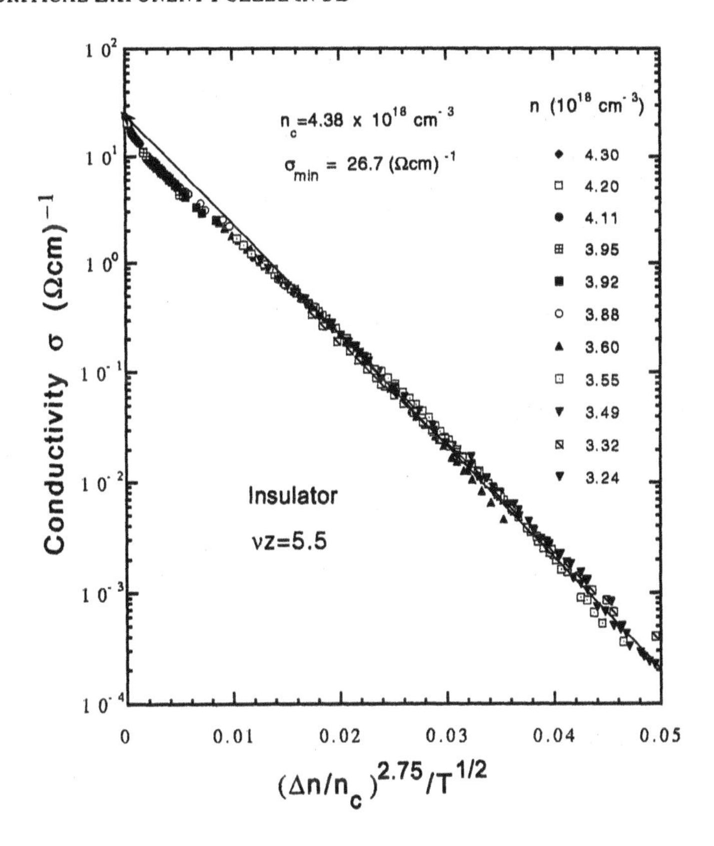


Figure 3.5: For different values of dopant concentration $n$ used here as the tuning parameter, the conductivity $\sigma$ of unstressed Si:B is shown on a logarithmic scale as a function of $(T^*/T)^{1/2}$; here $T^* \propto (\Delta S)^{z\nu}$.

dependence yields scaling on *both* sides of the transition for the conductivity of unstressed Si:B. Surprisingly, the full scaling that was obtained for the stressed samples appears not to hold for the unstressed case. On the other hand, as shown in Fig. 3.5, all the data obtained for insulating samples down to $0.75n_c$ (which is clearly outside the critical region) collapse onto a single curve if one chooses to plot $\sigma$ itself rather than $\sigma/T^x$ with $x = \mu/z\nu = 1/2$ or $1/3$. The conductivity for concentrations ranging from $0.75n_c$ to $n_c = 4.38 \times 10^{18}$ cm$^{-3}$ collapses onto a single curve; although three positive-slope samples (the lowest three curves of Fig. 3.1) are included that have generally been assumed to be on the metallic side of the transition, further careful work is required to determine whether the collapse holds or whether it breaks down very near the transition. The conductivity of unstressed Si:B on the insulating side of the transition is thus given by $\rho = \rho_0 F(T^*(n)/T)$ with $T^* \propto (n_c - n)^y$, $F(0) = 1$, and a prefactor $\rho_0$ that is independent of temperature $T$ and dopant concentration $n$; this implies a resistivity $\rho = \rho_0$ that is independent of temperature at the critical point $n = n_c$. Deep in the insulating phase the conductivity obeys exponentially activated Efros-Shklovskii variable-range hopping, $\rho = \rho_0 \exp[-(T^*/T)^{1/2}]$.

It is remarkable that, despite considerable effort over a period of decades, a complete and satisfactory understanding of the behavior of doped semiconductors near the metal-insulator transition has not yet emerged; this remains one of the most interesting and important open questions in condensed matter physics.

## 3.2 NOVEL PHENOMENA IN DILUTE 2D SYSTEMS: NEW PHYSICS OR OLD?

While we continue our efforts to understand the metal-insulator transition in three dimensions, new developments in disordered, dilute two-dimensional systems have opened an entirely new area of exciting physics, launched by the availability of silicon MOSFETs with unusually high mobilities fabricated in the (former) Soviet Union. These samples allowed measurements at substantially lower densities than had been accessible earlier, a regime where interaction energies are very large compared with the Fermi energy. The experimental findings have called into question our long-held belief [2, 3] that there is no metallic phase in two dimensions in the limit $T \to 0$. The strange and enigmatic behavior of dilute $2D$ systems [21] is illustrated in the next two figures.

The resistivity of a high-mobility silicon metal-oxide-semiconductor field-effect transistor (MOSFET) is shown in Fig. 3.6(a) as a function of density at different temperatures, and in Fig. 3.6(b) as a function of temperature at different densities [22]. The crossing point in frame (a) indicates a critical density $n_c$ below which the behavior is insulating, and above which the resistivity decreases with decreasing temperature, behavior that is normally associated with a metal (see frame (b)). The curves shown in Fig. 3.1(b) as a function of temperature $T$ can be collapsed onto two branches by applying a single scaling parameter $T_o$, a feature generally associated with quantum phase transitions. These claims first met with considerable skepticism, but were soon confirmed in MOSFETs obtained from different sources, and similar behavior was subsequently found for other two-dimensional systems (p-GaAs, n-GaAs, p-SiGe, etc.).

As is true for many other interesting open questions in condensed matter physics, both interactions and disorder play a role, and their relative importance is unclear: (i) the transition from insulating to metallic temperature-dependence occurs at very low electron (hole) densities ($\approx 10^{11}$ cm$^{-2}$ or lower), where interaction energies are much larger than kinetic energies; (ii) the resistivity is on the order of $h/e^2$, suggesting that disorder plays a role.

A second important feature of these dilute 2D systems is their dramatic response to external magnetic fields applied parallel to the plane of the electrons [23]. As shown in Fig. 3.7, the resistivity increases by several orders of magnitude with increasing field and saturates to a constant plateau value above a density-dependent magnetic field on the order of 2 or 3 Tesla. The magnetoresistance

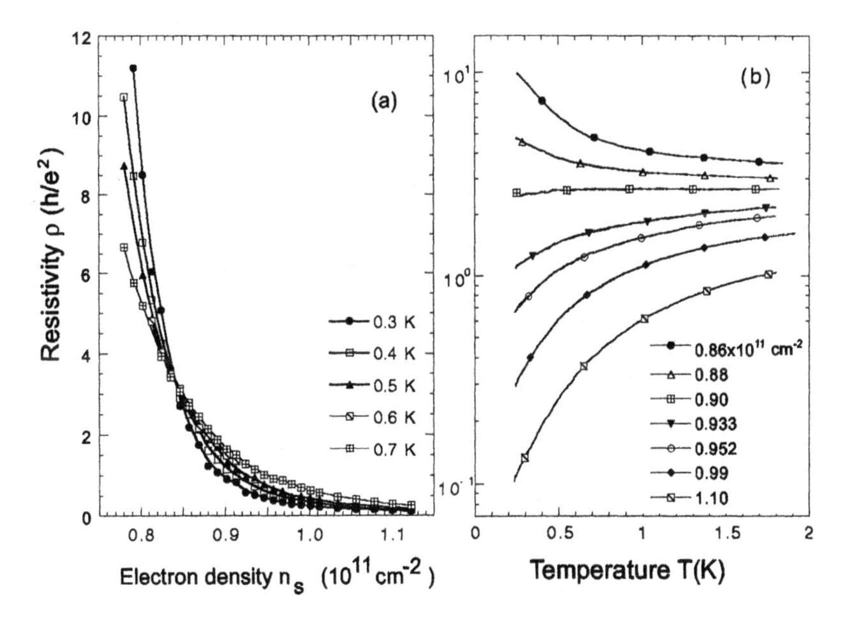


Figure 3.6: (a) Resistivity as a function of electron density for the two-dimensional system of electrons in a high-mobility silicon MOSFET; different curves correspond to different temperatures. (b) Resistivity as a function of temperature; here different curves are for different electron densities.

is larger at lower temperatures and in higher mobility samples. It should be noted that the curves of Fig. 3.7 span densities that have both metallic and insulating temperature dependence in zero field. The dramatic field-dependence thus appears to be a general feature of these dilute two-dimensional electron systems that is distinct from the temperature-dependence.

A lively debate has ensued concerning the significance of these findings: whether they represent fundamentally new physics or whether they can be explained by an extension of physics that is already understood. A view held by many is that these features signal a true zero-temperature quantum phase transition to a novel ground state at $T = 0$ (such as a "perfect" metal, a superconductor, a spin liquid, a Wigner glass, etc.) [21]. Others argue that the anomalous metallic behavior can be explained within a single-particle description in terms of a temperature-dependent Drude conductivity, and that "metallic" behavior is observed in a restricted range of temperatures so that localization prevails in the limit of zero temperature. Suggestions include scattering at charged traps, temperature-dependent screening, interband scattering, and thermal smearing of a percolation theshold [21].

The high mobilities that are now attainable in MOSFETs and heterostructures have opened a new area of investigation in low density 2D electron (hole) systems

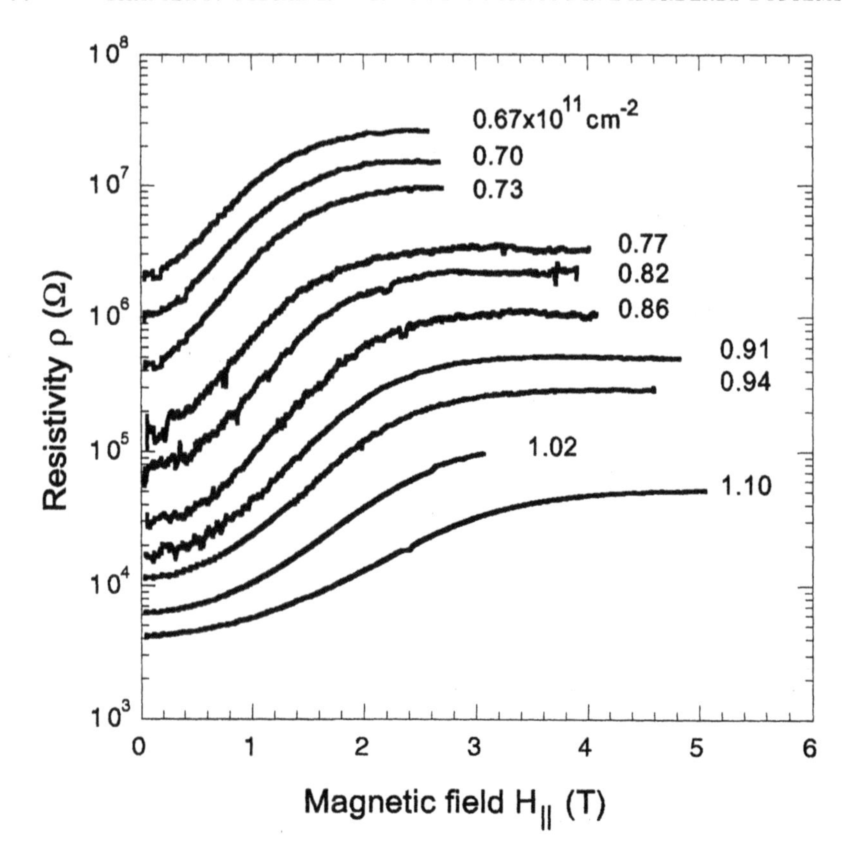


Figure 3.7: For different electron densities, the resistivity of a silicon MOSFET at 0.3 K as a function of magnetic field applied parallel to the plane of the 2D electron system. The top three curves are insulating in zero field while the lower curves are conducting.

where interactions are very strong. Regardless of how many-body effects will be incorporated into a full description in this regime, the behavior of these materials and the physics that is emerging are new and fascinating. My closing remark as an experimentalist is that much more experimental information is needed before serious progress can be made.

## ACKNOWLEGMENT

I thank the US Department of Energy for support under grant No. DE-FG02-84-ER45153.

## BIBLIOGRAPHY

[1] N. F. Mott, J. Non-Cryst. Solids **1**, 1 (1968).

[2] E. Abrahams, P. W. Anderson, D. C. Licciardello, and T. V. Ramakrishnan, Phys. Rev. Lett. **42**, 673 (1979).

[3] For a review and references, see P. A. Lee and T. V. Ramakrishnan, Rev. Mod. Phys. **57**, 287 (1985).

[4] W. L. McMillan and J. Mochel, Phys. Rev. Lett. **46**, 556 (1981).

[5] Hertel, D. J. Bishop, E. G. Spencer, J. M. Rowell and R. C. Dynes, Phys. Rev. Lett. **50**, 743 (1983); D. J. Bishop, E. G. Spencer, and R. C. Dynes, Solid State Electronics **28**, 73 (1985).

[6] M. A. Paalanen, T. F. Rosenbaum, G. A. Thomas, and R. N. Bhatt, Phys. Rev. Lett. **48**, 1284 (1982); G. A. Thomas, M. A. Paalanen, and T. F. Rosenbaum, Phys. Rev. B **27**, 3897 (1983).

[7] A. Mobius, *et al.*, J. Phys. C: Solid State Phys. **18**, 3337 (1985).

[8] J. G. Massey and Mark Lee, Phys. Rev. Lett. **75**, 4266 (1995); Phys. Rev. Lett. **77**, 3399 (1996).

[9] F. Wegner, Z. Phys. B **25**, 327 (1976).

[10] J. Chayes, L. Chayes, D. S. Fisher and T. Spencer, Phys. Rev. Lett. **57**, 2999 (1987).

[11] H. Stupp, M. Hornung, M. Lakner, O. Madel, and H. v. Lohneysen, Phys. Rev. Lett. **71**, 2634 (1993).

[12] See the comment by T. F. Rosenbaum, G. A. Thomas, and M. A. Paalanen, Phys. Rev. Lett. **72**, 2121 (1994), and the reply by H. Stupp, M. Hornung, M. Lakner, O. Madel, and H. v. Lohneysen, Phys. Rev. Lett. **72**, 2122 (1994).

[13] It is important to note that the data of the Bell group shown in Fig. 3.2 and of the Karlsruhe group shown in Fig. 3.3 differ substantially in numerical detail. This points to a difference between the two experiments associated with the samples, the stress, the method of measurement, or some other factor.

[14] K. M. Itoh, E. E. Haller, *et al.*, Phys. Rev. Lett. **77**, 4058 (1996).

[15] T. G. Castner, Phys. Rev. B **52**, 12 434 (1995).

[16] For reviews, see D. Belitz and T. A. Kirkpatick, Rev. Mod. Phys. **66**, 621 (1994); S. L. Sondhi, S. M. Girvin, J. P. Carini and D. Shahar, Rev. Mod. Phys. **69**, 315 (1997).

[17] D. Belitz and T. R. Kirkpatrick, Phys. Rev. B **52**, 13 922 (1995).

[18] S. Bogdanovich, M. P. Sarachik, and R. N. Bhatt, Phys. Rev. Lett. **82**, 137 (1999); Ann. Phys. (Leipzig) **8**, 639 (1999).

[19] S. Waffenschmidt, C. Pfleiderer, and H. v. Lohneysen, Phys. Rev. Lett. **83**, 3005 (1999).

[20] K. M. Itoh, M. Watanabe, Y. Ootuka, and E. E. Haller, Ann. Phys. (Leipzig) **8**, 631 (1999).

[21] For reviews and references see E. Abrahams, S. V. Kravchenko, and M. P. Sarachik, preprint cond-mat/0006055 (2000), to be published in Rev. Mod. Phys.; M. P. Sarachik and S. V. Kravchenko, Proc. Natl. Acad. Sci. USA **96**, 5900 (1999); Phys. Stat. Sol. **218**, 237 (2000).

[22] Kravchenko, S. V., Kravchenko, G. V., Furneaux, J. E., Pudalov, V. M. and D'Iorio, M. (1994) *Phys. Rev. B* **50**, 8039-8042; S. V. Kravchenko, W. E. Mason, G. E. Bowker, J. E. Furneaux, V. M. Pudalov, and M. D'Iorio, Phys. Rev. B **51**, 7038 (1996).

[23] D. Simonian, S. V. Kravchenko, M. P. Sarachik, and V. M. Pudalov, Phys. Rev. Lett. **79**, 2304 (1997); V. M. Pudalov, G. Brunthaler, A. Prinz, and G. Bauer, JETP Lett. **65**, 932 (1997).

## CHAPTER 4

# THE NATURE OF SUPERFLUID $^3$HE IN SILICA AEROGEL

D. D. OSHEROFF, B. I. BARKER[1], AND Y. LEE

Department of Physics, Stanford University
Stanford, CA 94305-4060

ABSTRACT

New NMR studies on liquid $^3$He confined in 98.2% porosity silica aerogel show that below a transition temperature which is suppressed below that of the bulk, the superfluid first forms in an equal spin pairing $A$-like state and then undergoes a first order phase transition to a $B$-like state at lower temperatures. The equilibrium $AB$ transition temperature at all pressures from 12 to 32 bars is within 0.97 $T_c$, but upon cooling the metastable $A$ phase can exist to temperatures as low as 0.75 $T_c$ at 32 bars. At the $AB$ transition on cooling, an abrupt drop in the liquid magnetization and rise in the NMR frequency are simultaneously observed. Values of the $B$-phase magnetization as well as estimates of the suppressed $A$ and $B$ phase energy gaps are presented.

## 4.1 INTRODUCTION

Liquid $^3$He in its normal state at ultra-low temperatures is perhaps the purest and most isotropic fluid known. Yet below about 2.5 mK it supports three anisotropic BCS states in which the atoms themselves form Cooper pairs. These are known to be $p$-wave states, and were the first unconventional BCS states discovered. While the degenerate Fermi fluid is highly isotropic and does not reside in a background lattice to break this high degree of symmetry, the ordered phases exhibit varying degrees of anisotropy. The energy gap of the high temperature $A$ phase actually vanishes along the $\mathbf{l}$ axis, which is parallel to the angular momenta of all the

[1]* Present Address: Department of Physics; University of California; Berkeley, CA 94720-7300. Email: bbarker@socrates.berkeley.edu

local Cooper pairs. The $B$ phase by contrast has an isotropic energy gap, but is characterized by a rotation of the spin coordinate system with respect to the orbital coordinate system about an arbitrary axis $\mathbf{n}$ by an angle $\cos^{-1}(-1/4) \simeq 104°$, necessary to minimize the dipole-dipole energy of the atoms in the Cooper pairs [1].

The difference in free energy between the $A$ and $B$ superfluid phases varies with sample pressure, temperature, and magnetic field. In zero field the $B$ phase is the only stable phase below about 21.5 bars pressure. Above this pressure, a wedge of $A$ phase is stable near $T_c$ which grows in width from zero at 21.5 bars to about 0.2 $T_c$ at the melting pressure ~34.4 bars. Because the $A$ phase has a higher susceptibility than the $B$ phase, magnetic fields tend to lower the free energy of the $A$ phase with respect to the $B$ phase, making the $A$ phase stable at $T_c$ at all pressures, and to lower temperatures in general. One can readily differentiate between these two states using NMR techniques, since the $A$ phase is an equal-spin-pairing state and has a nearly constant magnetic susceptibility approximately equal to that of the Fermi liquid, while Cooper pairs in the $B$ phase consist of all three spin triplet combinations and thus the $B$ phase has a susceptibility less than the Fermi liquid. In addition, both phases exhibit NMR frequency shifts which are state dependent and depend upon the direction of the relevant anisotropy axes with respect to an applied magnetic field. Because these anisotropy axes are not tied to any intrinsic symmetry of a host lattice, the superfluids readily exhibit liquid crystal like textures resulting from competing effects of surfaces and both magnetic and flow fields to orient the anisotropy axes.

The liquid crystal-like textures and unusual spin dynamical modes of the $^3$He superfluid phases are essentially unique to these fluids, however the depairing of these superfluids by non-magnetic impurities should be similar to that which must occur in other non-$s$-wave BCS states such as the high $T_c$ and heavy fermion superconductors. While it is impossible to dissolve atomic or molecular impurities in superfluid $^3$He, it has been shown that one can mimic the effects of impurities by placing $^3$He in a low density silica aerogel. Most studies have been carried out in 98.2% porosity aerogel. Here the gel strands have effective diameters of about 5 nm, compared with the zero temperature coherence length $\xi_0$, which varies from 20 nm near the melting pressure (34.4 bars) to 80 nm at zero pressure. Thus, the strands are too small for the superfluid textures to adjust to their presence, and if randomly oriented and distributed, should mimic the depairing effects of dissolved impurities. However, the response of the $^3$He superfluids to depairing by random impurities should be richer than that which can occur in the superconductors, because of the absence of a background lattice symmetry and the existence of multiple stable phases.

Following the application of aerogel to studies of helium by Chan *et al.* [2], Porto and Parpia [3] where the first to show that superfluidity in $^3$He could exist within 98% porosity aerogel. They used torsional oscillator techniques to measure the superfluid density and $T_c$ as a function of sample pressure, and did much work to better characterize their aerogel samples over the length scales relevant to superfluid $^3$He [4]. These measurements were not able to determine the nature of the stable phase, however, nor to see any evidence for a transition between superfluid phases. Sprague *et al.* [5, 6] then employed pulsed NMR to probe the nature of superfluid $^3$He in aerogel. Their conclusion was that, provided the aerogel surfaces were coated with localized $^3$He atoms, the only superfluid found at any pressure was an equal-spin-pairing $A$-like phase. However, when they replaced the two layers of localized $^3$He atoms adsorbed on the aerogel surfaces by adding 3.4 atomic layers of $^4$He, some of which must have been superfluid, they saw only a $B$-like phase [7]. We question the validity of the $A$-like phase identification, and revisit this issue near the end of this paper.

We have carried out two recent NMR studies of superfluid $^3$He in aerogel using continuous wave NMR techniques which are able to record more subtle aspects of the NMR absorption spectra and can do so with less disruption to the superfluids. In addition, we have made more systematic studies as a function of the nature of the aerogel surface adsorbate, including two localized $^3$He layers, one layer of localized $^3$He and one layer of localized $^4$He, and two mono-layers of localized $^4$He. The $^4$He is more tightly bound to the aerogel surfaces than is $^3$He, and hence $^4$He will remove the first and then both localized layers of $^3$He atoms. The $^3$He spins localized on the aerogel surfaces exhibit a Curie-Weiss magnetization with a positive Weiss constant of about 0.4 mK. Because of the large surface area of the aerogel, these localized spins possess a magnetization that exceeds that of the liquid by about a factor of two at 2 mK, rising to over a factor of seven at 1 mK. The presence of this large magnetization obscures the behavior of the superfluid in two ways. Sprague *et al.* [5] showed that exchange between the superfluid and localized layers is sufficient to produce a single NMR resonance in which the liquid frequency shift is considerably diminished by the magnetization of the localized spins:

$$\langle\omega\rangle = \frac{\langle\omega_\ell\rangle M_\ell + \langle\omega_s\rangle M_s}{M_\ell + M_s}. \tag{4.1}$$

Here the resonant frequency of the liquid and localized spins are $\omega_\ell$ and $\omega_s$, respectively. The liquid and solid magnetizations are $M_\ell$ and $M_s$, respectively. The magnetization of the localized spins changes so rapidly below the superfluid $T_c$ that it can mask any decrease in the magnetization of the superfluid, particularly if it were smaller than that found in the bulk. While one usually characterizes the temperature dependence of this magnetization as being Curie-Weiss in nature, this is only an approximation which is not very good for temperatures approaching the Weiss constant.

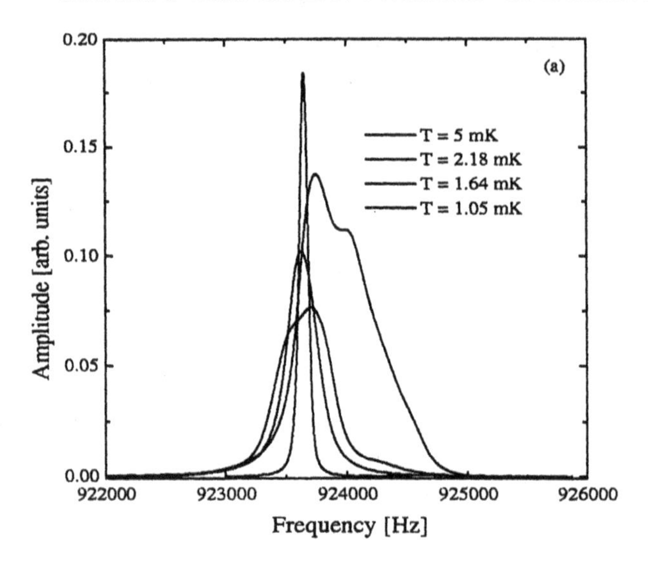


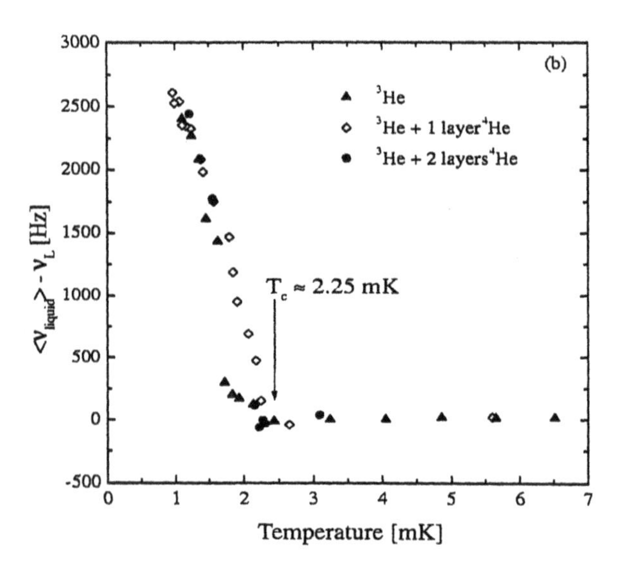


Figure 4.1: (a) Typical $^3$He NMR spectra from our first aerogel sample at the temperatures indicated. $T_c$ for this sample was 2.20 mK. Broader NMR traces correspond to lower the temperatures. Notice the abrupt change in the character of the spectra between 1.64 mK and 1.05 mK. (b) Typical average frequency shifts for superfluid $^3$He in aerogel sample 1 with various aerogel surface adsorbates, on cooling (solid symbols) and warming (open symbols). Notice that the frequency shifts below 1.6 mK all agree.

## 4.2 RESULTS OF THE FIRST STUDY

In the first of our studies [8], we employed a sample of aerogel grown by a rapid supercritical extraction process, rather different from the more conventional process in which the gelation is allowed to occur slowly after the precipitation of the silica from the precursor fluid [9]. The results of this study, all at 32 bar sample pressure and in a magnetic field of 28.2 mT, can be summarized as follows:

(1) As the sample is cooled slowly, there is a sharp onset temperature $T_c$ = 2.2 mK at which the NMR signal of the $^3$He in the aerogel begins to broaden rather dramatically from about 140 Hz (full width half maximum) just above $T_c$ to about 500 Hz at 1.64 mK. The average frequency shift at 1.64 mK was small compared to the line broadening. We assumed that the broadening was inhomogeneous, and resulted from an $A$-like superfluid in which the direction **l** was somehow pinned to inhomogeneities in the aerogel.

(2) In a range of temperatures between about 1.65 mK and 1.55 mK, and over a narrow temperature interval perhaps a 20 $\mu$K wide, the shape of the NMR spectrum changed dramatically to one with a substantially positive net frequency shift, and with almost none of the NMR signal below the Larmor frequency. While the peak of the NMR line was still close to the Larmor frequency, there was a long high frequency tail reminiscent of $B$ phase spectra in 'bulk' samples caused by competing orienting effects of the rotation axis by surfaces and the magnetic field [10]. In the $B$ phase the NMR frequency is given by the equation

$$\omega^2(T) = \omega_L^2 + \sin^2(\theta) \cdot \Omega_B^2(T). \tag{4.2}$$

Here $\Omega_B$ is the temperature dependent longitudinal resonant frequency which is field independent, $\omega_L$ is the Larmor frequency, and $\theta$ is the angle between the rotation axis $\mathbf{n}$, and the magnetic field $\mathbf{B}$. In the bulk, $\mathbf{n}$ is oriented parallel to $\mathbf{B}$. Depairing effects due to surfaces and a magnetic field will be minimized when $\mathbf{n}$ is aligned so that a rotation about $\mathbf{n}$ by $\cos^{-1}(-1/4)$ will rotate the surface normal vector into the magnetic field. (This effect was first suggested by P.W. Anderson to W.F. Brinkman who had just given a physics colloquium at Cambridge University in 1973 [11].) The $A$ phase NMR frequency depends upon the orientation of **l** with respect to the magnetic field, however in the $A$ phase the textural healing length for **l** to assume its bulk orientation normal to $\mathbf{B}$ is much shorter than the healing length for $\mathbf{n}$. Typical spectra obtained as a function of temperature with two layers of localized $^3$He on the aerogel (pure $^3$He in the cell) are shown in Fig. 4.1a. The data without localized $^3$He atoms on the aerogel were similar but the frequency shifts were larger owing the lower magnetization in the

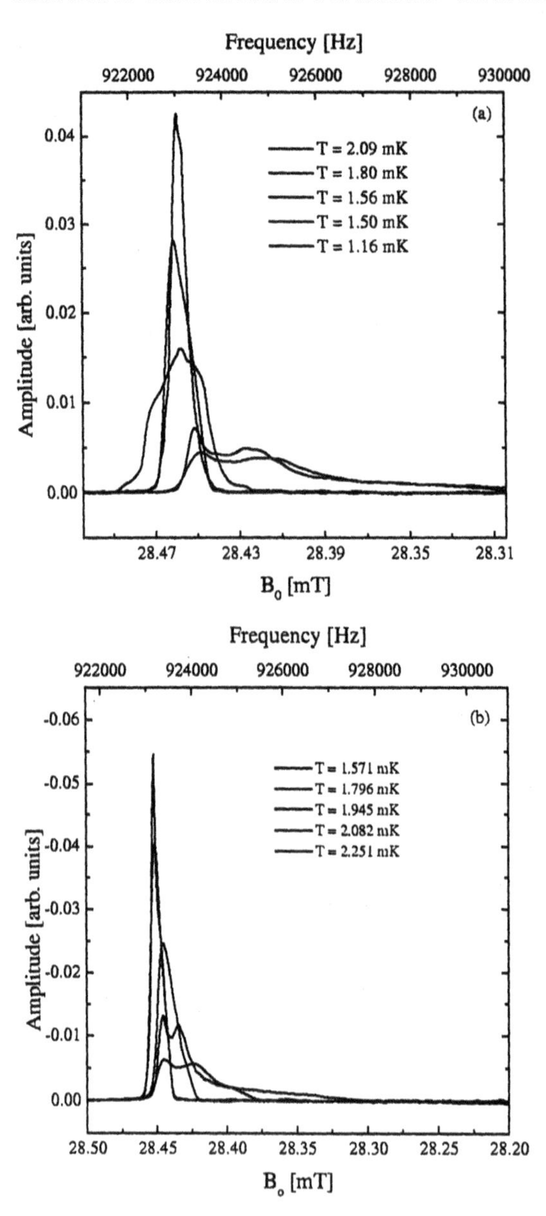


Figure 4.2: (a) Typical NMR spectra upon cooling for $^3$He in aerogel sample 2 with no localized $^3$He on surfaces. The lower the peak the lower the temperature. Notice the abrupt change in lineshape between 1.56 mK and 1.50 mK. (b) Typical NMR spectra upon warming for $^3$He in aerogel sample 2 with no localized $^3$He on the aerogel surfaces. The higher the peak, the higher the temperature. Notice the continuous variation upon warming.

denominator of Eq. 4.1. Such data from our second sample are shown in Fig. 4.2a and will be discussed below.

At this time, we did not interpret the sudden change in the NMR spectrum as evidence for a transition between different superfluid phases, but rather a depinning of the $A$ phase **l** vector from whatever weak pinning the aerogel might provide due to its inhomogeneities on length scales large compared to the coherence length. Magnetization measurements made with varying amounts of $^4$He covering the aerogel surfaces were inconclusive owing to tiny instabilities in the spectrometer, and were not interpreted as evidence for the existence of a $B$ phase. The NMR spectra upon cooling and warming showed a great deal of hysteresis, with the broad $A$ phase NMR line seen only upon cooling.

(3) The average frequency shifts in the liquid below 1.6 mK (after correcting for the presence of the localized $^3$He spins using Eq. 4.1) were found to be independent of the number of layers of localized $^3$He atoms adsorbed on the aerogel, as is shown in Fig. 4.1b. This suggested that the nature of the superfluid state did not depend in any way on the presence of localized $^3$He spins on the aerogel surfaces, in contrast with the results by Sprague *et al.* [7]. This was true even though the actual average frequency shifts observed depended dramatically on the magnetization of the localized $^3$He spins. It should be noted that most of the results of Sprague *et al.* were taken in a field of 111.7 mT, not the 28.2 mT field we used in this study, although in our second study we found that the behavior we discovered was similar at the higher field.

(4) At about 1 mK, it was found that the shape of the NMR spectrum depended upon the orientation of **H** in the plane normal to the axis of the cylindrical aerogel sample. This dependence did not depend upon the orientation of the field in which the sample was cooled, and suggested that the aerogel sample was not isotropic even on length scales of order 1 mm.

Once we found the dependence of the NMR spectrum upon magnetic field orientation in the plane normal to the symmetry axis of our sample, we became concerned that our aerogel sample was anomalous, owing to the differences in growth techniques employed. Rather than continuing measurements on this sample, we chose to make a second cell using an aerogel sample for which gelation had occurred at least sufficiently prior to hypercritical extraction of the remaining fluid to prevent convection during the extraction process. In addition, our second sample was shorter, resulting in significantly faster thermal equilibrium times. Both samples had a nominal porosity of 98.2%.

## 4.3 RESULTS OF THE SECOND STUDY

We soon found that the dependence of the NMR spectra upon field orientation was also exhibited by the second sample. It is possible that all aerogel, perhaps due to convection resulting from heat generation during silica precipitation, is slightly anisotropic. In addition, the broadening of the line observed upon cooling through $T_c$ was also observed in the second sample, although both $T_c$ and the temperature at which the abrupt change in lineshape were seen were slightly different; 2.1 mK vs. 2.2 mK and 1.5 mK vs 1.6 mK. The nominal porosities of the two samples were the same. This magnitude of variation in $T_c$ for nominally identical aerogel densities is typical.

For this sample, more detailed studies were made. Upon cooling below $T_c$ to about 1.5 mK we found that the average negative frequency shift grew linearly with $\tau = 1 - T/T_c$. By 1.7 mK and with almost no localized $^3$He on the aerogel, the average shift was -60 Hz. The sign of this small average frequency shift did vary with the rotation of the magnetic field normal to the axis of the aerogel sample. Generally, the behavior of the NMR spectra in this temperature range was quite similar to that found in our earlier aerogel sample, including the strong hysteresis seen above 1.5 mK upon warming and cooling. NMR spectra obtained with no localized $^3$He atoms on the aerogel for warming and cooling of the second sample showing this hysteresis are presented in Figs. 4.2a and 4.2b.

The key to recognizing that the changes at 1.5 mK were caused by a phase transition from an $A$-like phase to a $B$-like phase upon cooling was the measurement of the magnetization of the sample after replacing nearly all the $^3$He atoms localized on the aerogel surfaces with $^4$He atoms. Evidently there remained a small fraction of a monolayer of localized $^3$He, for the magnetization above $T_c$ obeyed the expression $M_{3He} = 5.06/T + 10.28$, where $T$ is in mK. This temperature dependent component was subtracted from the measured values below $T_c$ to obtain the magnetization of the liquid. Owing to the very broad NMR linewidths in this limit, integrating the NMR absorption signal was a difficult matter, and considerable time had to be spent to better stabilize the spectrometer. Ultimately, we obtained the magnetization and average frequency shift upon cooling shown in Fig. 4.3a. Note that the magnetization drops abruptly at about 1.6 mK, just as the average frequency shift jumps from a negative value to a positive value. This is precisely what one would expect if the sample converted from an $A$-like equal-spin-pairing phase to a $B$-like phase. However, upon warming, both the liquid magnetization and the frequency shift were continuous. Extrapolation to zero of the frequency shifts in Fig. 4.3b suggest $T_{AB}$ = 1.98 mK, while $T_c$ = 2.05 mK was established from observing the point at which the NMR spectra stop changing upon slow warming. Note that one cannot extrapolate the $B$ phase frequency shifts to zero upon warming to establish $T_c$ . Thus $\tau_{AB} \simeq 0.03$, compared to its bulk value at this pressure of 0.20. This suggests that the presence of the aerogel tends to stabilize the isotropic $B$ phase, as predicted by Thuneberg *et al.* [13].

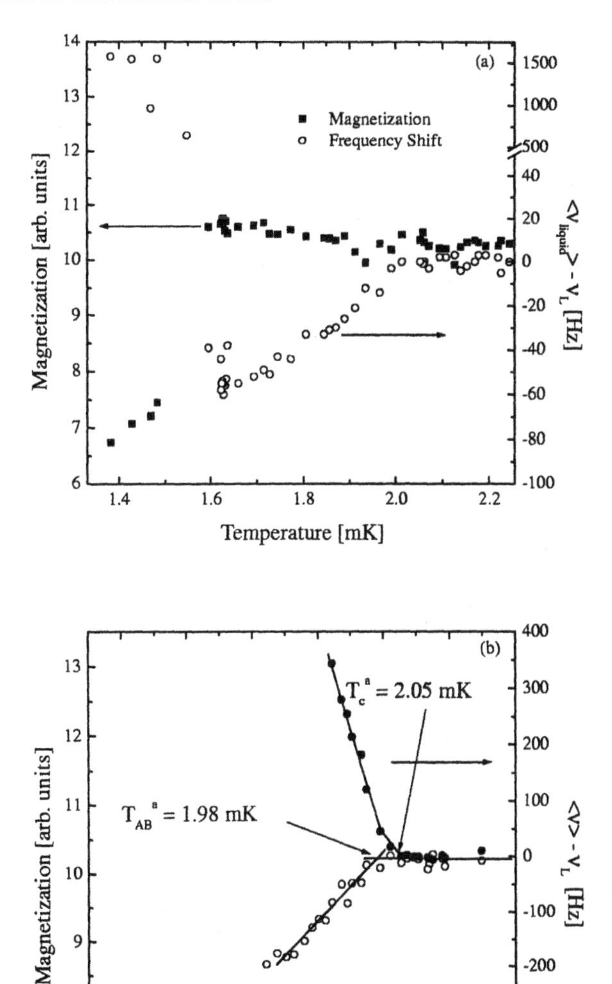


Figure 4.3: (a) Magnetization and frequency shift data for aerogel sample 2 with no localized $^3$He component upon cooling. Notice the simultaneous jump in the frequency and drop in the magnetization at about 1.6 mK, signaling a first order phase transition. (b) Magnetization and frequency shift data for aerogel sample 2 with no localized $^3$He component upon warming. Notice that the frequency shifts are now all positive, and the magnetization rises continuously up to $T_c$. Thus the transition at 1.6 mK must be supercooled.

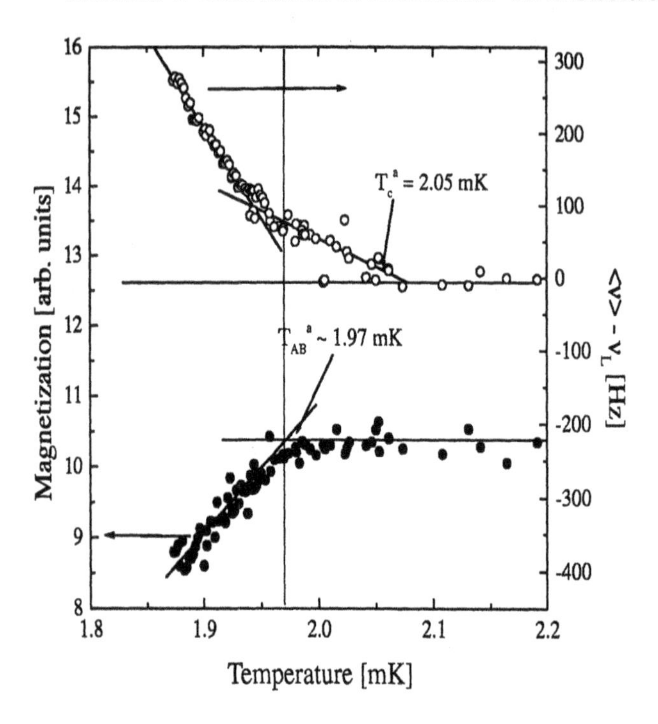


Figure 4.4: Magnetization and frequency shift data for aerogel sample 2 with no localized $^3$He component upon warming in a field of 113.6 mK. This is similar to the data in Fig. 4.3b, but now we can roughly measure the slope of the frequency shifts in the $A$ phase region near $T_c$.

To better observe the $A$ phase upon warming, we also made measurements in a field of 113.6 mT. Those data are shown in Fig. 4.4. Here we found $T_c$ = 2.06 mK upon extrapolation of the $A$-phase frequency shifts to zero. From extrapolation of the magnetization to the Fermi liquid value, $T_{AB}$ = 1.98 mK, and by determination of the vertex in the frequency shifts, $T_{AB}$ = 1.95 mK, suggesting that $\tau_{AB} \simeq 0.045$. We were also able to measure the slope of the frequency shift near $T_c$ in the $A$ phase (with limited accuracy). We found that $d(\nu_L^2)/dt \simeq 1.7 \times 10^{10}\text{Hz}^2$. This value should be proportional to $d\Delta^2/d\tau$, where $\Delta$ is the BCS energy gap. For bulk $A$ liquid at 32 bars Ahonen *et al.* [12] found $d(\nu_L^2)/dt \simeq 4.8 \times 10^{10}$ Hz$^2$. Thus, it appears that the $A$-phase gap in our aerogel is 0.50 $\Delta_{A-bulk}$.

There is a problem with the obvious assumption made above that one can obtain $\Omega_L^2(T)$ from the average frequency shift even for A phase in aerogel. This certainly is not true for the spectra we observe upon cooling. The frequency shift in the $A$ phase varies as $\sim \cos(2\phi)$, where $\phi$ is the average angle between $\mathbf{l}$ and the spin anisotropy axis $\mathbf{d}$ (the direction along which $m_s = 0$). Generally, $\mathbf{d}$ will be oriented normal to the magnetic field, and **l** will be either parallel or antiparallel to **d** in the bulk liquid. If one were to orient **l** randomly throughout the $^3$He sample, the average frequency shift would be minus one third of the maximum positive shift when **l** and **d** are parallel. To have zero average shift in a sample, with **d** perpendicular to the magnetic field, **l** would have to be uniformly distributed with $\phi$ varying from 0 to almost 70°. It seems difficult to understand how **l** and **d** could be parallel everywhere upon warming and so badly mis-oriented upon cooling. If there are indeed strong pinning sites in the aerogel for **l**, one would expect them to be important both on cooling and warming. Another possibility is that the sample goes through $T_c$ independently in numerous unconnected regions upon cooling, resulting in a myriad of topological singularities in the **l** texture. Upon warming, however, the orientation of **l** at the $A-B$ interface forms a template that prevents such topological singularities.

It is clearly desirable to obtain an independent estimate of the energy gap in our experiments. One can do so from an estimate of $\Omega_L^2(T)$ in the $B$ phase using the NMR spectra. However, we know that **n** is unlikely to be always perpendicular to the magnetic field. Indeed, the bulk orientation is parallel to the field, giving no frequency shift. If one carefully looks at NMR spectra in geometries confined on the mm distance scale, however, one usually finds a broad spectrum that does extend up to the shift indicated by $\Omega_L^2(T)$ for the $B$ phase. Figure 4.5a shows these values for our sample, compared to values calculated for 'bulk' $B$ phase from $\Omega_L^2(T)$, assuming **n** is normal to the magnetic field. Since $\Omega_L^2(T)$ in the $B$ phase depends inversely upon the $B$-phase susceptibility, which is likely to be different for $^3$He in aerogel than in bulk, we must use measured values of the susceptibility to determine the gap from $\Omega_L^2(T)$. For the second sample, typical susceptibility data are shown in Fig. 4.5b. Here we have corrected for the small temperature dependent magnetization measured above $T_c$ for a sample with nominally no localized $^3$He present. Notice that the decrease in the magnetization is only about half of what is seen in the bulk as a function of $\tau$. From these values, and the maximum frequency shifts we observed in the $B$-phase NMR spectra upon warming (shown in Fig. 4.5a), we find that the average $B$-phase gap at 32 bars is about 0.55 $\Delta_{B-bulk}$. This should be regarded as a lower limit to the gap, since we cannot insure that at some point in the NMR coil **n** was actually perpendicular to **B**. The fact that the two determinations of the gap agree fairly well suggests that the **l**-vector in the $A$ phase, upon warming, must be oriented nearly normal to the magnetic field.

One final question regarding our data is how the $A$ phase can supercool so far when the equilibrium $T_{AB}$ is so close to $T_c$. It is true that one can pin the $AB$ interface in the bulk nearly to $T=0$ by forcing it to propagate through an orifice of order 100 nm in diameter. The pinning of the $AB$ interface by an aerogel strand should be much weaker than that by a solid surface, but we have actually observed bulk $B$ phase outside our aerogel samples with supercooled $A$ phase in the aerogel near 1.6 mK.

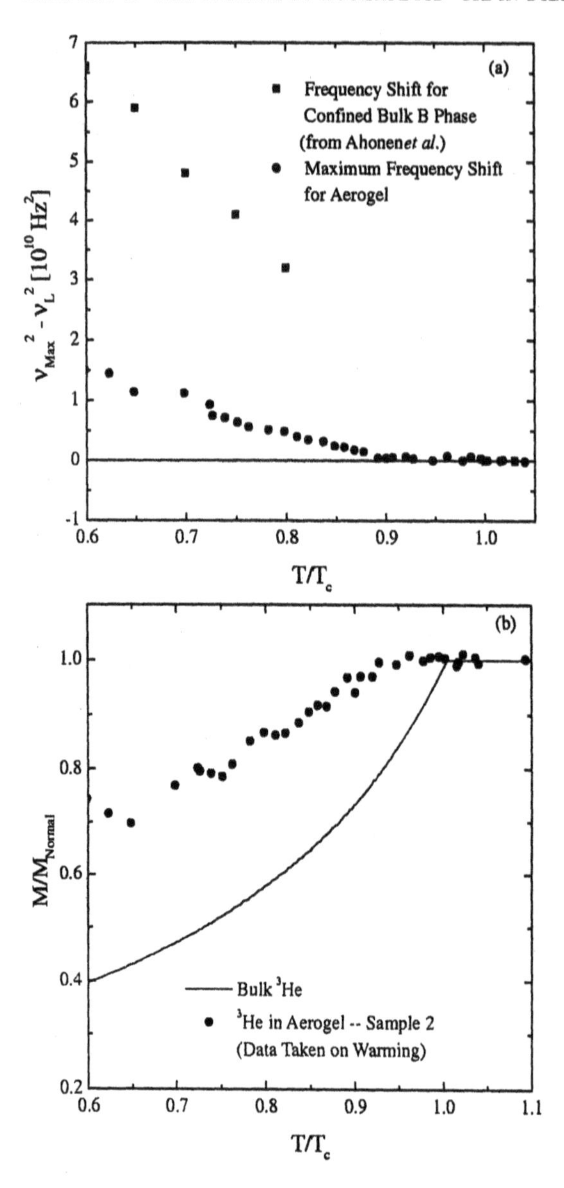


Figure 4.5: (a) Frequency shifts for the $B$ liquid in aerogel sample 2. The squares are data from bulk $B$ phase at about the same pressure (ref. 12). (b) Magnetization for the $B$ liquid in aerogel sample 2 measured upon warming. The solid line is the magnetization calculated for the $B$ phase using $F_o^a$ = -0.80, which fits existing data well below the bulk $T_{AB}$. The high value of $F_o^a$ compensates for the strong coupling effects, and the actual value is -0.75.

We have also taken data upon cooling at a number of pressures as low as 12 bars. In each case there was some evidence for a negative frequency shift just below $T_c$ , which can be interpreted as evidence for a narrow $A$ phase region. At 24 bars the evidence was much stronger, including with a clear transition from a very broad NMR line with both negative and positive shifts to a $B$-like NMR spectrum. At 16 bars and 12 bars only a small portion of the NMR spectral weight shifted to lower frequencies, and never below the lowest frequency where the Fermi liquid signal had spectral weight.

## 4.4 COMPARISON WITH PREVIOUS RESULTS

To date, ours is the only observation of an $A - B$ transition in superfluid $^3$He in aerogel to be reported. However, in the middle of our studies and quite independently, Alles *et al.* [14] carried out a simultaneous torsional oscillator and NMR study of $^3$He in both 98 % and 99 % porosity aerogel. The NMR was carried out in a field of only 5.1 mT, and the signal to noise was sufficiently low that those authors could not afford to remove the localized $^3$He atoms from the aerogel surfaces. However, from the shape of their NMR spectra they concluded that the phase within their aerogel had to be $B$-like at all sample pressures from about 5 to 29 bars in 98 % and 1.5 to 29 bars in 99 % aerogel. Although there was a large contribution to the magnetization coming from the localized $^3$He atoms, one can see a small feature in the magnetization vs. temperature suggesting a drop in the liquid susceptibility. In addition, the group found in the torsional oscillator experiments that $\rho_s$ did not depend upon the external magnetic field, suggesting a state with an isotropic gap. The measurements of $\rho_s$ would also have been independent of magnetic field, however, if there were an $A$-like phase present but with $\mathbf{l}$ distributed nearly isotropically as appears to be the case in our data upon cooling. Thus, if Alles *et al.* had measured samples like ours, they would not have seen an abrupt change in $\rho_s$ upon cooling because of the distribution of $\mathbf{l}$ throughout the sample, and on warming the region over which an $A$ phase would be stable would be so small that they probably would not have been able to see the change in $\rho_s$ on warming either.

The most difficult data to understand in light of our results and those by Alles *et al.* are the original studies by Sprague *et al.* [5, 6, 7]. While these studies were mostly carried out at about 112 mT, we see no evidence of significant differences in the behavior of the superfluid in our aerogel data between 28.2 mT and 111 mT. Sprague *et al.* [7] did, however, find clear evidence for a $B$-like phase in their experiment when they replaced the localized $^3$He atoms with $^4$He atoms.

It is important to know if the behavior of $^3$He in aerogel depends upon subtle variations in the aerogel which are not controlled in the growth process, or whether the identification by Sprague *et al.* with pure $^3$He in their cell was in error. There are a number of reasons to suspect that this might have been the case, but we cannot explain all their results on this basis. For example, these authors studied the average NMR shift as a function of $\phi$, the tipping angle [6]. For $A$ phase in the bulk, one finds a frequency shift that varies as $1/4 + 3/4 \cos\phi$. The authors report data consistent with this up to a tipping angle of only about $40^o$, with the frequency shift dropping rapidly to zero for larger tipping angles. According to Fomin [15], this is precisely what one would expect for a $B$-phase sample in which the $\mathbf{n}$ axis is not initially aligned parallel to the applied magnetic field. Sprague *et al.* also measured the average frequency shift as a function of magnetic field [7]. In the $A$ phase one expects that $\omega^2_{Liquid} - \omega^2_{Larmor}$ will be independent of magnetic field. However, these authors report that this quantity saturates at low temperatures at a lower value for higher magnetic fields. While this might result from heating due to vibrations in the magnetic field, the authors do not suggest this as a possibility. On the other hand, in $B$-phase textures it is a well-known result that $\omega^2_{Liquid} - \omega^2_{Larmor}$ decreases as the field increases. This occurs because the force tending to align $\mathbf{n}$ parallel to $\mathbf{B}$ in the bulk increases with increasing magnetic field, but the gradient energies do not.

Sprague *et al.* do report the presence of a negative frequency shift near $T_c$ in their NMR spectra, which appears superficially to be consistent with our data. They took their data only upon warming, however, where we saw no negative frequency shift. It might be possible that an occasional large angle NMR pulse could have perturbed their sample sufficiently to disperse the $\mathbf{l}$ texture, but we certainly have no way of knowing if this was the case. These authors attributed the small negative shift as coming from a small region of bulk $^3$He that coupled to their NMR coil, and defined $T_c$ for the $^3$He in the aerogel as being the onset of a positive frequency shift. They report a magnetic field dependence to this temperature, which is not seen at the bulk $A$-phase transition, and attribute it to interactions with the localized $^3$He atoms. While this is plausible, one generally finds that $T_{AB}$ depends quadratically upon applied magnetic field, in a manner similar to the field dependence to the transition reported. This suggests that the phase which Sprague *et al.* identified as the $A$ phase might indeed be the $B$ phase.

While the above arguments tend to support the conclusion that both experiments measured primarily $B$-phase NMR spectra upon warming, there are serious problems with this conclusion. We have measured values of $\omega(T)^2 - \omega_L^2$ in our samples at low temperatures that were consistent with those reported by Sprague *et al.* One would not expect this to be the case for $B$-phase samples, since the different magnetic fields used in the two experiments should have produced different $\mathbf{n}$ textures in the two experiments. For similar reasons, one would not expect the average frequency shift in the earlier experiments to be independent of magnetic field at intermediate temperatures, although this was indeed the case. We thus cannot suggest with any certainty that the very similar behavior that we have seen in two very different aerogel samples is in fact completely general.

## 4.5 CONCLUSIONS

There is no doubt that we have indeed seen an $A$ to $B$ phase transition in $^3$He confined in high porosity aerogel. The fact that we have actually seen very similar behavior in two aerogel samples grown by very different processes, over an extended range of sample pressures, and independent of the magnetic field and presence or absence of localized $^3$He atoms on the aerogel all suggests that this behavior is quite general. In addition, we have observed a very strange NMR signature of the $A$ phase upon cooling, which seems to suggest either **l**-domains or **l**-pinning within the aerogel, although a nearly bulk-like $A$ phase signal upon warming. This behavior suggests that numerous regions of low aerogel strand density may pass through $T_c$ on cooling independently. We also observe a high degree of supercooling of the $A$-like phase, even though bulk $B$ phase exists outside our aerogel sample elsewhere in the cell. This indicates that the $AB$ interface becomes pinned in the aerogel, probably to regions of high aerogel strand density. All these observations are qualitatively new and the only observation which seems to be explained by existing theory is the one suggesting that the presence of aerogel tends to stabilize the $B$ phase over the $A$ phase [13].

These should certainly not be the last NMR studies of superfluid $^3$He in aerogel. For example, it would be instructive to measure the $B$ phase texture in a parallel plate geometry as a function of applied magnetic field. Not only should this give a much better estimate of the energy gap than any obtained to date, but one could also measure the bending energies for $^3$He in aerogel, which certainly will be less than the bulk values. In the background of all such studies, however, lurks the realization that the distribution of silica strands in aerogel is not isotropic nor the density uniform, as has been shown so clearly by Porto and Parpia [4]. Yet almost any property of the superfluid measured so far varies nearly linearly just below $T_c$, suggesting that the aerogel variations are on a sufficiently small distance scale as to be averaged over by the superfluid. We encourage more theorists to make contributions to our understanding of this fascinating system.

We wish to thank L.W. Hrubesh and J.F. Poco who provided us with our aerogel samples, and Mike Cross for an occasional refresher on superfluid spin dynamics. This work was supported by NSF grant number DMR-940950, DMR-9971694, and a grant from the Japanese NEDO Foundation. The aerogel samples provided by Hrubesh and Poco were grown at the Lawrence Livermore National Laboratory under U.S. DoE contract W-7405-ENG-48.

## BIBLIOGRAPHY

[1] See, for example, A. Leggett, Rev. Mod. Phys. **47**, 331, (1975), or D. Vollhardt and P. Wlfle, *The Superfluid Phases of Helium 3* (Taylor and Francis, London, 1990).

[2] M.H.W. Chan *et al.*, Phys. Rev. Lett. **61**, 1950 (1988).

[3] J. Porto and J. Parpia, Phys. Rev. Lett. **74**, 4667 (1995).

[4] J. Porto and J. Parpia, J. Low Temp. Phys. **101**, 397 (1995).

[5] D. T. Sprague *et al.*, Phys. Rev. Lett **75**, 661 (1995).

[6] D. T. Sprague *et al.*, J. Low Temp. Phys. **101**, 185 (1995).

[7] D. T. Sprague *et al.*, Phys. Rev. Lett. **77**, 4568 (1996).

[8] B. I. Barker *et al.*, J. Low Temp. Phys. **113**, 635 (1998).

[9] J. Poco, P. Coronado, R. Pekala, and L. Hrubesh, Mat. Res. Soc. Symposium Proc. **431**, 297 (1996). This describes the growth technique for our first sample.

[10] H. Smith, W. F. Brinkman, and S. Engelsberg, Phys. Rev. B **15**, 199 (1976).

[11] W.F. Brinkman, *private communication*, 1974.

[12] A. I. Ahonen, M. Krusius, and M.A. Paalanen, J. Low Temp. Phys., **25**, 421 (1976).

[13] E.V. Thuneberg, S.K. Yip, M. Fogelstrom, and J.A. Sauls, Phys. Rev. Lett. **80**, 2861 (1998).

[14] H. Alles *et al.*, Phys. Rev. Lett. **83**, 1367 (1999).

[15] Igor Fomin, *private communication during the ULT99 Symposium, St. Petersburg, Russia.*

# CHAPTER 5

# RVB DESCRIPTION OF HIGH-$T_c$ SUPERCONDUCTORS

PATRICK A. LEE
Department of Physics, Massachusetts Institute of Technology
Cambridge, MA 02139

## ABSTRACT

We review the development of the RVB idea, focusing on the underdoped region of the phase diagram. The pseudogap phenomenon is interpreted as a staggered flux phase and recent calculations of staggered current correlators based on projected wave functions and exact diagonalization are cited to support this point of view.

## 5.1 INTRODUCTION

In January 1987, a few months after the discovery of high-$T_c$ superconductivity in the cuprates, Phil Anderson developed the idea of the resonating valence bond (RVB) while he was attending a conference in Bangalore. Our gathering to pay tribute to Phil for his 50 years of leadership in condensed matter physics also marks the thirteenth anniversary of the RVB idea and it is fitting to reflect on the progress that has been made since that time.

In his *Science* paper, [1] Anderson made two important points. The first is to identify the high-$T_c$ problem as the problem of doping into a Mott insulator. This identification was made at a time when it was not known whether the parent compound was Néel ordered, or even if it should be considered a Mott insulator. Today there is general agreement on the correctness of this identification: superconductivity arises by doping an antiferromagnetic Mott insulator and the parameter space for a basic model is reasonably clear. A large segment of the community believes that a one-band Hubbard or $t$-$J$ model is a good starting point to study this phenomena. A second point Anderson made was that the anomalous properties of the normal state is part and parcel of the phenomenon of high-$T_c$ superconductivity and he introduced the notion of RBV as a liquid of singlets which becomes superconducting when the holes become phase coherent. Since that time many experiments, particularly transport and ARPES [2], have shown that the normal state is indeed anomalous. This is particularly apparent in the underdoped region, where the pseudogap phenomenon clearly shows that a radical departure from conventional Fermi liquid physics is called for.

## 5.2 REVIEW OF RVB THEORY

The RVB notion was put on a more formal footing by Basharan, Zou and Anderson. [3] They started with the $t$-$J$ model

$$H = -tc^{\dagger}_{i\sigma}c_{j\sigma} + J\left(\vec{S}_i \cdot \vec{S}_j - \frac{1}{4}n_i n_j\right) \tag{5.1}$$

with the constraint of no double occupation. The exchange term is traditionally decoupled by $J\vec{S}_i \cdot \vec{S}_j \approx J\langle\vec{S}_i\rangle \cdot \vec{S}_j$ and the mean-field theory gives rise to antiferromagnetic ordering. Instead they used the slave boson method to enforce the constraint of no double occupation, i.e. $c_{i\sigma} = f_{i\sigma}b^{\dagger}_i$, subject to the constraint $\sum_\sigma f^{\dagger}_{i\sigma}f_{i\sigma} + b^{\dagger}_i b_i = 1$. The exchange term can be written in terms of $f_{i\sigma}$ in two forms $-J\left|\sum_\alpha f^{\dagger}_{i\sigma}f_{j\sigma}\right|^2$ or $-J\left(f^{\dagger}_{i\uparrow}f^{\dagger}_{j\downarrow} - f^{\dagger}_{i\downarrow}f^{\dagger}_{j\uparrow}\right)(f_{j\downarrow}f_{i\uparrow} - f_{j\uparrow}f_{i\downarrow})$ which invites the following novel mean-field decoupling

$$\begin{aligned} \chi_{ij} &= \sum_\alpha f^{\dagger}_{i\sigma}f_{j\sigma} \\ \Delta_{ij} &= f_{i\uparrow}f_{j\downarrow} - f_{i\downarrow}f_{j\uparrow}\ . \end{aligned} \tag{5.2}$$

These new mean fields capture the notion of singlet formation on the bond $(i, j)$ and I shall refer to the subsequent development following this line of approach rather broadly as RVB theory.

The mean-field phase diagram based on these decouplings take the schematic form shown in Fig. 5.1. [4, 5] As the temperature is lowered, $\chi_{ij} \neq 0$, so that the fermions now acquire an energy band and a Fermi surface. At a lower temperature, the fermions form a pairing state with $d$-wave symmetry. The bosons become essentially Bose condensed (with exponentially large correlation length with decreasing $T$) below a cross-over temperature $T^{(0)}_{BE} = 2\pi xt$. Below $T^{(0)}_{BE}$ the boson field can be treated as a c-number. In the overdoped region this gives rise to a Fermi liquid phase, similar to the theory of heavy fermion systems. In the intermediate doping range, the simultaneous presence of $\Delta_{ij}$ and $\langle b\rangle$ gives rise to a pairing order parameter for physical electrons $\langle c_{i\uparrow}c_{j\downarrow}\rangle$ which is of $d$-wave symmetry. Above $T^{(0)}_{BE}$ spin charge separation occurs at the mean-field level. In the pairing state a $d$-wave type gap occurs in the spin excitation spectrum, but not in the charge excitation, and it is natural to identify this as the spin gap phase.

Finally, the region IV in Fig. 5.1 is a non Fermi liquid state which may be referred to as a "strange metal."

We can go beyond mean-field to include fluctuations about the mean-field solution. The most important fluctuations are the phase fluctuations of the order parameter $\chi_{ij} = |\chi_{ij}|e^{i\theta_{ij}}$. Particles hopping around a plaquette acquire a phase related to $\theta_{ij}$, just like electrons in the presence of a magnetic flux. These low lying excitations are U(1) gauge fields.[6, 7] When coupled to the fermions and bosons they enforce the constraint locally, not just on average as in mean-field theory.

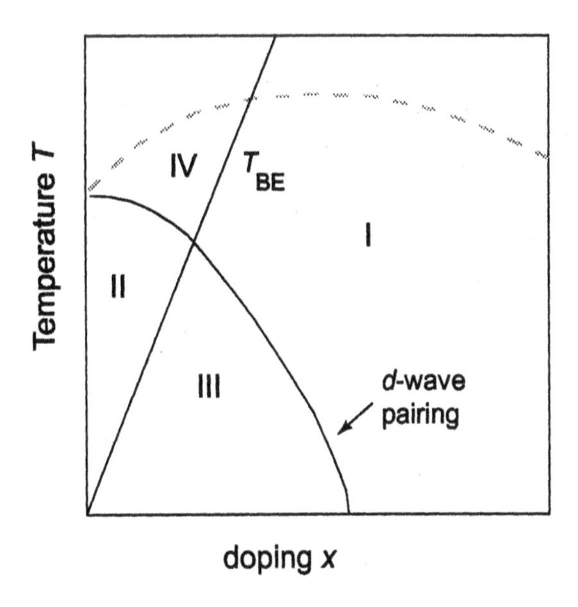


Figure 5.1: Schematic mean-field phase diagram of the $t$-$J$ model. Below the dashed line the uniform RVB order parameter $\chi_{ij}$ is nonzero. The mean-field pairing line below which $\Delta_{ij} \neq 0$ (dotted) and the Bose-condensation line (solid) divide the phase diagram into four regions. Region I is a Fermi-liquid phase, region II is the spin-gap phase, region III is the superconducting phase, and region IV is the strange metal phase.

It should be noted that $\chi_{ij}$, $\Delta_{ij}$, and $\langle b_i \rangle$ are not gauge invariant and would vanish upon gauge averaging. The only physical order parameter is the superconducting pairing $\langle c_{i\sigma} c_{j\sigma} \rangle$ and the only physical phase boundary is the transition to region III, the $d$-wave superconducting state. The rest of the phase diagram should be viewed as cross-over. Viewed in this light the schematic phase diagram captures many features of the high-$T_c$ phase diagram that have since been discovered, including the $d$-wave pairing state and the pseudogap region.

The theory outlined up to now is subject to a number of criticisms which I summarize as follows:

1. The undoped system is a Néel state, not an RVB state as implied by Fig. 5.1.

2. Mean-field theory enforces constraint only on average. This objection is partly mitigated by gauge theory which treats phase fluctuations at the Gaussian level and enforces the constraint locally.[6, 7] However, the gauge fluctuation is strong and indeed the coupling constant is infinite, since the gauge field started out as a field which enforces a constraint and has no restoring force. Gaussian fluctuation is then inadequate.

3. The scheme outlined above, which we shall referr to as the U(1) theory, is only one of many different ways to enforce the constraint. For example, Wen and Lee [8] introduced an SU(2) theory, which has a slightly different mean-field phase diagram. Recently Senthil and Fisher [9] have introduced a Z(2) scheme which enforces the constraint exactly at half-filling and fixes $f_\sigma^\dagger f_\sigma + b^\dagger b$ to be odd integers away from half-filling. Which of these schemes, if any, should we trust? If this question is to be ultimately decided by comparison with experiment, are there physical quantities we can compute which distinguish between different approaches?

Before attempting to answer these objections, I first summarize some of the salient features of the SU(2) formulation. At half-filling it was pointed out [10] that the action when formulated in terms of fermions possesses a hidden SU(2) symmetry distinct from the conventional SU(2) symmetry relating up and down spin. By introducing SU(2) doublets

$$\psi_\uparrow = \begin{pmatrix} f_\uparrow \\ f_\downarrow^\dagger \end{pmatrix} , \quad \psi_\downarrow = \begin{pmatrix} f_\downarrow \\ -f_\uparrow^\dagger \end{pmatrix} , \tag{5.3}$$

the action is invariant under the rotation $\psi \to g\psi$ where $g$ is a $2 \times 2$ SU(2) matrix. This is a consequence of the redundancy of the fermion representation: adding a spin-up fermion is the same as removing a spin-down fermion upon projection to the physical subspace. This symmetry explains why a variety of mean-field solutions have identical spectra and energies. These become *identical* states after enforcing the constraint by projection. Mean-field states of special interest are the $d$-wave pairing states and a variety of flux phases, where $\chi_{ij}$ is complex in such a way that the product around a plaquette gives a net phase $\Phi$ which is gauge invariant, corresponding to a gauge "magnetic flux." In particular, the staggered flux phase is one where the flux $\Phi$ takes + and – values on the $A$ and $B$ sublattice. This is illustraed in Fig. 5.2a. In mean-field theory the unit cell is doubled and the fermion dispersion is identical with that of the $d$-wave pairing state, with gap nodes at $(\pi/2, \pm\pi/2)$. When $\Phi = \pi$, translation symmetry is restored.

With doping the SU(2) symmetry is broken and out of the multitude of degenerate states ($\pi$ flux, $s+id$, $d$-state, etc.) the $d$-wave pairing state has the lowest free

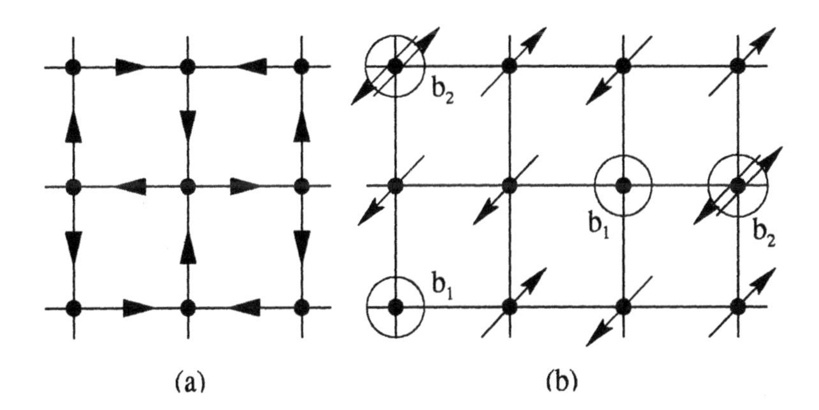


Figure 5.2: (a) Staggered-flux phase. Links with arrows correspond to $\theta_{ij} = \Phi/4$ in the direction of the arrow. (b) A typical configuration of the half-filled fermion state. Arrows denote fermions. Circled sites are physical holes which are spin singlets made up of either empty or two-fermion sites. $b_1$ and $b_2$ bosons are assigned to these respective sites.

energy, giving rise to the phase diagram shown in Fig. 5.1. The reason is that with doping, $\sum_\sigma f_{i\sigma}^\dagger f_{i\sigma} = 1 - b_i^\dagger b_i = 1 - x$ and the chemical potential moves from the particle-hole symmetric point $\mu = 0$. The flux phases are semiconductors (with gap nodes) and the chemical potential now moves away from the nodes, leaving a small pocket of Fermi surfaces. On the other hand, in a pairing state, the gaps are associated with the Fermi surface so that the nodal point remains, but shifted away from $(\pi/2\,,\,\pi/2)$ towards the origin. This state is clearly energetically more favorable, albeit by a small amount (of order $x^{3/2}$). Wen and Lee were motivated by the notion that for small $x$, fluctuations among these almost degenerate states may be important degrees of freedom overlooked in the U(1) mean-field theory. They introduced a new formulation of the constraint of no double occupation which preserves the SU(2) symmetry even away from half-filling. The trick is to introduce two bosons forming an SU(2) doublet

$$h = \begin{pmatrix} b_1 \\ b_2 \end{pmatrix} . \tag{5.4}$$

The physical electron is an SU(2) singlet formed out of the product of the fermion and boson doublet:

$$\begin{aligned} c_\uparrow &= \tfrac{1}{\sqrt{2}} h^\dagger \psi_\uparrow &=& \left(b_1^\dagger f_\uparrow + b_2^\dagger f_\downarrow^\dagger\right)/\sqrt{2} \\ c_\downarrow &= \tfrac{1}{\sqrt{2}} h^\dagger \psi_\downarrow &=& \left(b_1^\dagger f_\downarrow - b_2^\dagger f_\uparrow^\dagger\right)/\sqrt{2} \ . \end{aligned} \tag{5.5}$$

The physical Hilbert space is constrained to be the SU(2) singlet subspace. One of the constraints is that the $z$ component $\tau_3$ vanishes, i.e.,

$$\left\langle \sum_i f_{\sigma i}^\dagger f_{\sigma i} + b_{1i}^\dagger b_{1i} - b_{2i}^\dagger b_{2i} \right\rangle = 1 \ . \tag{5.6}$$

Note that in contrast to the U(1) constraint, this can be satisfied by keeping the fermions at half-filling, with $\langle b_1^\dagger b_1 \rangle = \langle b_2^\dagger b_2 \rangle = \frac{1}{2}x$. This is illustrated in Fig. 5.2b. A typical configuration of the half-filled fermion system has $1 - x$ singly-occupied sites and $\frac{1}{2}x$ empty sites and $\frac{1}{2}x$ doubly-occupied sites. The main observation is that since the fermions keep track of the spin degrees of freedom, the empty and doubly-occupied sites are both spin singlets, and have the correct spin quantum number to be a physical hole. Instead of throwing out these configuration as in the U(1) theory, we simply declare that these are all valid representations of a physical hole and attach a $b_1$ boson to the empty site and a $b_2$ boson to the doubly-occupied site. More accurately, the linear superposition of Eq.(5) is used.

The SU(2) mean-field phase diagram resembles the U(1) phase diagram shown in Fig. 5.1 except that the $\pi$ flux phase is no longer restricted to the $x = 0$ line, but bulges out to occupy a finite area of the phase diagram. More importantly, region II has a new interpretation. Instead of a pairing phase between the fermions, this phase now includes fluctuations of all other U(1) mean field states related by an SU(2) gauge rotation. To emphasize this distinction, we refer to region II as the staggered-flux phase. Note that unlike U(1) mean-field theory, the staggered-flux phase does not break translation symmetry in that it is related by a gauge rotation to the $d$-wave state, which clearly does not break translational symmetry. One of the interesting outcomes of the SU(2) theory is that by assuming an attraction between the fermions and bosons to form a bound state, we obtain Fermi surface segments, i.e., region of zero energy physical hole excitations, which are in agreement with ARPES experiments [8]

## 5.3 PROJECTED WAVEFUNCTIONS AND STAGGERED CURRENT FLUCTUATIONS

Now we are ready to address some of the criticisms listed earlier. Let us begin with objection (3). If gauge fluctuations were treated exactly, the U(1) and SU(2) formulations are both equivalent to the $t$-$J$ model. The reason for exploring different formulations is precisely that they have different mean-field solutions, and the hope is that one of them will be closer to reality. However, the mean-field solutions are highly approximate and the phase diagrams are not that different. Do we have any physical prediction which distinguishes one theory from the next? Motivated by the difficulty of treating the gauge field analytically beyond the Gaussian level (objection #2), we turn to numerical studies of projected wavefunctions. This field has a long history. For example, it has been known for a long time that if one takes a spin density wave state for the fermion (i.e., introduce a staggered magnetization mean-field) and performs what is called the Gutzwiller projection, i.e., project out all doubly-occupied configurations, one does not obtain a very good wavefunction.

On the other hand, projection of a $\pi$-flux phase does surprisingly well. [11] The best state is the combine staggered magnetization with some flux, either $\pi$-flux or staggered-flux and an excellent energy is achieved. [12, 13, 14] Thus a partial answer to objection #1 is that the $\pi$-flux phase is in some sense close to the Néel-ordered ground state of the quantum Heisenberg model. With doping, the best state is a projected $d$-wave state. [13] Not surprisingly, this state has long range pairing order after projection. Recently we calculated the current-current correlation function of this state [15]

$$c_j(k,\ell) = \langle j(k) j(\ell) \rangle \tag{5.7}$$

where $j(k)$ is the physical electron current on the bond $k$. The average current $\langle j(k) \rangle$ is obviously zero, but the correlator exhibits a staggered circulating pattern as shown in Fig. 5.3. (Note that the pattern is shifted by $\pi$ relative to a pattern constructed from the reference bond at the origin.) Within our numerical accuracy, this correlation decays as a power law and the decay is faster with increasing doping. Such a pattern is absent in the $d$-wave BCS state before projection, and is a result of the Gutzwiller projection.

We were motivated to look for the staggered pattern in the current-current correlation function because that is what we expect to find in the staggered flux phase. Consider a plaquette with a hole in one corner site (4) and spins on the other three corners (labelled 1–3). A hole hopping around the plaquette sees a wandering spin quantization axis from site to site and will pick up a Berry's phase $\Phi$ which is given by the solid angle subtended by the 3 spin directions. [16] The flux in the flux phases is designed to capture this piece of physics, as a hole hopping in the presence of a gauge "magnetic flux" will also pick up an Aharonov-Bohm phase.

More formally, it is known [16, 7] that if we define $Q_{ij} = \sum_\sigma f^\dagger_{i\sigma} f_{j\sigma}$, then

$$Im \left( \prod_\square : Q_{12} Q_{23} Q_{34} Q_{41} : \right) = \left( f^\dagger_{4\sigma} f_{4\sigma} \right) \vec{S}_1 \cdot \vec{S}_2 \times \vec{S}_3 + \text{permutations}. \tag{5.8}$$

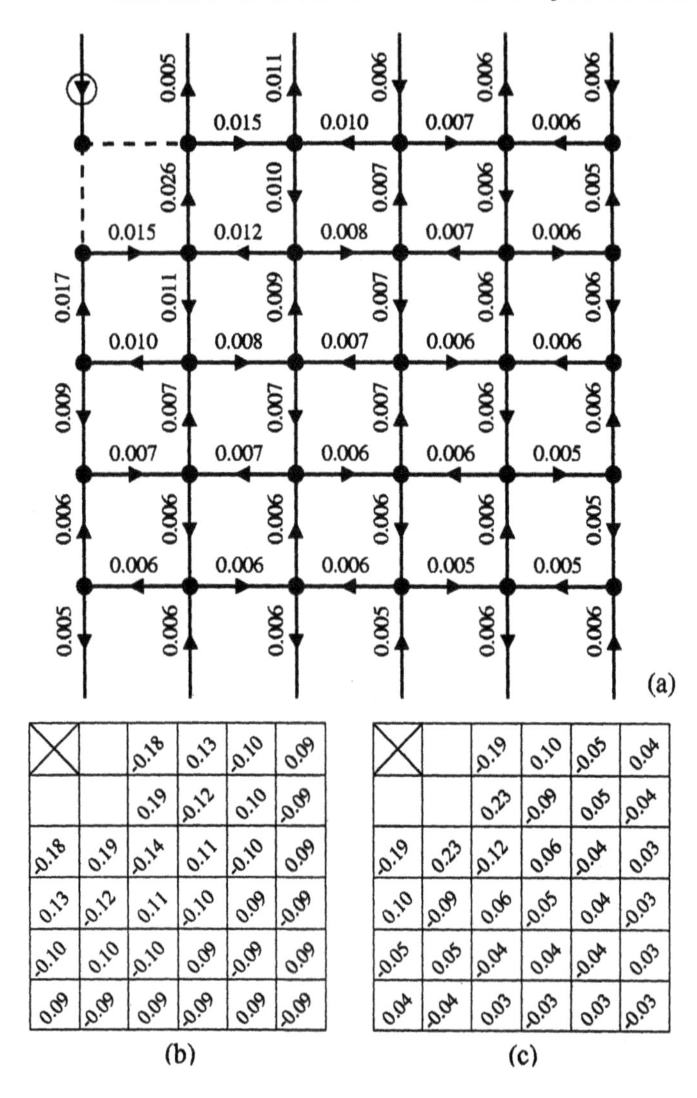


Figure 5.3: (a) Current-current correlations for 2 holes in the $10 \times 10$ lattice. Boundary conditions are periodic in one and antiperiodic in the other directions (the data are averaged over the two orientations). The number on a link is the correlation of the current on this link and of the current on the circled link divided by hole density. The arrows point in the direction of the postive correlations of the current. (b) The same data in the form of vorticity defined as the sum of the current around a plaquette. The number on a plaquette is the vorticity correlation divided by $x$ with the crossed plaquette. (c) Same as (b) for 10 holes in $10 \times 10$ lattice.

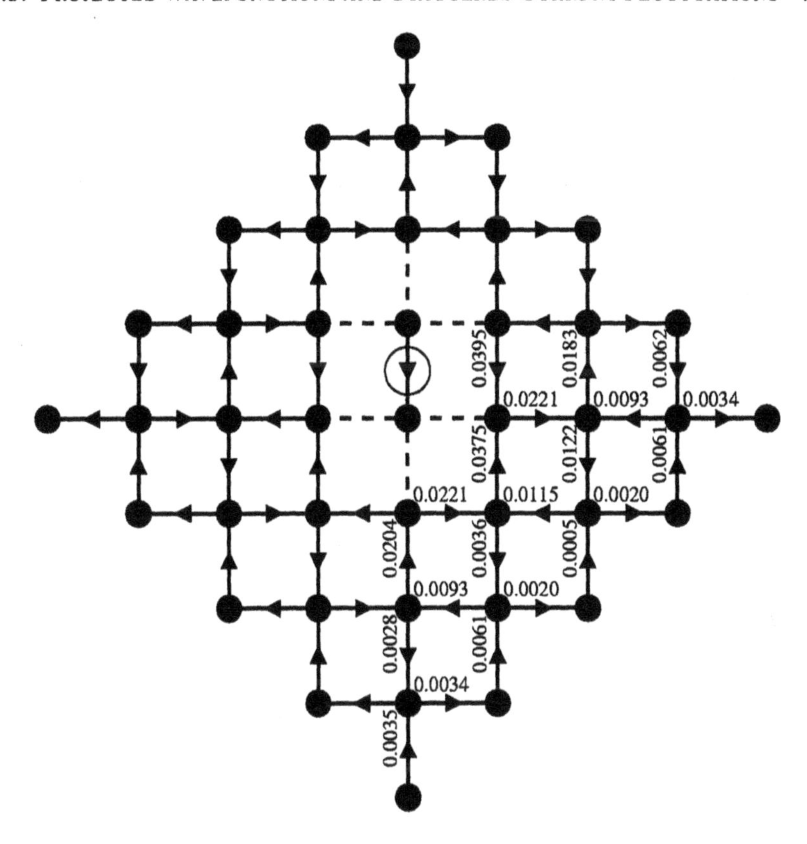


Figure 5.4: The current correator divided by the hole density ($c_j/x$) for the exact ground state of 2 holes in a 32-site $t$-$J$ model with $J/t = 0.3$ (from Leung [18]).

In the mean-field theory, the L.H.S. of Eq.(8) is $|\chi|^4 \sin \Phi$. Thus the flux is also related to the spin chirality defined as $K = \vec{S}_1 \cdot \vec{S}_2 \times \vec{S}_3$. This led us to calculate the spin chirality correlation for the projected $d$-wave state,

$$c_\chi(k,\ell) = \left\langle \vec{S}_1 \cdot \vec{S}_2 \times \vec{S}_3(k) \;\; \vec{S}_{1'} \cdot \vec{S}_{2'} \times \vec{S}_{3'}(\ell) \right\rangle \tag{5.9}$$

where $(k,\ell)$ labels plaquettes and the spins 1,2,3 form a triangle with a fixed orientation around plaquette $k$. We find that $c_\chi$ is very small and decays rapidly. On the other hand, we note that in Eq.(8), $f^\dagger_{4\sigma} f_{4\sigma} = 1 - b^\dagger b = 1 - n_H$ where $n_H$ is the hole density. We thus consider

$$c_{\chi H}(k,\ell) = \left\langle n_H(4) \vec{S}_1 \cdot \vec{S}_2 \times \vec{S}_3(k) \;\; n_4(4') \vec{S}_{1'} \cdot \vec{S}_{2'} \times \vec{S}_{3'} \right\rangle \quad . \tag{5.10}$$

We find that $c_{\chi H}$ shows power-law-like decay, very similar to the current-current correlation $c_j$. [17] These results tell us that the spin-chirality is associated with the presence of a hole. Each hole generates a staggered-chirality pattern around it which can be $+$ or $-$, and the correlation between two such patterns may be what is responsible for the current pattern.

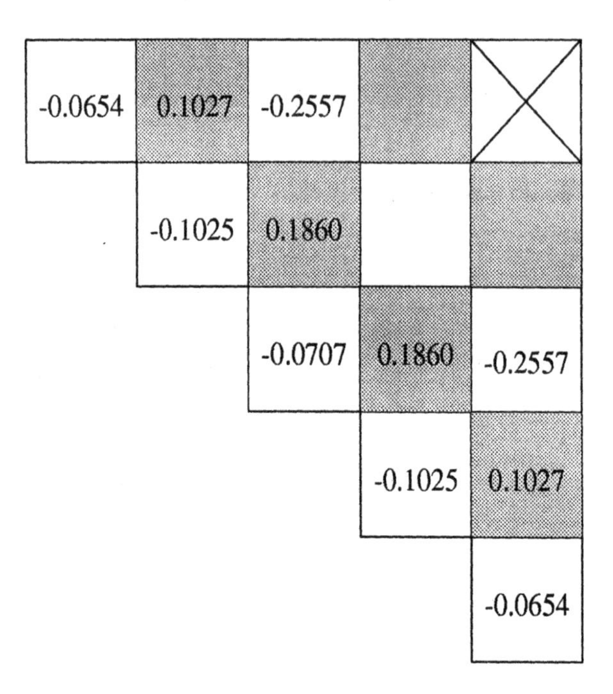


Figure 5.5: The vorticity correlator divided by $x$ from exact diagonalization (from Leung [18]). The numbers are to be compared with Figs. 5.3b or 5.3c and are of the same order of magnitude.

This kind of staggered pattern is totally unexpected in the U(1) mean-field theory, but is very naturally in the SU(2) theory. In fact, our original motivation was to construct a trial wavefunction by projecting staggered-flux mean-field state to the SU(2) singlet subspace. It is obvious that this is equivalent to projecting $d$-wave BCS state to the SU(2) singlet subspace, because the two states are related by SU(2) rotation before projection. It turns out that the latter state is also identical to the Gutzwiller projected $d$-wave state (for proof see ref. [15]). We believe the current-current correlator $c_j$ and the hole-chirality correlation $c_{\chi H}$ constitutes physical predictions which distinguishes the U(1) and SU(2) theory. Its existence will support the notion that the SU(2) thoery is a better starting point for the underdoped cuprates.

The above results are based on a projected wavefunction. We received some confirmation of these ideas by a study on the exact diagonalization of a 32-site sample with 2 holes. [18] The ground state is a $d$-wave pairing state and the current-current correlation shown in Figs. 5.4 and 5.5 indeed exhibits the staggered pattern we predicted. We suggest that such correlation should exist in the underdoped cuprates and indeed we propose that the correlation should survive above the superconducting transition temperature into the pseudogap state. Up to now it has been quite mysterious whether the pseudogap state can be characterized by any kind of quasi-long-range order. The correlators $c_j$ and $c_{\chi H}$ are natural candidates within the SU(2) theory. Unfortunately, detection of such correlators experimentally appears to be a daunting task.

## 5.4 CONCLUSION

The original RVB mean-field phase diagram dated from a dozen years ago. With the subsequent experimental data, it looks attractive as a starting point to describe the high-$T_c$ phase diagram. I think the recent work on projected wavefunctions and exact diagonalization that showsthe existence of staggered current fluctuation patterns lends strong support to the validity of this line of investigation, in that a novel correlator which is otherwise completely unexpected appears to play a role in the underdoped region. Hopefully, clever experiments and future numerical work can lead to further development of the viewpoint launched by Phil Anderson thirteen years ago.

## ACKNOWLEDGMENTS

I am thankful to Xiao-Gang Wen and Dmitri Ivanov for collaboration on the work reported here and to Pakwo Leung for permission to reproduce Figures 5.4 and 5.5. This work is supported by NSF under the MSREC program, DMR98–08941.

## BIBLIOGRAPHY

[1] P.W. Anderson, Science **235** 1196 (1987).

[2] See the article by J.C. Campuzano in this volume.

[3] G. Baskaran, Z. Zou, and P.W. Anderson, Solid State Commun. **63** 973 (1987).

[4] G. Kotliar and J. Liu, Phys. Rev. B **38** 5142 (1988).

[5] Y. Suzumura, Y. Hasegawa, and H. Fukyuama, J. Phys. Soc. Jpn. **57** 2768 (1988).

[6] L. Ioffe and A.I. Larkin, Phys. Rev. B **39** 8988 (1989).

[7] P.A. Lee and N. Nagaosa, Phys. Rev. B **46** 5621 (1992).

[8] X.G. Wen and P.A. Lee, Phys. Rev. Lett. **76** 503 (1996).

[9] T. Senthil and M. Fisher, article in this volume.

[10] I. Affleck, Z. Zou, T. Hsu, and P.W. Anderson, Phys. Rev. B **38**, 745 (1988); E. Dagotto, E. Fradkin, and A. Moreo, Phys. Rev. B **38** 2926 (1988).

[11] T.C. Hsu, Phys. Rev. B **41** 11379 (1989).

[12] C. Gros, Ann. Phys. **189** 53 (1989).

[13] H. Yokoyama and M. Ogata, J. Phys. Soc. Jpn. **65** 3615 (1996).

[14] D.V. Dmitriev, V. Ya. Krivnov, Y.N. Likhachev, and A.A. Orchinnikov, Fiz. Tr. Tela. **38** 397 (1996) [Sov. Phys. Solid State Physics].

[15] D. Ivanov, P.A. Lee, and X.G. Wen, Phys. Rev. Lett. **84** 3958 (2000).

[16] X.G. Wen, F. Wilczek, and A. Zee, Phys. Rev. B **39** 11413 (1989).

[17] Guobin Sha, D. Ivanov, P. Lee, and X.G Wen, *unpublished.*

[18] P.W. Leung, *preprint.*

# Chapter 6

# Angle-Resolved Photoemission Results in Cuprates

J. C. Campuzano

Department of Physics, University of Illinois at Chicago, IL 60607
and Argonne National Laboratory

## Abstract

The two-dimensional nature of the high temperature superconductors allows the determination of the energy-momentum relationship of electronic states by angle resolved photoemission (ARPES). Furthermore, the shape of the ARPES spectra provides information on the many-body interactions so prevalent in these materials. In this paper we describe the results obtained by our group on the nature of the normal state, the effect of superconductivity on the photoemission lineshape.

## 6.1 Introduction

As first suggested by Anderson soon after the high temperature superconductors (HTSCs) were discovered [1], Angle Resolved Photoemission (ARPES) has played a major role in elucidating the electronic structure of the high temperature superconductors. This comes about because of the fortunate confluence of three factors: the quasi two-dimensional nature of these materials, the large energy scales involved, and the improvements in instrumental resolution. We have shown that under these circumstances ARPES can yield the spectral function [2],

$$I(\mathbf{k},\omega) = I_0(\mathbf{k};\nu;\mathbf{A})f(\omega)A(\mathbf{k},\omega) \tag{6.1}$$

for a quasi-two-dimensional system, assuming the validity of the impulse approximation. Here $\mathbf{k}$ is the in-plane momentum, $\omega$ is the energy of the initial state measured relative to the chemical potential, $f(\omega) = 1/[\exp(\omega/T) + 1]$ is the Fermi function and the one-particle spectral function (Fig. 6.1)

$$A(\mathbf{k},\omega) = (-1/\pi)\Im m G(\mathbf{k},\omega + i0^+). \tag{6.2}$$

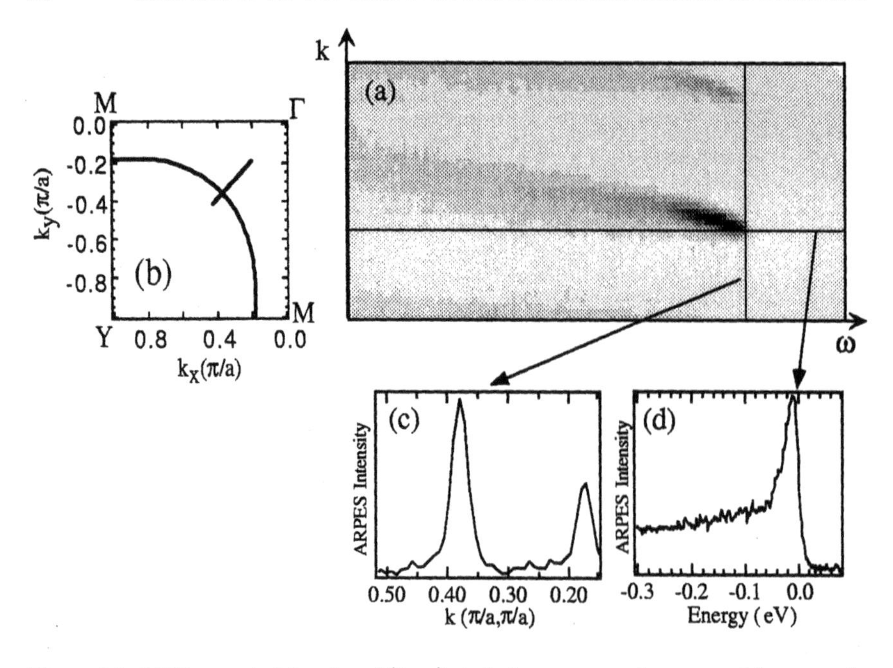


Figure 6.1: (a) The spectral function $A(\mathbf{k}, \omega)$ plotted as a gray scale measured for a sample described in [3], along the zone diagonal shown in (b). Panel (c) shows $A(\mathbf{k}, \omega = 0)$, and panel (d) shows $A(\mathbf{k} = k_F, \omega)$

The prefactor $I_0$ is proportional to the dipole matrix element $|M_{fi}|^2$ and thus a function of $\mathbf{k}$ and of the incident photon energy $\hbar\nu$ and polarization $\mathbf{A}$.

The measurement of $A(\mathbf{k}, \omega)$ provides a direct window into the many-body interactions. We believe this has resulted in a qualitative change in thinking about ARPES data and its analysis. Along with this have come a variety of new physics results which have shed very important new light on the high-$T_c$ superconductors. Here we will focus on our results regarding questions raised by Anderson, concerning the nature of the electronic excitations in the normal state, and whether these states can be described using the concepts of existing theories, or whether a new description of matter will have to be found. We then explore the nature of the superconducting state. First, let us briefly review what sort of electronic structure we can expect from the HTSCs.

## 6.2 Nature of the electronic states

We will mostly study $Bi_{1.6}Pb_{0.4}Sr_2CuO_6$ (Bi2201) and $Bi_2Sr_2CaCu_2O_{8+x}$ (Bi2212). These materials are most suited to ARPES studies for two important reasons: 1) They are highly 2-dimensional, and therefore only the momentum parallel to the $CuO_2$ planes is relevant, a quantity easily determined by ARPES, and 2) They contain two adjacent planes of BiO, which are van der Waals bonded to each other, without charge exchange between the planes. Since ARPES is a surface-sensitive probe, this implies that when we cleave a crystal to make a fresh surface, the potential at the surface is not altered, and no surface states are formed to complicate the interpretation of the measurements. The measured electronic structure is representative of that of the bulk. We have check this assertion by comparing, whenever possible, our results to those given by bulk probes.

Even though the structure of these quaternary compounds is quite complicated, only a few weeks after their discovery, Anderson suggested that we need concern ourselves with only one set of states, the ones resulting from the hybridization of Cu $3d_{x^2-y^2}$ to O $2p$ orbitals [4], since these would be the only states crossing the Fermi energy, and furthermore, these states have a large separation of order 1eV from the nearest occupied states. Figure 6.2 shows an energy distribution curve obtained over a large energy range at the $(\pi, 0)$ point of the Brillouin zone. There are roughly three features observable in this spectrum: a peak at ~6 eV binding energy, which corresponds to the most strongly binding orbital. There is another peak near zero binding energy, corresponding to the most antibonding combination of orbitals. And in between, there are a series of peaks, not all resolvable, corresponding to the non-bonding combinations, plus all the states from all the other atoms in the structure. One can see that the state near the Fermi energy is separated by nearly 1 eV from the rest of the states. Therefore, as predicted by Anderson, all the action, as far as the low-energy physics is concerned, takes place in this single band.

## 6.3 EXCITATIONS AT THE FERMI SURFACE

We now examine in detail the question of the existence of quasiparticles at the Fermi surface. It is now well established that the cuprate superconductors are $d$-wave superconductors with points $N$ (see inset in Fig. 6.3a) characterized by nodes in the energy gap [5, 6], and points $A$ at the zone edge where the gap is maximum. Figure 6.3a shows ARPES spectra in the normal state along the Fermi surface of a Bi2212 ($T_c$=89K) sample. The linewidths are very large, even in this optimally doped sample. Curve 8 in Fig. 6.3a shows the lineshape at point $A$ of the Brillouin zone. Remembering that half the spectral function is cut off by the Fermi function, we see that the width of this peak is ~200 meV, or 2000 K, much larger than the temperature of the sample, i.e. the width is intrinsic. It is very important to emphasize that the large linewidths observed in ARPES are not extrinsic, or artifacts of any analysis. As we will see below, when quasiparticles do exist (for $T \ll T_c$) they are clearly seen in the experiment.

One expects transport properties to be dominated by states with the largest dispersion, those near the nodal point $N$. We can make a rough estimate of the electron lifetime near the nodes using the uncertainty principle $\Delta E \Delta t \geq \hbar/2$ to be of order $4 \times 10^{-15}$ s. We can further estimate the electron velocity from the

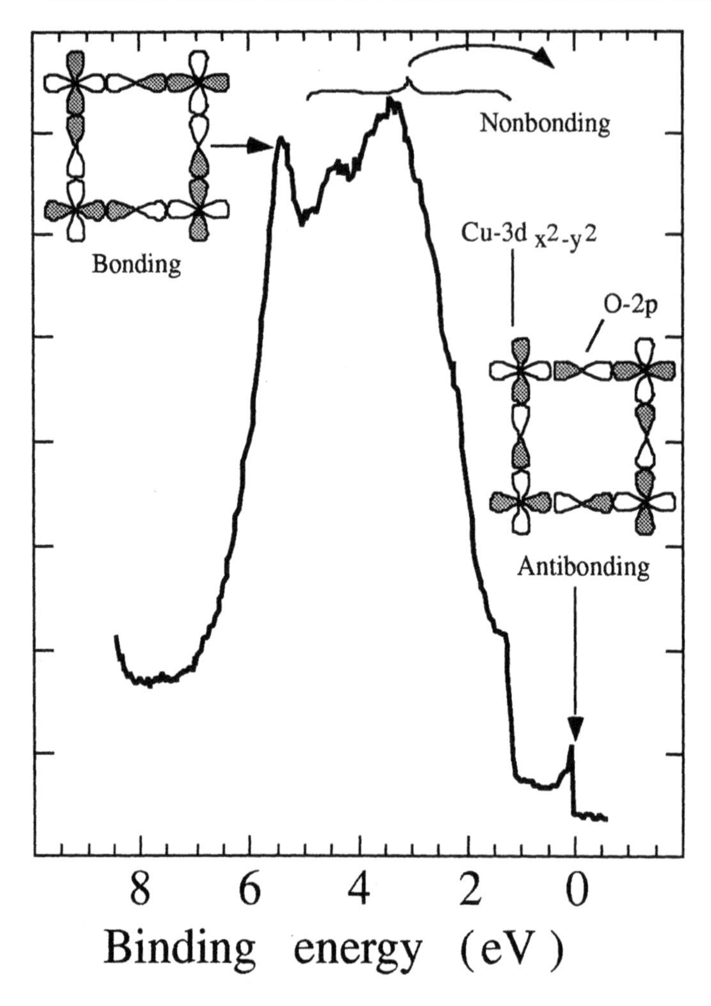


Figure 6.2: Energy distribution curve of the valence band at $(\pi, 0)$ in Bi2212-OD88K obtained at $\hbar\nu$ =22 eV. Note the 1 eV separation between the state at $E_F$ and the rest of the valence band.

measured dispersion to be of order $2.5 \times 10^7$ cm/s, which yields a mean free path is of order 10Å. One concludes that the electron travels a distance of only three lattice spacings before it decays. This is a very incoherent electron indeed. We must therefore conclude that the normal state spectral function is extremely broad, and *there are no well defined quasiparticles above* $T_c$! This anomalous normal state spectrum implies a breakdown of Fermi liquid theory, as also suggested by numerous transport experiments. But perhaps the most striking evidence of this anomalous state is provided by a more detailed look at the spectral function

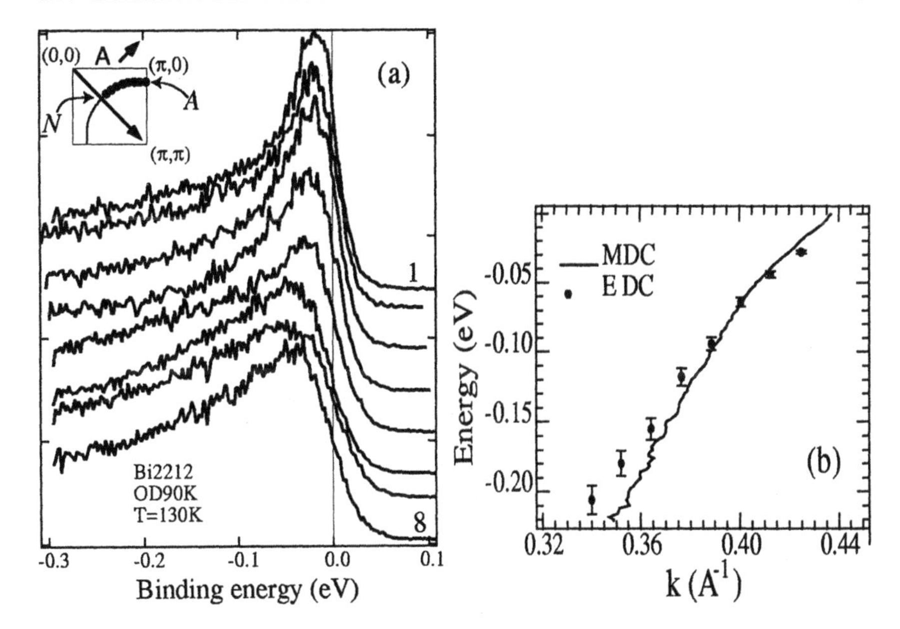


Figure 6.3: (a) EDC's from an optimally doped Bi2212 sample ($T_c$ =90K, obtained at $T$ =130K and $\hbar\nu$ =22 eV along the Fermi surface at points indicated in the inset. (b) Comparison of the MDC and EDC dispersion from the spectral function shown in Fig. 6.1.

(Eq. 6.2), which we write as:

$$A(\mathbf{k},\omega) = \frac{1}{\pi} \frac{|\Sigma''(\mathbf{k},\omega)|}{[\omega - \epsilon_{\mathbf{k}} - \Sigma'(\mathbf{k},\omega)]^2 + [\Sigma''(\mathbf{k},\omega)]^2}. \tag{6.3}$$

where the self-energy $\Sigma = \Sigma' + i\Sigma''$ and $\epsilon_{\mathbf{k}}$ is the bare dispersion. For $\mathbf{k}$ near $k_F$, and varying normal to the Fermi surface (shown in Fig. 6.1b), we may write $\epsilon_{\mathbf{k}} \simeq v_F^0(k - k_F)$, where both $k_F(\theta)$ and the bare Fermi velocity $v_F^0(\theta)$ depend in general on the angle $\theta$ along the Fermi surface.

In Fig. 6.3b we plot the dispersion of the spectral peak above $T_c$ obtained from constant-$\mathbf{k}$ scans (energy distribution curves or EDCs), and the peak in momentum obtained from constant-$\omega$ scans (momentum distribution curves or MDCs) [7] from the data of Fig. 6.3a. We find that the EDC and MDC peak dispersions are very different, a consequence of the $\omega$ dependence of $\Sigma$. To see this, we note from Eq. (1.3) that the MDC at fixed $\omega$ is a Lorentzian centered at $k = k_F + [\omega - \Sigma'(\omega)]/v_F^0$, with a width (HWHM) $W_M = |\Sigma''(\omega)|/v_F^0$, *provided* (i) $\Sigma$ is essentially independent [8] of $k$ normal to the Fermi surface, *and* (ii) the dipole matrix elements do not vary significantly with $k$ over the range of interest. That these two conditions are fulfilled can be seen by the nearly Lorentzian MDC lineshape of the data plotted in Fig. 6.1c.

On the other hand, the EDC at fixed $\mathbf{k}$ (Fig. 6.1d) has a non-Lorentzian lineshape reflecting the non-trivial $\omega$-dependence of $\Sigma''$, in addition to the Fermi cutoff at low energies. Thus, the EDC peak is *not* given by the pole condition $[\omega - v_F^0(k - k_F) - \Sigma'(\omega)] = 0$, but also involves $\Sigma''$, and therefore we do not have a Fermi liquid. The large discrepancy between the dispersion of the MDC and EDC is not expected to occur in a Fermi liquid. A Fermi liquid has a correction $\omega^3/\varepsilon^2_{F_{eff}}$ to the pole condition, which would be undetectable by ARPES for any $\varepsilon_{F_{eff}}$ greater than the energies in Fig. 6.3b.

As there is some evidence from transport that more Fermi liquid-like behavior develops for heavily overdoped materials, the question arises whether ARPES sees evidence for normal state quasiparticles in that case. The top curve of Fig. 6.4a shows a spectrum for a highly overdoped Bi2201 sample ($T_c$ = 4 K) at 20 K. Indeed, the spectral peak is much narrower than the optimally doped sample, consistent with more Fermi liquid-like behavior. However, it should be pointed out that even this highly overdoped system still does not behave as a conventional Fermi liquid. In Fig. 6.4b we plot the EDC's dispersing along the $(0,0) \to (\pi,0)$ direction, at points indicated in the inset. It was first noticed by Anderson that the spectra appear to "peel off" from a common curve given by the state closest to the chemical potential, a behavior one would *not* expect from a Fermi liquid.

## 6.4 QUASIPARTICLES IN THE SUPERCONDUCTING STATE

Upon lowering the temperature below $T_c$, the spectrum undergoes remarkable $T$-dependent changes in the line-shape at the antinode $A$, as can be seen by comparing curves (1) and (4) in Fig. 6.5a, which may be understood as follows. For $T < T_c$ the SC gap opens up and spectral weight at $\mathbf{k}_F$ shifts from $\omega = 0$ (in the normal state) to either side of it, of which only the occupied side ($\omega < 0$) is probed by ARPES. At the lowest temperature the EDC peak *is* the peak of the spectral function (unlike the normal state) since the Fermi function has now become sharper and spectral weight has moved down to below the gap energy.

A remarkable feature of the data is the sharpening of the peak with decreasing $T$ in the SC state. This is *not* a "BCS pile up" in the density of states, a description frequently used in the early literature, since we are not measuring a DOS. With a rapid decrease in linewidth below $T_c$, the only way the conserved area sum rule [2] can be satisfied is by having a large rise in intensity. The decrease in the linewidth below $T_c$ is a consequence of the the SC gap leading to a suppression of electron-electron scattering which was responsible for the large linewidth above $T_c$. *Thus coherent quasiparticle excitations do exist for* $T \ll T_c$.

Similarly, the nodal points $N$ exhibit quaisparticles in the superconducting state. Figure 6.5b shows a spectrum in the superconducting state, where a near resolution limited peak appears [9]. In addition, a break clearly separates the coherent quasiparticle part of the spectral function from the incoherent part, as indicated by the up-arrow [10]. A significant point is that in the vicinity of the $A$ point, the break in the spectra appears exactly at $T_c$ [11]. We will show below

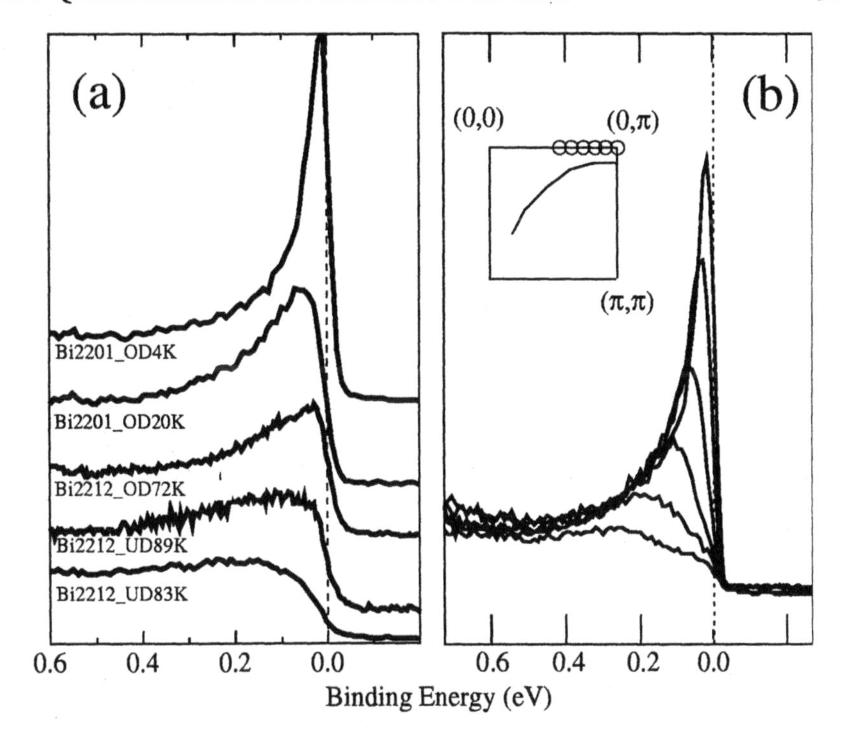


Figure 6.4: a) Variation of the ARPES lineshape in the BSCCO family as a function of doping. b) EDC's dispersing along the $(0,0) \rightarrow (\pi,0)$ direction, corresponding to the points marked in the inset for a Bi2201 sample with $T_c = 4$ K, obtained at $T = 30$ K at the $(\pi,0)$ point.

that the onset of superconductivity introduces a new energy scale, which leads to a reduction of the scattering rate of electrons over an energy range larger than the maximum superconducting gap [12] throughout the zone, thus allowing quasiparticles to exist even at the nodal points $N$. Although these data were obtained in the superconducting state, because they are at the node of the $d$-wave state, they are in fact gapless.

Figure 6.6a shows the temperature dependence of the lineshape at point $N$. In the normal state the trailing edge of the spectral peak smoothly evolves into an incoherent tail going to high binding energy. As the temperature is lowered below $T_c$, a break develops which separates the coherent quasiparticle peak from the incoherent tail. It is important to note that quasiparticles exist all along the Fermi surface (Fig. 6.6b).

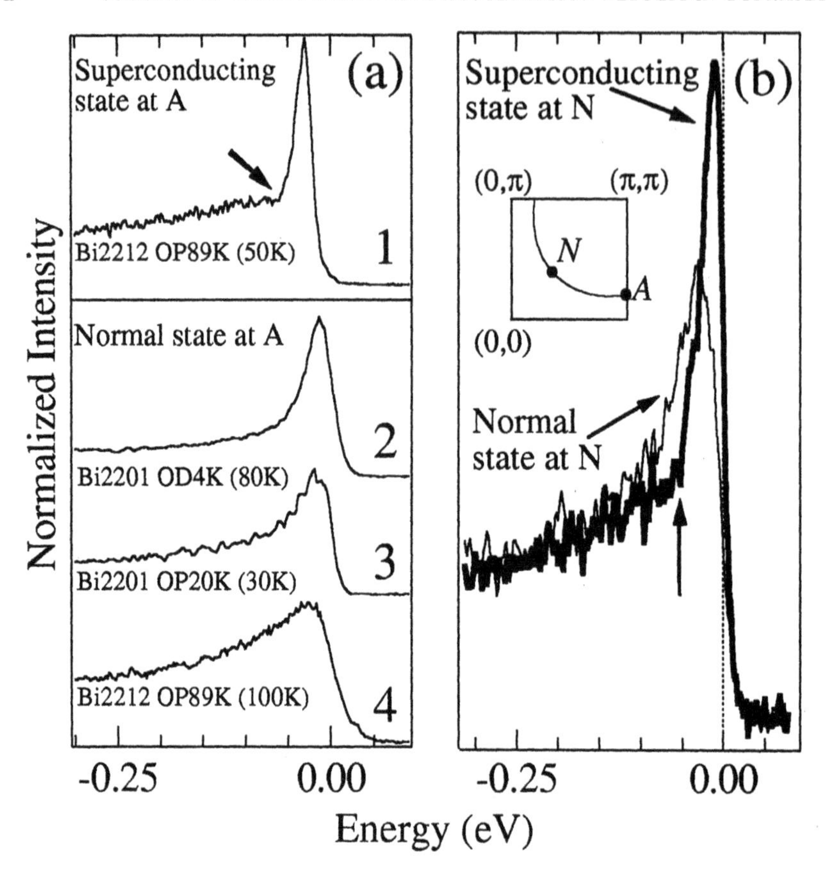


Figure 6.5: (a) ARPES spectra at point $A$ of the zone (inset in Fig. 6.1), with spectra labeled by doping (OD for overdoped, OP for optimal doped) and onset $T_c$. The temperature is given in parenthesis. Note the break in the high resolution spectrum at low temperature marked by an arrow. (b) ARPES spectra of Bi2212 ($T_c$=89K) in the normal and superconducting states at point $N$.

## 6.5 NEW ENERGY SCALES IN THE SUPERCONDUCTING STATE

One of the most striking features of the high-temperature superconductors is the linear dependence of the resistivity on temperature in the normal state. This behavior has been attributed to the presence of a quantum critical point, where the only relevant energy scale is the temperature [13]. However, new energy scales become manifest below $T_c$ due to the appearance of the superconducting gap and resulting collective excitations, as we now show. Thus, one does in general not expect response functions to continue to exhibit $\omega/T$ scaling in the superconducting state. The effect of the these new scales on the ARPES spectral

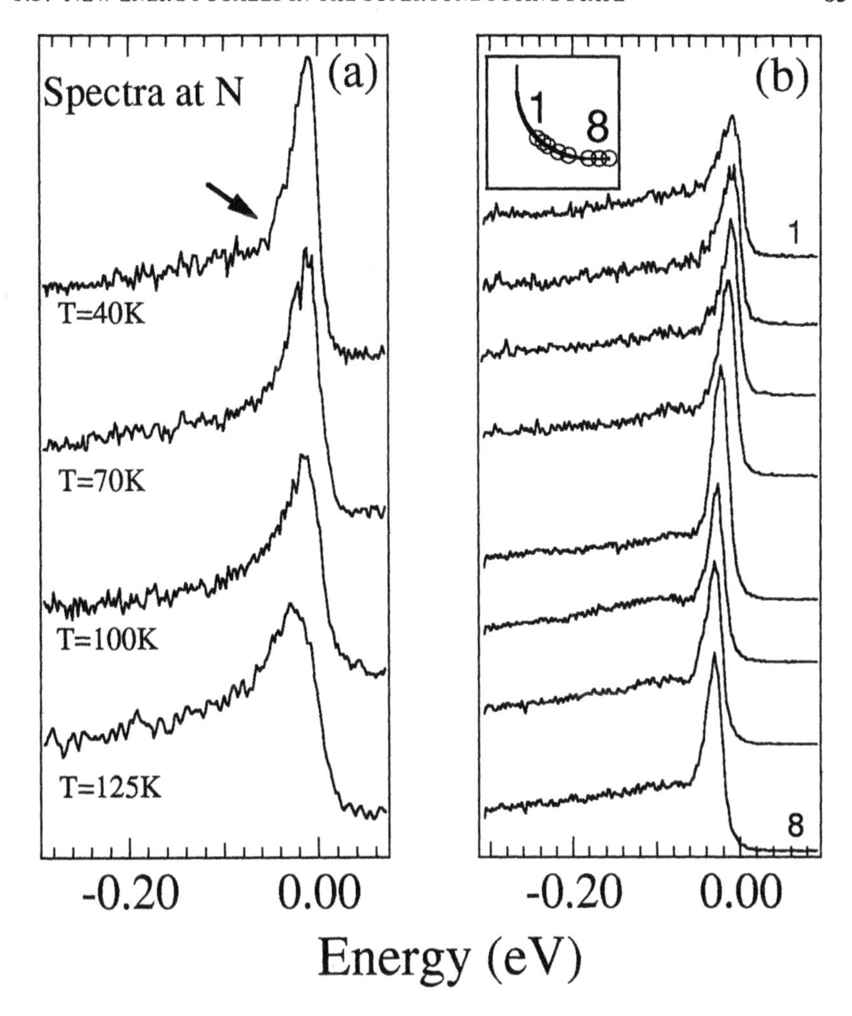


Figure 6.6: (a) Temperature dependence of the peak at $N$. (b) Comparison of the superconducting state spectra ($T$ =40K) along the Fermi surface for a few points indicated in the inset.

function below $T_c$ have been well studied near the $(\pi, 0)$ point of the zone [12, 14, 15].

From the arguments presented in Sec. 3, we see that it is much simpler to interpret the MDC peak positions, and consequently we will now focus on the change in the MDC dispersion going from the normal (N) to the superconducting (S) state shown in Fig. 6.7a. The striking feature of Fig. 6.7a is the development of a kink in the dispersion below $T_c$. We now present an analysis of this effect

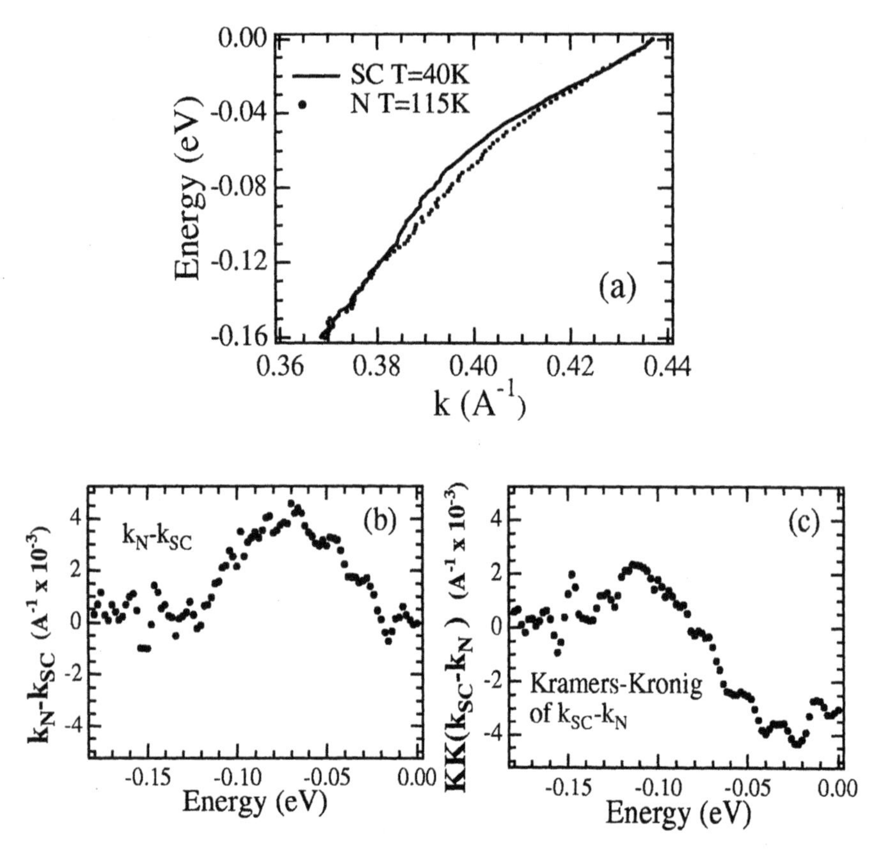


Figure 6.7: ARPES data along the nodal direction ($\Gamma Y$) from Fig. 6.1. (a) MDC dispersions at $h\nu$ =28 eV in the superconducting state (solid line) and normal state (dashed line). (b) change in MDC dispersion from (a). (c) Kramers-Krönig transform of (b).

that uses only the fundamental concepts of many-body theory, and therefore does not rely on any fittings to understand its origin.

At fixed $\omega$, let the dispersion change from $k_N$ to $k_S$ as in Fig. 6.7a. Since the MDC lineshape is Lorentzian, the peak in the spectral function is given by the pole condition, which in turn yileds $v_F^0(k_N - k_S) = \Sigma'_S(\omega) - \Sigma'_N(\omega)$ on going from the normal to the superconducting states. We thus directly obtain the change in real part of $\Sigma$, plotted in Fig. 6.7b. The Kramers-Krönig transformation of $\Sigma'_S - \Sigma'_N$ then yields $\Sigma''_S - \Sigma''_N$, plotted in Fig. 6.7c, which shows that $|\Sigma''_S|$ is smaller than $|\Sigma''_N|$ at low energies. Note that because the change in $|\Sigma'_N|$ goes to zero at both ends of the curve in Fig. 6.7a, the Kramers-Krönig transformation can be applied without further assumptions.

We can now compare the change in $|\Sigma''_N|$ obtained from the change in dis-

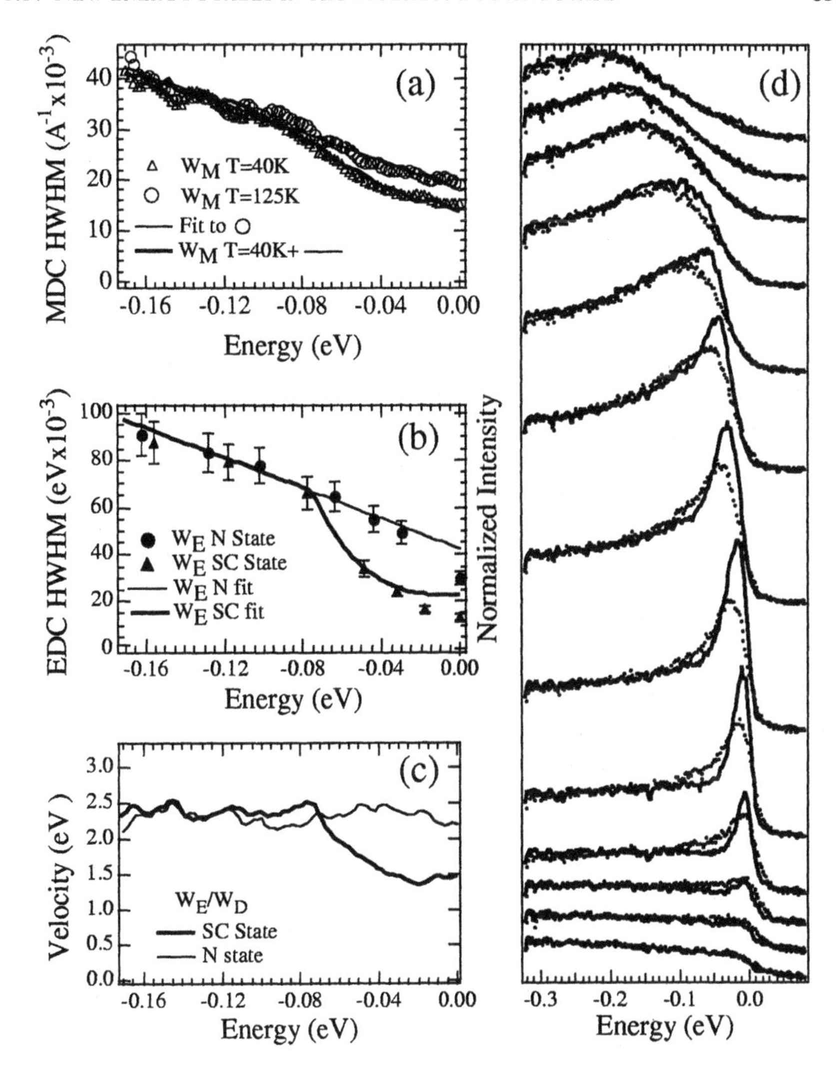


Figure 6.8: (a) Comparison of changes in $\Sigma''$ obtained directly from the MDC widths (HWHM) to the one obtained from the dispersion in Fig. 6.7a by using the Kramers-Krönig transform. (b) HWHM data obtained from EDCs in (d) (Ref. [16]). (c) Renormalized velocity obtained from dividing (b) by (a). Note that the numeric values of $v$ may be slightly different from the ones obtained from the dispersion because the MDC and EDC widths are affected differently by the experimental resolution. (d) EDC's in the normal (dotted) and superconducting (full lines) states along the $(\pi, \pi)$ direction in the vicinity of the Fermi crossing.

persion to the change in $W_M = |\Sigma''|/v_F^0$ estimated directly from the MDC Lorentzian linewidths in the data, as shown in Fig. 6.8a. The normal-state curve was obtained from a linear fit to the corresponding MDC width data points in Fig. 6.8a, and then the difference from Fig. 6.7c was added to it to generate the low temperature curve. We are thus able to make a quantitative connection between the appearance of a kink in the (MDC) dispersion below $T_c$ and a drop in the low-energy scattering rate in the superconducting state relative to the normal state. We again emphasize that we have estimated these $T$-dependent changes in the complex self-energy without making fits to the spectral (EDC) lineshape, thus avoiding the problem of modeling the $\omega$ depedence of $\Sigma$ and the extrinsic background).

Alternatively, one can also analyze the data by starting with the linewidths and deducing changes in the dispersion. The ratio of the EDC and MDC widths is given by $W_E/W_M = v_F^0/[1 - \partial\Sigma'/\partial\omega] = v_F$, the renormalized Fermi velocity, shown in Fig. 6.8b. One can see a sudden drop in velocity around 80 meV, from $\approx$ 2.5 eVÅabove to $\approx$ 1.5 eVÅbelow 80 meV. This results in a sharp kink in the dispersion, which can be directly seen in the raw data shown in Figs. 6.1 and 6.9. The changes in the self-energy are evident in the raw spectra shown in Fig. 6.8d.

Notice that we have completed a full circle. From the change in dispersion of the constant-$\omega$ cuts (MDCs) of $A(\mathbf{k}, \omega)$, we derived the change in the real self-energy, whose Kramers-Krönig transform yields the change in the imaginary self-energy. This agrees with the change in width of the constant-$\omega$ cuts (MDCs) of $A(\mathbf{k}, \omega)$, which together with the change in width of the constant-$\mathbf{k}$ cuts (EDCs) of $A(\mathbf{k}, \omega)$, yields the change in dispersion. This remarkable connection between different aspects of the data reveals the full power of analyticity and causality in many-body theory.

The reduction in $\Sigma''$ below $T_c$ reflects a drop in the scattering rate. We now ask ourselves how this drop in the scattering rate evolves as we move along the Fermi surface. Away from the node, a quantitative analysis (like the one above) becomes more complicated [17]. Here, we will simply look at the evolution of the dispersion. In Fig. 6.9, we plot raw (2D) data as obtained from our detector for a series of cuts parallel to the $MY$ direction (left panels). We start from the bottom row that corresponds to a cut close to the node and reveals the same kink described above. As we move towards $(\pi, 0)$, the dispersion kink (shown as crosses and dots) becomes more pronounced, and at around $k_x$=0.55 develops into a break separating the faster-dispersing, high-energy part of the spectrum from the slower-dispersing, low-energy part. This break leads to the appearance of two features in the EDCs, shown in the right panels of Fig. 6.9. Further, towards $(\pi, 0)$, the low-energy feature (the quasiparticle peak) becomes almost dispersionless. At the $(\pi, 0)$ point, this break effect becomes the most pronounced, giving rise to the well-known peak-dip-hump structure.

We note that the evolution of the dispersion from kink to break occurs continuously, at the exact same energy of 80 meV. We must therefore conclude that they are in fact of the same origin. As discussed earlier, the most natural explanation of this effect near $(\pi, 0)$ is in terms of electrons interacting with a collective mode

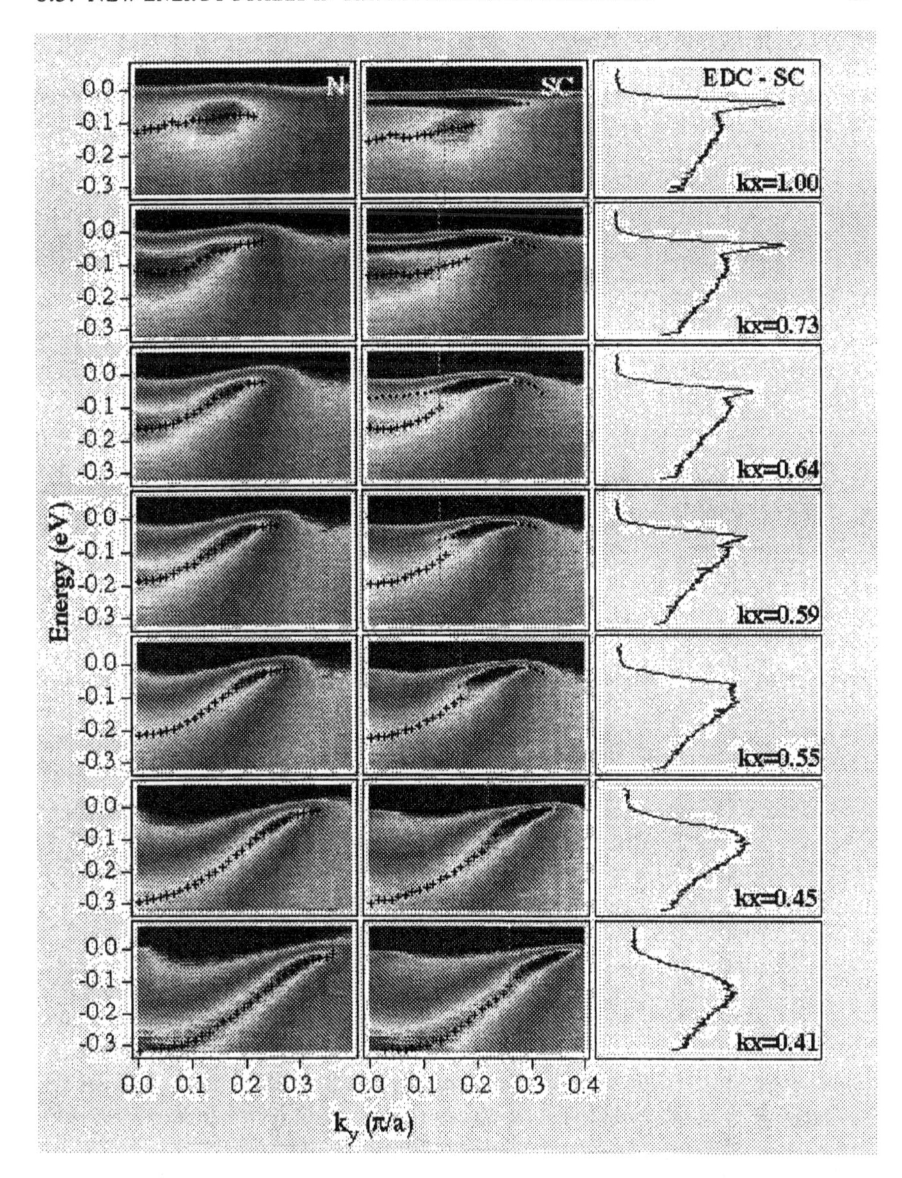


Figure 6.9: Left panels: Normal-state ARPES intensity plots along selected cuts (shown in zone inset in Fig. 6.1). EDC peak positions are indicated by crosses. Middle panels: Superconducting state intensity plots at the same cuts as in the left panels. Crosses indicate positions of peaks that are the same as in the normal state, and dots show peak positions that change in the superconducting state. Right panels: EDCs at locations marked by the dotted vertical lines in the middle panels.

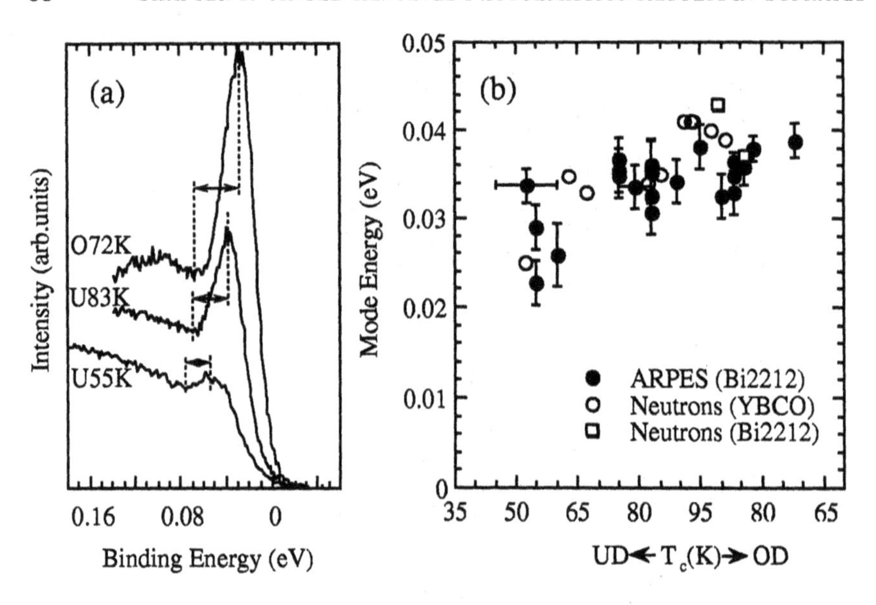


Figure 6.10: FIG. 5. Doping dependence of the mode energy: (a) Spectra at $(\pi, 0)$ showing the decrease in the energy separation of the peak and dip with underdoping (from Ref. [15]. ( b) Doping dependence of the collective mode energy inferred from ARPES together with that inferred from neutron data (for the latter, YBCO results as compiled in Ref. [19], Bi2212 results of Refs. [20, 21, 22]).

[14] below $T_c$, whose energy and characteristic wavevector matches [15] the sharp magnetic resonance observed directly by neutron scattering. Since the break evolves smoothly from the kink, and sets in at the same energy, we are led to the conclusion that the same collective mode effect dominates the superconducting lineshape at all points in the zone, including the node [18]. In essence, there is a suppression of the low energy scattering rate due to the finite energy of the mode.

Since the neutron mode is characterized by a $(\pi, \pi)$ wavevector, one would expect its effect on the lineshape to be much stronger at points in the zone that are spanned by $(\pi, \pi)$. This is indeed the case, as can be seen in Fig. 6.9. Near the node, the drop in the scattering rate is such that it causes a kink in the dispersion. But upon approaching the $(\pi, 0)$ point, the drop in $\Sigma''$ is so drastic as to produce a dip in the spectral function. At this point, we can only put bounds on how fast $\Sigma''$ drops. Near the node it drops close to $\omega^3$ (see Fig. 6.8b), while near $(\pi, 0)$ it must drop at least as fast as $\omega^6$ [23]. The energy at which the drop occurs is consistent with our previous work, since it is seen in the EDC at an energy equal to the sum [24] of the mode energy $\Omega_0$ and the maximum gap $\Delta_0$, both of which are $\approx$40 meV for optimally doped samples. However ARPES and neutron scattering give the same number for the mode energy not only for optimal doping, but continue to agree as the doping is reduced, as shown in Fig. 6.10 [15]. It is the presence of this finite-energy scale that leads to deviations from the normal-state quantum critical scaling below $T_c$.

In summary, we have discussed some of the salient features of the ARPES spectra of the cuprates, which indicate a rather unusual normal state, as long proposed by Anderson. We have also shown that ARPES is able to measure the spectral function in such exquisite detail that rather quantitative statements can be made regarding the many-body aspects of the problem. Finally, we have shown that new energy scales appear in the superconducting state, which give rise to large changes in the nature of the electron states between the normal and superconducting states.

To conclude, since the work presented here is the result of a most congenial collaboration, I would like to acknowledge my collaborators (in alphabetical order) H. Ding, H. Fretwell, P. Guptasarma, D. Hinks, K. Kadowaki, A. Kaminski, J. Mesot, T. Mochiku, M. Norman, M. Randeria, T. Sato, T. Takahashi, M. Takano, T. Takeuchi, and T. Yokoya, thank them for embarking on this adventure with me, and the National Science Foundation and the Department of Energy for making it possible.

## BIBLIOGRAPHY

[1] P.W. Anderson, Science **235**, 1196 (1987).

[2] M. Randeria *et al.*, Phys. Rev. Lett. **74**, 4951 (1995).

[3] The optimally doped $Bi_2Sr_2CaCu_2O_{8+\delta}$($T_c$=90K) samples were grown using the floating zone method. The samples were mounted with either the $\Gamma X$ or $\Gamma M$ axis parallel to the photon polarization vector, and cleaved in situ at pressures less than $5 \cdot 10^{-11}$ Torr. We used a Scienta SES 200 electron analyzer with energy resolution of 16 meV and a momentum resolution of 0.0097 $Å^{-1}$. Measurements were carried out at the Synchrotron Radiation Center in Madison WI, on the U1 undulator beamline supplying $10^{12}$ photons/sec.

[4] P. W. Anderson, *The Theory of Superconductivity in the High $T_c$ Cuprates* (Princeton Univ. Press, Princeton, 1997).

[5] Z.-X. Shen and D. S. Dessau, Phys. Rep. **253**, 1 (1995).

[6] J. C. Campuzano *et al.*, in *The Gap Symmetry and Fluctuations in High-$T_c$ Superconductors*, eds. J. Bok *et al.* (Plenum, New York, 1998), p. 229.

[7] T. Valla *et al.*, Science **285**, 2110 (1999).

[8] Note that: (1) a linear dependence of $\Sigma'$ on $k - k_F$ can be simply absorbed into the definition of $v_F^0$; (2) $\Sigma$ does have an implicit dependence on $\theta$.

[9] Note the node itself is a single point, so makes a contribution of measure zero to the spectrum.

[10] The separation of the spectral function into a quasiparticle piece and an incoherent part can be justified from the breakup of the Green's function into corresponding singular and regular parts. See P. Nozieres, *Theory of interacting Fermi systems* (Addison-Wesley, Reading, 1964).

[11] M. R. Norman *et al.*, Phys. Rev. B **57**, R11093 (1998).

[12] M. R. Norman *et al.*, Phys. Rev. Lett. **79**, 3506 (1997).

[13] B. Batlogg and C. M. Varma, Physics World, Feb., p. 33 (2000).

[14] M. R. Norman and H. Ding, Phys. Rev. B **57**, R11089 (1998).

[15] J.C. Campuzano *et al.*, Phys. Rev. Lett. **83**, 3709 (1999).

[16] A. Kaminski *et al.*, Phys. Rev. Lett. **84**, 1788 (2000).

[17] Such an analysis would need to include (1) the non-zero superconducting gap which affects $\Sigma$ through a pairing contribution; and (2) the quadratic dispersion about the $\Gamma M$ symmetry line, which is close to $E_F$ near $M = (\pi, 0)$.

[18] Similar kinks have been seen in ARPES in normal metals due to phonons, see M. Hengsberger *et al.*, Phys. Rev. Lett. **83**, 592 (1999); T. Valla *et al.*, *ibid.*, 2085 (1999).

[19] Neutron results are reviewed by P. Bourges, in The Gap Symmetry and Fluctuations in High Tc Superconductors, edited by J. Bok et al. (Plenum, New York, 1998), p. 349.

[20] H. A. Mook, F. Dogan, and B. C. Chakoumakos, e-print cond-mat/9811100.

[21] H. F. Fong *et al.*, Nature **398**, 588 (1999).

[22] P. Dai et al., Phys. Rev. Lett. **77**, 5425 (1996).

[23] M. R. Norman *et al.*, Phys. Rev. B **60**, 7585 (1999).

[24] As discussed in ref. [15], the different doping dependences of $\Omega_0$ and $\Delta_0$ can be used to discriminate between $2\Delta_0$ and $\Omega_0 + \Delta_0$ which happen to coincide near optimality.

# CHAPTER 7

# SPIN EXCITATIONS IN COPPER OXIDE SUPERCONDUCTORS

BERNHARD KEIMER

Max-Planck-Institut für Festkörperforschung, 70569 Stuttgart,Germany [1], and
Department of Physics; Princeton University, Princeton, NJ 08544, U.S.A.

ABSTRACT

We review recent results of neutron scattering studies of the high-temperature superconductors $La_{2-x}Sr_xCuO_4$, $YBa_2Cu_3O_{6+x}$, and $Bi_2Sr_2CaCu_2O_{8+\delta}$, focusing on the magnetic resonance peak that dominates the spectrum in the latter two compounds.

## 7.1 INTRODUCTION

Magnetic spectroscopy with neutrons has played a key role in elucidating the microscopic properties of high temperature superconductors. Neutrons are a weakly interacting probe entirely free of surface effects. The differential scattering cross section is simply proportional to the wave-vector and energy dependent susceptibility $\chi''(q,\omega)$, a quantity that is highly sensitive to Coulomb correlations which in turn are undoubtedly at the root of many of the unusual physical properties of the cuprates. Further, the neutron flux at a typical research reactor source peaks at energies comparable to the superconducting gap in these materials. Neutron scattering can hence be used efficiently to test theories of the superconductivity in the cuprates in a highly specific manner. As first pointed out by P.W. Anderson [1], the neutron cross section in the superconducting state even contains terms that are sensitive to the symmetry of the pairing wave function. The downside of the weak interaction of the neutron spin with electronic moments is that large crystals are needed in order to obtain a sufficiently large signal. This constraint is particularly pertinent in the cuprates which start out as spin-$\frac{1}{2}$ antiferromagnetic

[1] Present Address

insulators when undoped. The inelastic cross section is proportional to $S^2$ and is further reduced by about a factor of two due to zero-point fluctuations for $S = \frac{1}{2}$. In the metallic regime these fluctuations are enhanced and the neutron signal is further weakened.

The number of cuprates amenable to investigation by neutron spectroscopy is therefore limited by the state-of-the-art in crystal growth, a limitation that has proven to be severe. Superconducting cuprates are complex materials containing at least four different elements, and the phase relations are often unfavorable for crystal growth. Some particularly interesting materials contain toxic and volatile elements such as Tl and Hg, others are bound mostly by weak van-der-Waals forces perpendicular to the copper oxide planes (which contain the strongest bonds and provide rigidity to the structure) and therefore exhibit an extremely anisotropic growth habit. The result of all of these constraints is that until recently, more than a dozen years after the discovery of high temperature superconductivity, only two families of cuprates had been investigated by magnetic neutron spectroscopy: $La_{2-x}Sr_xCuO_4$ and $YBa_2Cu_3O_{6+x}$. Due to the different crystal structures and doping mechanisms of these two cuprates families, the influence of soft lattice fluctuations, disorder and magnetic interlayer interactions on $\chi''(q, \omega)$ is still a matter of debate. The magnetic spectra of $La_{2-x}Sr_xCuO_4$ and $YBa_2Cu_3O_{6+x}$, though similar at some level, exhibit a variety of important differences, and it is still largely unknown which features should be considered generic to the cuprates. Our strategy will be to systematically overcome the technical constraints in crystal growth and explore different cuprate families by neutron spectroscopy. We have started with $Bi_2Sr_2CaCu_2O_{8+\delta}$ [2], which is particularly interesting because it is the system of choice for another important momentum and energy resolved probe, angle-resolved photoemission spectroscopy (ARPES), a technique that is capable of measuring the single-electron spectral function on a comparable energy scale. This opens up the possibility of comparing the data obtained by both techniques on a quantitative basis.

## 7.2 NEUTRON SPECTROSCOPY RESULTS IN CUPRATES

Before describing the latest results, we review the present state of affairs in neutron spectroscopy of $La_{2-x}Sr_xCuO_4$ and $YBa_2Cu_3O_{6+x}$. For $x \sim 0$, both materials are $S = \frac{1}{2}$ antiferromagnetic insulators with predominantly two-dimensional exchange interactions (nearest-neighbor superexchange $J_{\parallel} \sim 100$ meV within the copper oxide layers). Despite the similar $J_{\parallel}$, the Néel temperatures in both materials are rather different: $T_N \sim 325$ K [3] in $La_2CuO_4$ and $T_N \sim 515$ K in $YBa_2Cu_3O_6$ [4]. This is because a purely two-dimensional Heisenberg system does not order at $T > 0$, and the interlayer interactions that ultimately drive the transition to long range order are rather different in the two systems, being almost frustrated in $La_2CuO_4$ but unfrustrated in $YBa_2Cu_3O_6$. Moreover, and more importantly, the spin system of $YBa_2Cu_3O_6$ consists of weakly coupled bilayer units with strong intra-unit exchange interactions ($J_\perp \sim 10$ meV [5]) whereas the unit in $La_2CuO_4$ comprises only a single layer. These strong intra-bilayer interactions are manifest throughout the phase diagram of $YBa_2Cu_3O_{6+x}$, and they are an important possible origin of differences between the two systems. As a (certainly inexact) one-dimensional analogy, consider the very different behaviors of a single chain and a two-leg ladder.

Doping also proceeds quite differently in the two materials. In $La_{2-x}Sr_xCuO_4$, the Sr content can be varied continuously. At low Sr concentrations the system remains insulating at zero temperature, but the magnetic correlation length becomes finite. The temperature and doping dependence of the correlation length in this regime [6] can be best described in models [7] in which the holes segregate into disordered one-dimensional (1D), meandering structures that act as domain boundaries for finite-size, locally Néel ordered segments. The early work was performed on crystals that had been grown in Pt crucibles, and a small number of Pt impurities were dissolved in the crystal. Recently, crystals grown by the crucible-free floating zone technique were shown [8] to exhibit a superstructure in which the segregated holes form well ordered "stripes" in the (1,1) direction of the copper oxide layers. It appears that in this (nonmetallic) regime of the phase diagram, defects can easily pin and disorder the stripe superstructure.

In $YBa_2Cu_3O_{6+x}$, holes are introduced by filling interstitials in an initially oxygen-free $Cu^{1+}$-layer by oxygen atoms which eventually donate holes into the layers. At low oxygen concentration, however, the holes only convert nonmagnetic $Cu^{1+}$ to $Cu^{2+}$, thus introducing disorder into the *interlayer* coupling. As the oxygen concentration is raised, this layer organizes itself into CuO chains, holes are donated into the layers, and the material becomes metallic. This chain-formation process is difficult to control, and homogeneous $YBa_2Cu_3O_{6+x}$ samples with low $T_N$ or low $T_c$ near the insulator-metal boundary are notoriously hard to prepare. Once the chains are formed, however, metallic $YBa_2Cu_3O_{6+x}$ can be prepared in *stoichiometric* form, that is, without disorder from dopant ions. In particular, fully oxygenated $YBa_2Cu_3O_7$ is entirely "self-doped" and hence (in principle) free of any disorder. In underdoped $YBa_2Cu_3O_{6+x}$, finite-size chain segments may introduce disorder (though on a larger length scale than in $La_{2-x}Sr_xCuO_4$). Even this rather mild form of disorder can be annealed out through heat treatments that result in the formation of well ordered chain superstructures [10], in particular the so-called "orthorhombic-II" structure at $x = 0.5$ in which every other chain is empty. Finally, our $YBa_2Cu_3O_{6+x}$ crystals are prepared by crucible-free techniques so that extrinsic defects are not incorporated into the crystals. The absence, or near-absence, of disorder in $YBa_2Cu_3O_{6+x}$ contrasts sharply with the situation in $La_{2-x}Sr_xCuO_4$ where disorder due to Sr acceptors is always present.

## 7.3 STRIPE FORMATION IN $LA_{2-x}SR_xCUO_4$

Finally, $La_{2-x}Sr_xCuO_4$ and $YBa_2Cu_3O_{6+x}$ also differ in their low-frequency lattice dynamics. Due to steric effects, *i.e.* a mismatch between the natural lattice parameters of the copper oxide layers and of the intervening charge reservoir layers, the $CuO_2$ layers are buckled in most cuprates. In $YBa_2Cu_3O_{6+x}$ and many other cuprate families, the buckling distortion is frozen in at some high temperature, and the associated optical phonon modes have high energies in the experimentally relevant temperature range. In $La_{2-x}Sr_xCuO_4$, the tetragonal-to-orthorhombic transition where the buckling sets in occurs at much lower temperatures, below room temperature near optimal doping. Even below this transition, soft phonons remain that can drive additional, lower temperature structural transitions into orthorhombic or tetragonal phases with larger unit cells [11]. The material can be nudged into these phases by isoelectronic substitutions of La (*e.g.*, by Nd). In the low-temperature tetragonal phase of Nd-substituted $La_{2-x}Sr_xCuO_4$ with $x \sim \frac{1}{8}$ (*i.e.*, in the superconducting regime of the phase diagram), Tranquada *et al.* [12] first observed a statically ordered state which was interpreted as arising from charge stripes, this time parallel to (1,0) or (0,1), acting as antiphase domain boundaries between antiferromagnetically ordered regions. The neutron signal arising from the charge order itself is weak, and the most salient signature of charge stripes is an associated spin density modulation that manifests itself as four well-defined incommensurate peaks at $\mathbf{Q} = (\pi(1 \pm \delta), \pi)$ and $\mathbf{Q} = (\pi, \pi(1 \pm \delta))$ (in square lattice notation with unit lattice constant). The interpretation of the neutron data in terms of one-dimensional domains is supported by transport measurements that show a vanishing Hall effect in the ordered phase.

Recently, ordered (1,0) stripes were also observed in Nd-free $La_{2-x}Sr_xCuO_4$ with $x$ close to $\frac{1}{8}$ [9]. In a previous experiment, low-energy fluctuations in orthorhombic, optimally-doped $La_{2-x}Sr_xCuO_4$, with sharp features in momentum space very similar to those in the statically ordered materials were observed [13, 14]. These can hence be best understood in a picture in which they are due to fluctuating stripe domains. The low temperature lattice instability, which is intrinsically unrelated to the electronic structure, can thus at least be held responsible for pinning the soft charge-stripe fluctuations in some variants of $La_{2-x}Sr_xCuO_4$. By continuity, one may wonder to what degree the accidental proximity to this lattice instability is a necessary precondition for the soft charge-stripe fluctuations observed throughout the phase diagram of this particular cuprate family. A closely related, and much more important, question is whether the stripe order and/or fluctuations in turn are a necessary precondition for high temperature superconductivity, or whether they are actually detrimental to superconductivity. After all, the maximum superconducting transition temperature of $La_{2-x}Sr_xCuO_4$ ($\sim$ 40 K) is low compared to other high-$T_c$ compounds, even the "single-layer" material $Tl_2Ba_2CuO_{6+\delta}$. These questions can only be answered by comparing the spin dynamics of $La_{2-x}Sr_xCuO_4$ to that of other cuprates where low-temperature lattice instabilities of this type are not present.

## 7.4 MAGNETIC RESONANCE PEAK IN CUPRATES

Neutron scattering experiments on the $YBa_2Cu_3O_{6+x}$ system have, however, revealed excitations that are peaked at $\mathbf{Q}_{AF} = (\pi, \pi)$, the ordering wave vector of the 2D antiferromagnetic (AF) state observed when the doping level is reduced to zero. In particular, the commensurate "resonance peak" at $\mathbf{Q} = (\pi, \pi)$ that dominates the spectrum in the superconducting state [17, 18, 1, 19, 22, 2] is difficult to reconcile with scenarios based on fluctuating 1D domains incommensurate with the host lattice. Recently, an incommensurate pattern with a four-fold symmetry reminiscent of $La_{2-x}Sr_xCuO_4$ has also been discovered in some constant-energy cuts of the magnetic spectrum of underdoped $YBa_2Cu_3O_{6.6}$, which was taken as experimental support for stripe-based scenarios of superconductivity [15, 16]. Even more recently, however, we have conducted a neutron scattering study of near-optimally doped $YBa_2Cu_3O_{6.85}$ ($T_c$ = 89 K) demonstrating that (unlike in $La_{2-x}Sr_xCuO_4$) the incommensurate pattern appears only below $T_c$ [23]. The data of Fig. 7.1 show clearly that both the resonance peak and the incommensurate signal set on sharply at $T_c$. These data are consistent with 2D theories (not invoking stripes) that predict a downward dispersion of the magnetic resonance peak. At high energies, above the resonance peak, the spin excitations in the underdoped materials exhibit a different, *upward* dispersion reminiscent of spin waves in the AF insulator. The overall shape of the spin excitations in the superconducting state therefore resembles an "hour glass" in an energy-momentum diagram. It should be noted, however, that by far the largest intensity belongs to the commensurate resonance peak at the "neck" of the hour glass. Low-energy magnetic excitations in the normal state are commensurate and centered at $\mathbf{Q} = (\pi, \pi)$ at all energies investigated.

The resonance peak has thus far not been observed in $La_{2-x}Sr_xCuO_4$. The most likely reason is the effect of disorder due to Sr acceptors already mentioned above. In fact, experiments in which disorder was introduced into optimally doped $YBa_2Cu_3O_{6+x}$ in a controlled fashion have demonstrated that the resonance peak is extraordinarily sensitive to minute amounts (1% or less) of Zn [20] and Ni [21] impurities. Nevertheless, the best way to make sure that this unusual excitation is not stabilized by chemical or structural peculiarities unique to $YBa_2Cu_3O_{6+x}$ (*e.g.*, the CuO chains) is to investigate the spin dynamics of other high-$T_c$ compounds. We have begun with $Bi_2Sr_2CaCu_2O_{8+\delta}$, a bilayer superconductor with maximum $T_c$ around 91 K that does not contain copper-oxide building blocks other than the layers. Doping proceeds by donation of holes from the Bi-O layers into the $CuO_2$ layers. This is because of an incommensurate modulation with a noninteger Bi/O ratio that also represents a significant potential source of (phason or amplitudon) disorder. While not much is known about

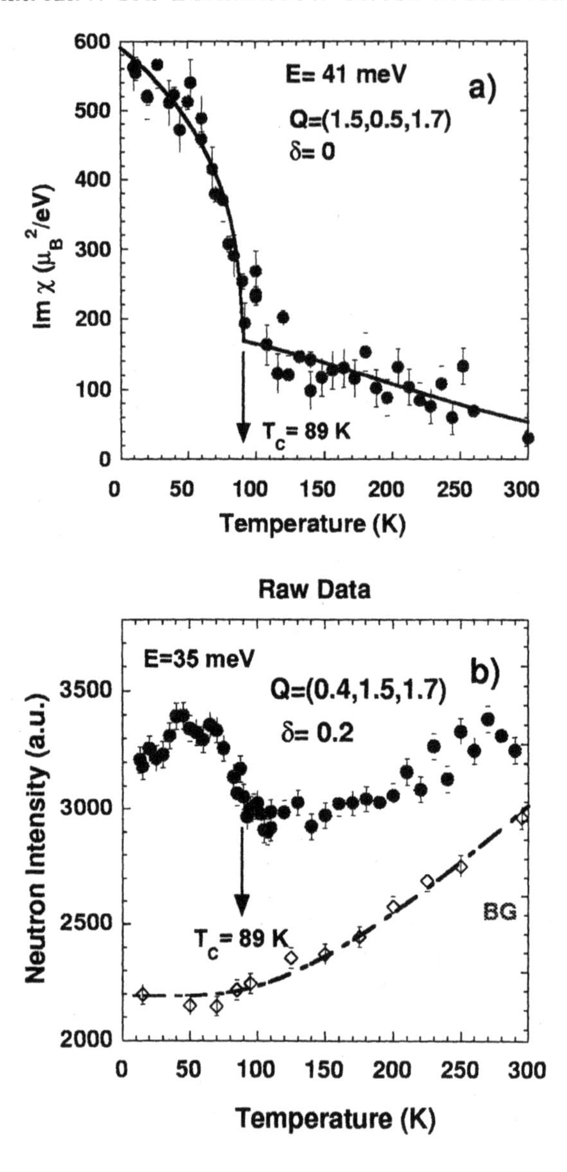


Figure 7.1: (a) Temperature dependence of spin susceptibility in absolute units at the resonance energy $E_r$= 41 meV of near-optimally doped $YBa_2Cu_3O_{6.85}$ ($T_c$ = 89 K). (b) Temperature dependence of the neutron intensity at E = 35 meV (below the resonance peak) and at the incommensurate wave vector for which the magnetic intensity is maximum at this energy (full circles). The open squares represent the background.

disorder in the Bi-O layer, one may *a priori* expect $Bi_2Sr_2CaCu_2O_{8+\delta}$ to be intermediate between $La_{2-x}Sr_xCuO_4$ and $YBa_2Cu_3O_{6+x}$ in terms of the influence of disorder on the electronic properties. This expectation is in fact confirmed by our experiments [2] on a large single crystal of optimally doped $Bi_2Sr_2CaCu_2O_{8+\delta}$. These do show a magnetic resonance peak whose behavior is nearly identical to that of the resonance peak in optimally doped $YBa_2Cu_3O_{6+x}$: The magnetic signal is localized in momentum space around the commensurate wave vector $\mathbf{Q} = (\pi, \pi)$, centered at a characteristic energy of 43 meV, and present only in the superconducting state. However, the resonance peak is substantially broader in energy than in optimally doped $YBa_2Cu_3O_{6+x}$ which signals a reduced lifetime of the mode, probably due to scattering from defects. As the limiting factor in these neutron scattering studies is the signal-to-background ratio, with the background mostly due to single-phonon and multiphonon events, the magnetic peak would rapidly become unobservable if the lifetime reduction continued with increased disorder; this is probably what happens in $La_{2-x}Sr_xCuO_4$.

## 7.5 ORIGIN OF THE RESONANCE PEAK

Since the resonance peak is common to two compounds with rather different chemical composition and lattice structure, we can now be sure that it is a phenomenon generic to the high-$T_c$ materials. The only feature shared by both $YBa_2Cu_3O_{6+x}$ and $Bi_2Sr_2CaCu_2O_{8+\delta}$ is the bilayer structure, and obviously the next step in our program is to investigate a clean single-layer material. Meanwhile, we have used the newly available $Bi_2Sr_2CaCu_2O_{8+\delta}$ system to answer an important open question concerning the dependence of the mode energy on the superconducting transition temperature and the doping level. This question is important because two rivaling models for the resonance peak predict different behavior in this regard. Specifically, theoretical interpretations of the resonance peak fall into two categories, attributing the mode to instabilities in the particle-particle [30] and particle-hole [31] channels, respectively.

In the former model, the magnetic resonance peak is identified with the so-called $\pi$-excitation that can be visualized as a spin triplet pair of electrons with center of mass momentum $(\pi, \pi)$ and the same relative wave function as a $d$-wave Cooper pair. The sharp resonance observed in the neutron measurements can be ascribed to resonant scattering of Cooper pairs into $\pi$-pairs. In this model, the intensity of the magnetic resonance peak is controlled by the magnitude of the $d$-wave order parameter, whereas the mode energy depends on hole doping only. The interpretation of the resonance mode is embedded into the more general SO(5) model linking antiferromagnetism and $d$-wave superconductivity. The SO(5) symmetry is exact at the quantum critical point linking both phases, and the mode energy is a quantitative measure of the departure from this situation in the superconducting phase. Unless different physics intervenes, it should therefore monotonically increase with doping.

The competing model [31] attributes the magnetic resonance peak to an excitonic particle-hole bound state that is pulled below the continuum of spin flip excitations by magnetic interactions. As this resonant excitation is a consequence of a delicate interplay between the superconducting gap, the shape of the Fermi surface in the normal state, and the antiferromagnetic spin correlations, its energy should be very sensitive to the magnitude of the gap. At least in the optimally doped and overdoped regimes of the phase diagram, where "pseudogap" effects are not important, the gap is expected to scale with $T_c$.

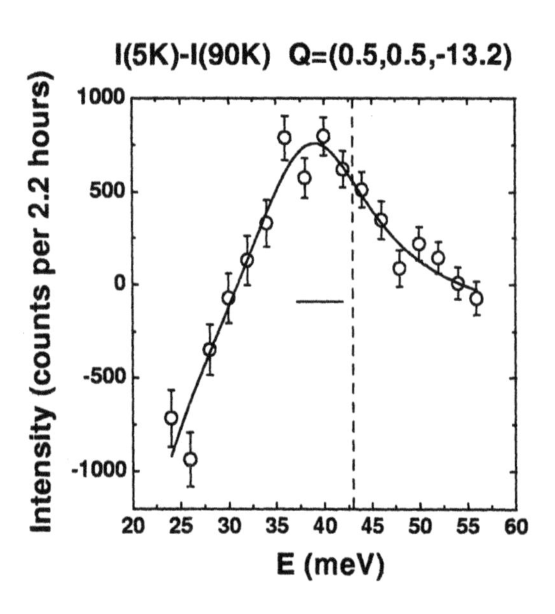


Figure 7.2: Difference spectrum of the neutron intensities at T = 5 K ($\prec T_c$) and T = 90 K ($\succ T_c$), at a wave vector equivalent to $\mathbf{Q} = (\pi, \pi)$ in overdoped $Bi_2Sr_2CaCu_2O_{8+\delta}$ ($T_c = 83$ K) The line indicates the resonance energy in an optimally doped sample of the same material.

In pure $YBa_2Cu_3O_{6+x}$, only the underdoped and optimally doped regimes are experimentally accessible, where $T_c$ is monotonically related to the hole doping. (Overdoping is possible by Ca-substitution, but the sample preparation is difficult and comes at the expense of introducing additional disorder.) Complications from "pseudogap" effects aside, our observation of a decrease of the mode energy upon approaching the antiferromagnetic phase is thus compatible with both scenarios. A clean separation is only possible in the overdoped state where the monotonic relation no longer holds and $T_c$ begins to fall upon increasing the doping level.

Figure 7.2 shows the low temperature magnetic spectrum of an overdoped $Bi_2Sr_2CaCu_2O_{8+\delta}$ sample with $T_c = 83$ K taken at wave vector $(\pi, \pi)$. Clearly, the resonance peak is shifted downward from its position in the optimally doped state (43 meV). These data are an essential complement to an extensive data set on

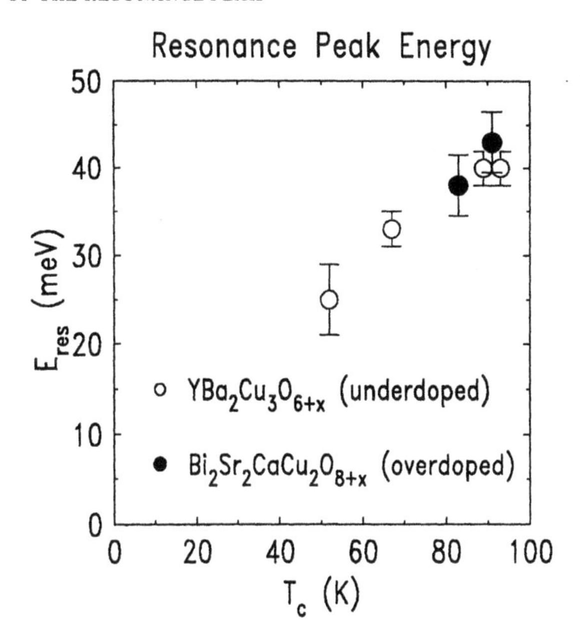


Figure 7.3: A synopsis of the resonance peak energy $E_{res}$ in $YBa_2Cu_3O_{6+x}$ and $Bi_2Sr_2CaCu_2O_{8+\delta}$ (Ref. [2, 22]), plotted as a function of the superconducting transition temperature $T_c$.

the resonance peak in underdoped $YBa_2Cu_3O_{6+x}$. A representative subset [22] is shown in Fig. 7.3 along with the presently available data on $Bi_2Sr_2CaCu_2O_{8+\delta}$. While we did not confirm the linear relationship between the mode energy and the doping level predicted by the SO(5) model, Fig. 7.3 suggests that the parameter controlling the mode energy is actually the transition temperature $T_c$. This is the behavior expected on the basis of the simplest possible interpretation of the resonance peak [1], first pointed out to us by P.W. Anderson, that attributes it to a particle-hole excitation across the *d*-wave superconducting gap (which in turn is proportional to $T_c$). Qualitatively, an analogous relationship is also expected to hold in more elaborate models incorporating interactions between the particle and hole in the final state. While this identification is fraught with uncertainty in the underdoped regime, where non-Fermi-liquid physics complicates the interpretation of spectroscopic data, there is general agreement that a simple Fermi-liquid based approach is more appropriate for the overdoped materials. This consideration adds to the significance of the data presented here. Of course, a direct measurement of the superconducting gap in our $Bi_2Sr_2CaCu_2O_{8+\delta}$ samples is ultimately required as a confirmation.

## 7.6 CONCLUSION

In conclusion, we have recently taken a big step forward in an ongoing effort to refine our understanding of spin excitations in the cuprates, especially the magnetic resonance peak. The results reported here will also help confirm and extend promising attempts to develop a unified phenomenology of magnetic and charge spectroscopies of the cuprates [24, 25, 26, 27] and to account for the superconducting condensation energy [28, 29].

## ACKNOWLEDGMENTS

We gratefully acknowledge I.A. Aksay, J. Bossy, P. Bourges, H.F. Fong, G.D. Gu, H. He, A.Ivanov, N. Koshizuka, B. Liang, C.T. Lin, D.L. Milius, L. P. Regnault, E. Schoenherr, and Y. Sidis for collaboration. The work at Princeton University was supported by the National Science Foundation under DMR-9809483.

## BIBLIOGRAPHY

[1] H.F. Fong *et al.*, Phys. Rev. Lett. **75**, 316 (1995); Phys. Rev. B **54**, 6708 (1996).

[2] H.F. Fong *et al.*, Nature **398**, 588 (1999); H. He *et al.*, cond-mat/0002013.

[3] B. Keimer *et al.*, Phys. Rev. B **45**, 7430 (1992).

[4] J.M. Tranquada, G. Shirane, B. Keimer, S. Shamoto and M. Sato, Phys. Rev. B **40**, 4503 (1989).

[5] D. Reznik *et al.*, Phys. Rev. B **53**, R14741 (1996).

[6] B. Keimer *et al.*, Phys. Rev. B **46**, 14034 (1992).

[7] See e.g. V.J. Emery, and S.A. Kivelson, Physica C **209**, 597, (1993).

[8] S. Wakimoto *et al.*, Phys. Rev. B **60**, R769 (1999).

[9] H. Kimura *et al.*, J. Phys. Chem. Solids **60**, 1067 (1999).

[10] N.H. Andersen *et al.*, Physica C **318**: 259 (1999).

[11] B. Keimer *et al.*, Z. Phys. B **91**, 373 (1993).

[12] J.M. Tranquada, *et al*, Nature, **375**, 561, (1995).

[13] T.E. Mason, *et al.*, Phys. Rev. Lett., **68**, 1414 (1992).

[14] T.R. Thurston,*et al.*, Phys. Rev. B, **46**, 9128 (1992).

[15] H.A. Mook, *et al.*, Nature, **395**, 580, (1998).

[16] M. Arai, *et al.*, Phys. Rev. Lett. **83** 608 (1999).

[17] J. Rossat-Mignod *et al.*, Physica C **185-189**, 86 (1991).

[18] H.A. Mook *et al.*, Phys. Rev. Lett. **70**, 3490 (1993).

[19] P. Bourges *et al.*, Phys. Rev. B **53**, 876 (1996).

[20] H.F. Fong *et al.*, Phys. Rev. Lett. **82**, 1939 (1999).

[21] Y. Sidis *et al.*, cond-mat/9912214; Phys. Rev. Lett., *in press.*

[22] H.F. Fong *et al.*, Phys. Rev. B **61**, 14773 (2000).

[23] P. Bourges *et al.*, Science **288**, 1234 (2000).

[24] Z.X. Shen and J.R. Schrieffer, Phys. Rev. Lett. **78**, 1771 (1997).

[25] J.C. Campuzano *et al.*, Phys. Rev. Lett. **83**, 3709 (1999); M.R. Norman and H. Ding, Phys. Rev. B **57**, R11089 (1998).

[26] D. Munzar, C. Bernhard, and M. Cardona, Physica C **318**, 547 (1999).

[27] J.P. Carbotte, E. Schachinger, and D.N. Basov, Nature **401**, 354 (1999).

[28] D.J. Scalapino and S.R. White, Phys. Rev. B **58,** 8222 (1998).

[29] E. Demler and S.C. Zhang, Nature **396**, 733 (1998).

[30] E. Demler and S.C. Zhang, Phys. Rev. Lett. **75**, 4126 (1995); E. Demler, H. Kohno, and S.C. Zhang, Phys. Rev. B **58**, 5719 (1998).

[31] See, *e.g.*, D.L. Liu, Y. Zha and K. Levin, Phys. Rev. Lett. **75**, 4130 (1995); L. Yin, S. Chakravarty and P.W. Anderson, Phys. Rev. Lett. **78**, 3559 (1997); A.A. Abrikosov, Phys. Rev. B **57**, 8656 (1998); D.K. Morr and D. Pines, Phys. Rev. Lett. **81**, 1086 (1998); J. Brinckmann and P.A. Lee, Phys. Rev. Lett. **82**, 2915 (1999); S. Sachdev, C. Buragohain, and M. Vojta, Science **286**, 2479 (1999); F. Onufrieva and P. Pfeuty, cond-mat/9903097.

## CHAPTER 8

# ANDERSON'S THEORY OF HIGH-$T_c$ SUPERCONDUCTIVITY

G. BASKARAN

Institute of Mathematical Sciences
C. I. T. Campus, Madras 600 113, India

### ABSTRACT

The discovery by Bednorz and Müller [1] marks a turning point in the history of superconductivity and strongly correlated electron systems. In 1987, Anderson proposed a theory for high-$T_c$ superconductivity, and followed up with several insightful papers. Looking back at these developments, I discuss the main ideas of Anderson which, I believe, provided (in 1987) the correct mechanism of cuprate superconductviity and a theoretical frame-work that may be crucial for future developments.

## 8.1 INTRODUCTION

It gives me great pleasure to speak at this meeting honoring Phil Anderson for 50 years of contributions. For the last 12 years Anderson has been working mostly on high-$T_c$ superconductivity [2]. A lot on this subject has been written by Anderson (alone and with collaborators), and his papers have drawn differing opinions from the physics community. The present article is a "stock-taking" of what happened during 1987. I am doubly pleased to describe the achievements of his group during this period, from the perspective of a close collaborator.

In the following sections, I discuss and attempt to justify a view, perhaps not universally held, that Anderson solved at a certain level, the problem of the electronic mechanism of superconductivity in these materials during 1987. I will not talk about Anderson's continuing efforts during the last 12 years, that have laid new foundations for the field of strongly correlated insulators, metals and superconductors, by providing key notions, as well as new many-body theory methodology. What was done in 1987 was to provide the framework for a theory, which of necessity, could not be adequately filled out at that early date, because of the lack of adequate data and theoretical tools. Thus, the papers of that time are an interesting mixture of sound insights leading to correct conclusions, and unsupported conjectures which surprisingly, missed their mark only occasionally (but were only sometimes properly labeled as guesswork). In the Appendix I provide a pedagogical summary of some aspects of RVB theory.

## 8.2 THE RVB THEORY OF 1987

The first paper of Anderson in Science [3] had immediate impact. It laid a solid *conceptual foundation* and suggested an *electronic mechanism* that was appreciated instantly because of its naturalness and novelty. The remaining three papers that followed [4, 5, 6] were built on this foundation, and were to some extent summarized in his Varenna school notes, also of 1987. These papers realized Anderson's suggestions in a natural mean-field-approximation scheme. They led to a way of going beyond this approximation by using an induced local U(1) gauge symmetry, and derived an 'RVB Ginzburg-Landau functional' that interpolated between the Mott insulating and the superconducting state. These four papers provided the foundation for an electronic mechanism of superconductivity and solved the basic issue of the mechanism of high-$T_c$ superconductivity to the extent I elaborate below. Table I may be helpful in contrasting the conventional BCS theory of superconductivity and Anderson's electronic mechanism-based theory.

The key results of Anderson's 1987 Science [3] paper are: i) the suggestion that $La_2CuO_4$ (LSCO) is a Mott insulator whose low temperature state is a quantum spin liquid with resonating neutral singlets, ii) the suggestion of neutral spin-1/2 excitations with a possible pseudo-fermi surface, iii) the implicit suggestion that this *Mott insulating state, rather than the conventional conducting fermi-liquid state, is the appropriate reference state for the description of superconductivity*, iv) the idea that doping delocalizes the singlets together with their charges and leads to high-$T_c$ superconductivity, v) identification of the one-band, large-$U$ (positive) Hubbard model as the relevant model, and vi) the discovery of a deep mathematical and physical connection/continuity between the spin-liquid ground state of a Mott insulator and a superconducting Gutzwiller projected BCS state. This made the conversion of a Mott insulator into a superconductor seemingly easy, avoiding issues of the normal state.

The parting of ways from a traditional BCS approach of building pair condensation in a fermi liquid reference state is perhaps the most novel aspect of Anderson's theory. Many theories that followed did not take advantage of the singlet correlations of the Mott insulating state in producing the superconducting state at low doping. Some of them concentrated on antiferromagnetic correlations and/or charge order correlations that actually compete with superconductivity at low energies.

Another aspect of this paper that has not been adequately appreciated is the so-called 'RVB' wave function:

$$|RVB;\phi\rangle \equiv P_G(\sum_{ij} \phi_{ij} b^\dagger_{ij})^{\frac{N}{2}}|0\rangle. \tag{8.1}$$

This is a remarkable wave function from the variational point-of-view. We start with an empty lattice and condense $\frac{N}{2}$ singlet pairs, all in the same momentum state (often zero), and remove the double occupancy on any site by a Gutzwiller projection. The pair function $\phi_{ij}$ is the variational function in this theory. This RVB wave function, for various choices of the pair function $\phi_{ij}$, describes

i) the short-range RVB state introduced by Anderson and Fazekas, and later studied in detail by Kivelson *et al.* and Sutherland,

ii) an RVB state with a pseudo Fermi Surface,

iii) an RVB state with 'point' Fermi Surfaces,

iv) the Kalmayer-Laughlin chiral spin-liquid state,

v) antiferromagnetic Mott insulator,

vi) $d$-wave and $s$-wave superconducting states,

vii) the ground state of the Haldane-Shastry Hamiltonian in $1d$,

viii) states with charge and spin stripe correlations, and

ix) the nodal liquid.

The physically motivated Gutzwiller projection does wonders; it enhances antiferromagnetic correlations and even introduces strong chiral correlations, in the process of reducing double-occupancy fluctuations.

In a sense, this is the 'Laughlin wave function' for cuprates - a thing that many seem to desire. Ironically, it even reproduces the Kalmayer-Laughlin wave function (a cousin of Laughlin's FQHE wave function) of the chiral spin-liquid state:

$$|RVB;\phi\rangle \equiv P_G(\sum_{ij} \phi_{ij} b^\dagger_{ij})^{\frac{N}{2}}|0\rangle \rightarrow \Pi_i(z_i - z_j)^2, \tag{8.2}$$

where $z_i = x_i + iy_i$ denotes the $2d$ lattice co-ordinates of a down-spin electron.

Following the first paper, there were three others on the issue of single-layer superconductivity from Anderson and his collaborators. The second paper by Anderson, Zou and myself [4] presented an RVB mean- field solution to the $t-J$ model. This paper fleshed out many aspects of Anderson's 1987 Science paper, providing a calculational framework to compute some important properties. The salient results of this work are: i) it provided a novel way of studying the Mott

| Theory | Refer. State | Mechanism | Methods | Excitations | Phenomen. Theory |
|---|---|---|---|---|---|
| BCS | Fermi Liquid | Exchange of Bose quanta | BCS, Nambu-Gorkov, Eliashberg, BdG, etc. | Bogoliubov QP | GL |
| RVB | Mott Insulator (Spin-Liquid State) | Super Exchange, Kinetic-energy frustration | RVB MFT, RVB Gauge Theory, RVB-BdG | Neutral Fermions? | RVB-GL |

Table 8.1: Comparision between conventional BCS theory and 1987 RVB theory. BCS = Bardeen-Cooper-Schrieffer theory, RVB = Resonating Valence Bond theory, QP = quasiparticles, GL = Ginzburg-Landau theory, BdG = Bogoliubov de Gennes Equations, MFT = mean-field-theory.

insulator in a framework called RVB mean field theory and found neutral spin-half fermions with a pseudo fermi surface, ii) it found a stable superconducting state (with extended-$s$ symmetry) for the doped Mott insulator, iii) it discussed the importance of preformed pairs and phase fluctuations at low temperatures in the underdoped regime, and iv) it provided a scale of transition temperatures in terms of the superexchange constant $J$ by viewing the superexchange interaction as the analogue of the BCS interaction term.

The above paper recognized that phase fluctuations arising from the elimination of double occupancy fluctuations should reduce the superconducting $T_c$ to zero in the Mott insulating case, and to a very small value in the underdoped regime. The importance of pre-formed pairs above $T_c$ in the underdoped regime and quantum phase fluctuations was suggested - these ideas/predictions, which predated detailed experiments, have matured into notions such as the spin-gap phase [8], and phase-fluctuation dominated bad-metal phase [9], with help from a variety of experiments performed after 1988.

It bears repeating that this paper constitutes an unequivocal prediction of the spin pseudogap phenomenon, long before it was noticed experimentally. Oddly enough, none of the authors seems to have realised this until much later.

This, like the other papers, was incorrect in detail while fruitful in principle. In this first mean-field theory, the extended-$s$ solution was overemphasized. Interestingly, Anderson chose (mainly based on then available experiments) to stick to the extended-$s$ solution for quite some time. With hindsight it is clear, both from theory and experiment, that $d$-wave is a natural RVB solution and the observation of $d$-wave symmetry does not discredit Anderson's electronic mechanism. In a year or so Affleck *et al.* had proposed the "$s + id$" solution, which was shown to be lower in energy than "$d$" alone, and Shastry demonstrated that this was the same as Laughlin's "flux phase". It was then shown by Hsu that the "$s+id$" (or the equivalent "flux phase") is in turn unstable at half-filling to the formation of antiferromagnetic ordering. In hindsight, it is clear that Anderson's focus on "extended $s$" was a tactical mistake, though it leads to good results for transport properties in the optimally doped normal state (and this was the reason for emphasizing it.)

The RVB mean field theory also paved the way to the construction of quasi-particles with fractional quantum numbers such as the spinon. If one works in a valence bond basis, for example, creating a spinon at a site, is a complicated non-local operation. In this approach, it is simply done by the local operation of creating, in the mean-field RVB state (that lives in an enlarged Hilbert space), a particle-hole pair followed by a Gutzwiller projection:

$$\zeta^\dagger_{i\sigma}\zeta^\dagger_{j\sigma'}|RVB\rangle \equiv P_G c^\dagger_{i\sigma} c_{i-\sigma'}|mRVB\rangle \tag{8.3}$$

where $|mRVB\rangle = (\sum_{ij} \phi_{ij} b^\dagger_{ij})^{\frac{N}{2}}|0\rangle$ is the unprojected RVB mean-field solution.

Using the above construction, non-trivial excitations such as the spinon of the Haldane-Shastry model in $1d$ and the Kalmayer-Laughlin model in $2d$ can be easily constructed.

Going beyond mean-field theory in a strongly correlated system was not an easy task. This was attempted in the third paper by Anderson and the present author [5], through the hypothesis of a local U(1) symmetry in the Mott insulating case. (This symmetry for the Mott insulator was exploited much earlier in a different context by Kohn.)

This paper was the progenitor of an extensive literature on gauge RVB theories involving many groups around the world, which is still actively growing. If there is ever to be a successful formal theory, it may well proceed along gauge theory lines such as those set out here. But more physical shortcuts such as Anderson's theory of the optimally doped normal state may provide useful insights, or even be superior. The conclusions of this paper were the following: i) It discussed the Mott insulator (Heisenberg antiferromagnet) in terms of electron variables rather than the spin-half magnetic moment operators, ii) it proposed the local U(1) symmetry and developed a functional integral formalism in terms of the complex link or RVB fields $\Delta_{ij}$ and emphasized the importance of Elitzur's theorem [11], iii) it found two types of spinon excitations: a topological one induced by the 'RVB magnetic charge' in 2D (to be called $\pi$ flux in later developments) and a non-topological one. The $\pi$ flux induced spinon was rediscovered by Sutherland, Rokhsar, Read and Chakrabarty and Kivelson in different forms. iv) Since it is a theory involving fermions coupled to gauge fields in a gauge invariant fashion, it conjectured the presence of a topological term on general grounds, later suggested as a Hopf term by Dzhyaloshinski, Polyakov and Wiegman [12] and as a Chern

Simons term by Zou [13], v) from the calculational point of view it derived, in an approximation, a RVB Ginzburg-Landau functional, that also unified the undoped and doped Mott insulator. It is important to elaborate the last point. The generalized Ginzburg-Landau action that was derived had the form

$$\begin{aligned} S_{rvb} \approx \quad & a_{rvb} \sum\nolimits_{\langle ijkl \rangle} \Delta^*_{ij} \Delta_{jk} \Delta^*_{kl} \Delta_{li} + h.c. + \\ & + a_2 \sum\nolimits_{\langle ij \rangle} |\Delta_{ij}|^2 + a_4 \sum\nolimits_{\langle ij \rangle} |\Delta_{ij}|^4 \\ & + b_2 \sum\nolimits_{\langle ij, jk \rangle} \Delta^*_{ij} \Delta_{jl} + h.c. \\ & + b'_2 \sum\nolimits_{\langle ij, kl \rangle} \Delta^*_{ij} \Delta_{kl} + h.c. \end{aligned} \tag{8.4}$$

Here $\Delta_{ij}$ is the pairing amplitude in the bond $ij$ and the various coefficients were determined using high-temperature series expansion in static approximations. Later work by Muller-Hartman's group [14] improved the estimation of these coefficients and studied the phase diagram. One can also go beyond the static approximation to include the induced dynamics of the amplitude field $\Delta_{ij}$.

The first term in the above action represents the '*memory of the Mott insulator*'. This contains the resonance or correlated motion of two singlets without charge transport, as is evident from the local U(1) symmetry of this term. This term in some sense represents the spin dynamics. The last two terms (which vanish when the doping $x$ tends to zero) represent the delocalization of singlets involving charge transport, as is evident from the global U(1) symmetry of these terms, and would only exist in the superconductor where the link variable has a definite phase.

The physics of superconductivity in the underdoped regime is seen as an interesting competition between the above two terms. It is a quantum effect in which the fully frustrated one electron kinetic energy in the Mott insulating state is partly regained through pair delocalization.

We can use the above RVB-GL action and do an integration over the phases $\theta_{ij}$ of the link field $\Delta_{ij} \equiv |\Delta_0| e^{i\theta_{ij}}$ to get the partition function. Elitzur's theorem ensures that the superconducting transition temperature $T_c$ will vanish at zero doping. A later paper by Nakamura and Matsui [15] used the above action and integrated over phase field $\theta_{ij}$ numerically, and found superconducting transition temperatures that were reasonable.

The above RVB-GL free energy was obtained through high-temperature expansion - a crude way of integrating over high frequency or performing a renormalization operation. The full low-energy effective action should contain the coupling between the link variable and the underlying fermions, and is given by

$$S_F = \sum_{\langle ij \rangle} t_{ij} c^\dagger_{i\sigma} c_{j\sigma} + h.c. - J \sum_{\langle ij \rangle} \Delta_{ij} b^\dagger_{ij} + h.c., \tag{8.5}$$

where $b_{ij} \equiv \frac{1}{\sqrt{2}}(c_{i\uparrow} c_{j\downarrow} - c_{i\downarrow} c_{j\uparrow})$ is the bond singlet operator.

The RVB-GL action Eq. 8.4 together with the fermion terms Eq. 8.5 is very general. For example, it reproduces, after a continuum approximation in the spin-gap region, the action suggested recently by Balents, Nayak and Fisher [55] on phenomenological grounds. Many authors, such as Wen, Lee, Nagaosa, Ioffe, Larkin, Wilczek, Zee, Senthil, Nayak, Fisher and Dung-Hai Lee have contributed to the elaboration of these fundamental ideas.

In the fourth paper, Anderson and collaborators [6] tried to incorporate various new experimental results (obtained during Jan-June '87) and some related important theoretical developments due to Kivelson, Rokhsar and Sethna [14]. They attempted a global understanding of the phase diagram including antiferromagnetic order and the tetragonal to orthorhombic structural transitions. This paper contains:

i) the argument that in the presence of the robust singlet correlations in the ground state, even a small doping can destroy the experimentally observed long-range antiferromagnetic order. This justifies the idea that a spin-liquid state rather than the fragile long-range antiferromagnetically ordered state is a meaningful starting point.

ii) At low doping, while the true superconducting phase transition is reduced to a small value, the ensuing RVB correlations (preformed singlets) around the RVB mean-field theory transition temperature can cause structural phase transitions provided the valence bond amplitudes also develop some relative phasing of the different bonds. This was suggested to lead to 'spin dependent overlap charges' that can explain the tetragonal-to-orthorhombic, or 'twitch' transition.

iii) the spin-half neutral fermion excitation was christened the 'spinon' (in an earlier paper Kivelson *et al.* named the spinless charged hole the 'holon'.) and the real-space argument of Kivelson *et al.* as well as the soliton doping phenomenon was given a **k**-space picture, implicitly suggesting a spin-charge decoupling at the fermi surface.

iv) there was an attempt to connect the 'charge-$e$ holon condensation' picture of Kivelson *et al.* to the RVB $2e$ superconductivity mechanism by arguing that 'holons are book keeping devices that keep track of valence bond (charge $2e$) delocalization, and eventually the off-diagonal long-range order (ODLRO) will occur in the electron-pair amplitude'.

This paper seems rather a mixed bag of sound physical ideas and speculations some of which have still not been settled, and some of which may be wrong. The connection to structural transitions does not carry over to other cuprates and may be a false scent.

Many of the ideas developed in 1987 were summarized in Anderson's lectures at Varenna in summer '87, which present the underlying physics and the BZA theory, but end with an overhasty attempt to guess at a full formal theory.

## 8.3 THREE APECTS OF ANDERSON'S 1987 MECHANISM

I will try to explain three theoretical aspects of Anderson's 1987 theory that, in my opinion, have been quite successful (more so than perhaps generally appreciated).

**i) Phonons are not major actors and pairing is not due to exchange of a boson.** After Bednorz and Müller's discovery, theorists had roughly two options. One was to build on a fermi liquid foundation and look for an additional contribution to a phonon-mediated mechanism of superconductivity, focusing only on the doped $CuO_2$ system. The other was to look for a new mechanism of superconductivity. Prior to the discovery of $YBa_2Cu_3O_{7-x}$ (YBCO), when the maximum observed $T_c$ ($\sim$ 35 K in LSCO) was near the limit set by the phonon mechanism, Anderson identified strong electron correlation to be primarily responsible for cuprate superconductivity. In a conversation around 20 Dec 1986, Anderson told me that "it is all a spin-half Mott phenomenon". It is nearly 12 years, and most of the community has accepted Anderson's suggestion that it is an electron mechanism and not a phonon mechanism, thanks to a variety of experimental results which came after Anderson's 1987 Science paper. A close connection between the superconducting condensation energy and the change in one-electron kinetic energy/super-exchange energy, as seen by experiments, has ruled out pairing as arising from an exchange of an 'elementary quantum'.

**ii) Mott Insulator as a Reference State** Recent ARPES measurements [20] have brought out a striking similarity, between the Mott insulator and the underdoped low $T_c$ materials, in the behavior of their normal-state spectral functions at low energies. Moreover, the long-range antiferromagnetic order, a fragile low-energy triplet condensation phenomenon [21] unlike the robust singlet correlations, has little signature in the details of the spectral functions. There seems to be a continuity between the normal state of the underdoped metal and the finite-temperature Mott insulating state with enhanced singlet correlations. In this sense the unorthodox view of a Mott insulator with preformed singlets as a reference state to discuss superconductivity has turned out to be experimentally relevant and correct.

The above, in turn, supports the correctness of identifying electron repulsion as the most important of energy scales.

**iii) The correct $tJ$ Model** After many efforts starting from the insightful work of Zhang and Rice [22], the validity of the one-band large-$U$ Hubbard model for a single $CuO_2$ layer may be considered established. In the relevant energy scales, ARPES also shows a single band, per $CuO_2$ layer.

In defense of the formalism:

**i) RVB mean field theory** Variational studies and numerical studies give enhanced $d$-wave pairing correlations in the $tJ$ or the large-$U$ Hubbard model. It is important to realize that the numerical study of finite fermionic systems can be misleading when one is looking for spontaneous symmetry breaking in the presence of gapless fermionic excitations in the ordered state.

I believe the reason why numerical attempts [23] fail to find strong signals of ODLRO may arise from this. Numerical work gives clear evidence for the pseudogap, but of course $T_c$ is much smaller. The historical record demonstrates that numerical workers can overestimate their accuracy on subtle points like these.

The RVB mean-field theory, in spite of its approximate nature, gives solutions exhibiting superconductivity and also gives the correct relative energies of the extended-$s$ [4] and $d$-wave [10] states. Quantitative comparison with experiments based on RVB theory have been attempted by several workers, notably Fukuyama [26] and collaborators.

The RVB mean field scenario provides a roughly correct phase diagram that agrees with experiment. Experiments, rather than numerical simulations, are the best support for the RVB mean field theory, at the moment.

**ii) Beyond Mean Field Theory**

The first gauge theory for cuprate [5], had an important difference compared to later developments. The entire discussion, including the Mott insulating state, was in terms of electron variables only. (The holon variable as a slave boson was introduced later by Zou and Anderson [27]). The first theory had the advantage that in the effective theory that was derived, there were no holon fields or gauge fields. The degrees of freedom present in the final action were the link pair amplitudes and the electron variables. This theory had local U(1) invariance only in the Mott insulating end. The doped Mott insulator had only global U(1) invariance, as it should be since we are dealing with moving electrons.

The slave boson-based gauge theory [28], on the other hand, has local U(1) gauge invariance even in the doped situation. It is interesting that all the technical difficulties faced by many authors, including Wiegman, Larkin, Ioffe, Patrick Lee, Nagaosa, Wen [29] and myself, for the last many years seem to stem from the the fact that *we have to work very hard to bring back the global U(1) gauge invariance to discuss the physics in terms of the Cooper pair variables and electrons.* Dung-Hai Lee [30] has overcome some major technical difficulties recently, by a conscious attempt to find a final effective theory with global U(1) invariance in terms of Cooper pair variables and the fermions. It is in this sense also, that the final action of Dung-Hai Lee and that of Balents, Nayak and Fisher are similar in spirit and content to the 1987 RVB-GL action.

From this point of view also the RVB-GL action captures the essential physics of quantum fluctuations arising from the strong on-site repulsion that is neglected in the RVB mean field theory.

**iii) Superconducting $T_c$**

As far as I know, a quantitatively satisfactory way of treating quantum fluctuations in RVB (aside from the recent analytic work of Dung-Hai Lee) is the early numerical work of Matsui and Nakamura [15], who numerically integrate over the phases of the link gauge fields in the RVB Ginzburg-Landau action (no gaussian approximation). In the underdoped region, they were able to get a reduction of the mean-field $T_c$'s versus $x$ that is close to the experimental values. This approach has not been pushed further.

## 8.4 SOME COMMENTS ON ANDERSON'S SOLUTION

The characteristics of Anderson's approach are amply illustrated in a chapter called 'The Central Dogma' that Anderson wrote around 1990 for his then-evolving book on high-$T_c$ [2].

The 'central dogmas' constrain the overall structure of any description of the actual mechanism. Anderson argues that "$\cdots$ they are empirical generalizations so direct and pervasive that it seems perverse or frivolous to ignore them. $\cdots$ As in molecular biology, there is enough irrelevant complexity that an unwitting theorist may never reach the neighborhood of the actual problem, even though he is working along a line which is widely represented in the literature. Understanding high-$T_c$ involves not one, but a multiplicity of steps, and it is vital to provide a map through the maze of alternative paths, almost all of which can be eliminated by simple logic using basic and well-founded experiments or theoretical reasonings, of a sort which should be immediately persuasive."

Briefly, the central dogmas are: i) All the relevant carriers of both spin and electricity reside in the $CuO_2$ planes and derive from hybridized $O_{2p}$ - $Cud_{x^2-y^2}$ orbitals which dominate the binding in these compounds. ii) Magnetism and high-$T_c$ are closely related, in a very specific sense: i.e. the electrons which exhibit magnetism are the same as the charge carriers. iii) The dominant interactions are repulsive and their energy scales are all large. iv) The 'normal' metal above $T_c$ is the solution of the plane one-band problem resulting from dogma III, and is not a Fermi liquid, in the sense $Z = 0$. But it retains the Fermi surface satisfying the Luttinger theorem, at least in the highest $T_c$ materials. We call this a (tomographic) Luttinger liquid. v) The above state is strictly two-dimensional, and coherent transport in the third direction is blocked. ... this two- dimensional state is not superconducting and has no major interaction tending to make it so, at least near the usual $T_c$. vi) Interlayer hopping together with the 'confinement' [31] of Dogma V is either the mechanism of or at least a major contributor to the superconducting condensation energy.

Some of these dogmas have stood up well, despite initial opposition. For example, the relevance of the single band model took nearly 5 years to be accepted by the majority of the community, starting from the efforts of Zhang and Rice.

It is interesting that dogma V questioned the original RVB theory of single layer superconductivity developed during 1987 and attempted to replace it by the inter-layer pair tunneling mechanism of Wheatley, Hsu and Anderson [32]. Recent experiments [33] have now shown that dogma V has turned out to be not correct on a quantitative level for the Tl- and Hg-based one-layer cuprates. In this sense, one of the dogmas has been either seriously questioned or falsified by experiments, but at the same time brought back the original single-layer RVB mechanism as a viable alternative.

However, the merit of the interlayer pair tunneling idea is that it made one think about the meaning of both in-plane and out-of-plane kinetic-energy driven mechanisms. It is remarkable that in several systems, including LSCO, there is a change in kinetic energy across the superconducting phase transition as seen, for example, in the data of Basov [34]. The change supports Anderson and Hirsch's idea of a connection between the superconducting condensation energy and the frustrated electron kinetic energy. The interlayer pair-tunneling idea is important in understanding the orgin of the 41 meV magnetic resonance in the superconducting state of YBCO.

## 8.5 APPENDIX

This appendix provides a brief, pedagogical summary of the basic ingredients of the RVB theory. We describe a) the meaning of the Mott insulator, b) the notion of an effective Hamiltonian, c) the proliferation of singlets in cuprates and RVB, and d) the meaning of induced, local-gauge symmetry.

a) **The Mott Insulator** We will take $La_2CuO_4$ as the prototype. Experiment and quantum chemistry tells us that there is a single, non-degenerate orbital on each copper site that determines the low energy electrical and magnetic properties. It is the $3d_{x^2-y^2}$ orbital hybridized with the surrounding oxygen $2p$ orbitals. In pure $La_2CuO_4$, there is only one valence electron per Cu site, so we have a one-band, tight-binding model on a square lattice representing the electron dynamics in the cuprate planes. The kinetic energy of the electrons is:

$$K.E. = -t \sum_{\langle ij \rangle} c_{i\sigma}^{\dagger} c_{j\sigma} + h.c. = -t \sum_{k} (\cos k_x + \cos k_y) c_{k\sigma}^{\dagger} c_{k\sigma}. \qquad (8.6)$$

With the hopping matrix element of $t \approx 0.25eV$, we get a band-width $W \approx 2eV$. It is a narrow band, compared to the Coulomb energy scale of 5 to 8 eV in the $3d$ states.

In the absence of Coulomb interaction, we have a half-filled band (one electron per site), which implies a metal. Experiments tell us that pure $La_2CuO_4$ is a good insulator with a gap of about 2 eV. This is where the fascinating phenomenon of a Mott insulator enters. It tells us that the band picture breaks down when you have strong Coulomb repulsion. While Coulomb interaction is long-ranged, a little thought tells us that, in converting a half-filled band metal into a Mott insulator, the most important term is the on-site repulsion between two electrons. And the Hubbard Hamiltonian just contains that:

$$H_{hub} = -t \sum_{\langle ij \rangle} c_{i\sigma}^{\dagger} c_{j\sigma} + h.c. + U \sum n_{i\uparrow} n_{i\downarrow} \qquad (8.7)$$

For the cuprates $U \approx 5eV$, which exceeds the band-width $W \approx 2eV$. Let us do a thought experiment. We increase $\frac{U}{t}$ from zero to a large value. When $U = 0$, we have a half-filled band - a good metal. When $U$ is very large compared to the band-width, the electrons avoid double occupancy and so cannot independently delocalize into Bloch states. Apart from some weak quantum mechanical charge fluctuations, each site contains exactly one electron with a high probability ($\sim 1 - (\frac{t}{U})^2$). Hence, the possibility of low-energy charge transport is precluded. It is an insulator with a charge gap of $\sim U - W$. This is called a Mott insulator.

If one thinks in terms of wave functions and other physical properties, the metallic state and Mott insulating state *are not adiabatically connected* to each other. So in our thought experiment we should encounter a phase transition. This is called a Mott transition. We should keep this in mind, because the conducting state in cuprates is obtained by adding carriers to the Mott insulating state, rather than adding charges to a half-filled band metal. There exists a fundamental difference between a 'doped Mott insulator' and a 'doped band metal', at least at small doping.

b) **Effective Hamiltonians**

It is clear intuitively that even though current-carrying low-energy excitations are absent in a Mott insulator, low-energy spin-fluctuation excitations are present. For example for $t = 0$, the ground state contains precisely one electron per site. But there is a large spin degeneracy $= 2^N$, where $N$ is the number of sites. The lowest energy current-carrying state has a gap of $U$ above the ground state energy. As we increase $t$ from a small value, we expect the $2^N$-fold spin degeneracy to be lifted, and the 'charge gap' or the Mott-Hubbard gap to decrease. All these can be clearly seen in a perturbation derivation of an 'effective Hamiltonian', which is actually a process of renormalization.

The basic idea in the effective Hamiltonian approach is to focus on the relevant 'low energy' part of the Hilbert space and take care of the virtual excursions to 'high energy' part of the Hilbert space through effective or induced interactions. The final Hamiltonian is defined in a reduced Hilbert space.

For the half-filled Hubbard model, we treat the $t$ term as a perturbation and see how the spin degeneracy of the $t = 0$ ground state gets lifted. To derive an effective Hamiltonian correct to order $(\frac{t}{U})^2$ it is sufficient to concentrate on two neighboring sites. The physics of Kramer's superexchange or Anderson's kinetic exchange applies in a natural way here.

Consider two neighboring Cu sites. In the absence of the hopping term $t$, the $2^2$ degenerate ground states are

$$|\uparrow,\downarrow\rangle, \quad |\downarrow,\uparrow\rangle, \quad |\uparrow,\uparrow\rangle, \quad |\downarrow,\downarrow\rangle.$$

In terms of the total spins of the two sites, the four degenerate states form a singlet or a triplet. There are the two excited states (of energy $U$),

$$|\uparrow\downarrow,0\rangle, \quad |0,\uparrow\downarrow\rangle. \tag{8.8}$$

They are both spin singlets. At finite $t$, the $2^2$-fold spin degeneracy of the ground state is lifted, through virtual mixing to the 'excited singlet states' involving double occupancy (Eq. 8.8). A simple 2nd-order perturbation theory gives us an energy reduction of $-\frac{2t^2}{U}$ for the singlet states and zero energy reduction for the triplet states.

All these are summarized nicely in the effective Hamiltonian obtained in a 2nd-order perturbation treatment of the kinetic energy (hopping) part of the Hubbard model:

$$H_{hub} \rightarrow -\frac{2t^2}{U} \sum_{\langle ij \rangle} \tau_{ij} \tau_{ji} \tag{8.9}$$

Here $\tau_{ij} \equiv \sum_\sigma c^\dagger_{i\sigma} c_{j\sigma}$. This Hamiltonian is defined in the restricted Hilbert space containing no double occupancy $n_{i\uparrow} + n_{i\downarrow} = 1$. In this subspace the Hamiltonian becomes the famous Heisenberg Hamiltonian:

$$H = -J \sum_{\langle ij \rangle} \tau_{ij} \tau_{ji} = J \sum_{\langle ij \rangle} (\mathbf{S}_i \cdot \mathbf{S}_j - \frac{1}{4}), \tag{8.10}$$

When we move away from half-filling by removal of electrons, as in doped $La_{2-x}Sr_xCuO_4$, the double occupancy restriction continues to be important up to optimal doping. Thus, the effective Hamiltonian is the same Heisenberg model, but containing the important (restricted) kinetic energy term that allows an electron to hop to a neighboring site (only) when it is empty. The corresponding model is called the $t-J$ model:

$$H_{t-J} = -t \sum_{\langle ij \rangle} c^\dagger_{i\sigma} c_{j\sigma} + h.c. - J \sum_{\langle ij \rangle} (\mathbf{S}_i \cdot \mathbf{S}_j - \frac{1}{4}), \tag{8.11}$$

with the double occupancy constraint $n_{i\uparrow} + n_{i\downarrow} \neq 2$.

The $t-J$ model in $2d$ has occupied the attention of a large body of condensed matter theorists.

c) **Proliferation of singlets and RVB**

Let us first consider the Mott insulator. The Heisenberg Hamiltonian may be rewritten in yet another suggestive fashion, in terms of the bond singlet operators $b^\dagger_{ij} \equiv \frac{1}{\sqrt{2}}(c^\dagger_{i\uparrow} c^\dagger_{j\downarrow} - c^\dagger_{i\downarrow} c^\dagger_{j\uparrow})$, as

$$H = -J \sum_{\langle ij \rangle} b^\dagger_{ij} b_{ij}, \quad \text{with} \quad n_{i\uparrow} + n_{i\downarrow} = 1. \tag{8.12}$$

We have used the important identity

$$(\mathbf{S}_i \cdot \mathbf{S}_j - \frac{1}{4} n_i n_j) \equiv b^\dagger_{ij} b_{ij}. \tag{8.13}$$

The singlet number operator $b_{ij}^{\dagger}b_{ij}$ has an eigenvalue of 0 or 1 in our single-occupancy subspace. In view of the above identity, in a translationally invariant ground state, the quantity that is maximized, consistent with the lattice structure, is the valence bond amplitude. *Thus, in the nearest-neighbor Heisenberg models, minimization of the ground state energy is synonymous with maximization of the strength of nearest-neighbor singlet bonds.* I have referred to this as the valence-bond-amplitude maximization hypothesis (VBAM) in a recent paper [35].

In a quantum spin system, there are many ways in which the valence bond amplitudes can be maximized, for example by forming singlets among nearest neighbor electrons. A nearest neighbor valence bond basis in which we pair electrons of neighboring sites into singlets in all possible ways is a good example where the valence bond amplitude is maximized on all the singlet bonds. Anderson suggested in 1973 that if we form a coherent superposition of these short-range valence bond states, the energy is further reduced, making it a good variational state for $2d$ frustrated spin systems such as the triangular lattice. These states are called RVB or a quantum spin-liquid states.

A famous example at the molecular level is benzene. Here we have the six $p_z$ orbitals of the six carbon atoms forming a 'half-filled band' through $p - \pi$ bonding. The band width $W = 4eV$ is smaller than the Hubbard $U \simeq 8$ eV. It is a mini Mott insulator! The famous in-phase superposition of the two valence bond states is a good approximation to the ground state.

Below, I give another heuristic picture for the proliferation of singlets in the doped and undoped cuprates. In a free fermi gas, any short range singlet correlation contained in the ground state is a consequence of the Pauli Principle rather than interactions. In the large-$U$ Hubbard model, accepted as a good model close to half-filling, every elementary collision between two electrons tries to establish essentially a nearest-neighbor singlet correlation (a valence bond). That is, the virtual transitions to doubly occupied state on a given copper site that takes place over a time scale of $\frac{\hbar}{U}$ lowers the energy (compared to the $U = \infty$ case) by the super exchange energy $J \approx \frac{2t^2}{U}$ and stabilizes the spin singlet state rather than a triplet state. Elementary collisions, in addition to the Pauli principle induce singlet correlations. Close to half-filling, elementary two-body collisions are more frequent than free hopping of charges. *The on site collision induced valence bond proliferation* is at the heart of the RVB theory.

Even though diagonal in terms of the number operators $b_{ij}^{\dagger}b_{ij}$, the Heisenberg Hamiltonian is not really diagonalized, as the number operators themselves do not commute whenever one of the sites coincide:

$$[b_{ij}^{\dagger}b_{ij}, b_{jk}^{\dagger}b_{jk}] = i(\mathbf{S}_i \times \mathbf{S}_k) \cdot \mathbf{S}_j \tag{8.14}$$

This non-commutativity propagates an irreducible minimum of bond triplet fluctuations in the ground state by making the ground state average $\langle b_{ij}^{\dagger}b_{ij}\rangle_G < 1$. In fact, $\langle b_{ij}^{\dagger}b_{ij}\rangle_G \approx 0.6391, 0.5846$ and $= 0.5$ respectively for 1d chain, 2d square lattice and infinite $d$ hyper-cubic lattice. In the $1d$ chain, the bond-triplet fluctuation is finite and manages to produce an algebraic AFM order. In $2d$, it increases further and is believed to produce true long-range order. In infinite $d$, it increase even further to produce perfect Néel order. The chiral operator appearing on the right hand side of Eq. 8.14 also tells us that chiral fluctuations are also induced in the ground state.

A long-range magnetic order, when it occurs in the ground state, is an inevitable consequence of VBAM in the presence of the constraints provided by the lattice structure and the above commutation relation. Thus in a hyper cubic lattice for $d \geq 2$

$$\text{VBAM + geometrical constraints} \Rightarrow \text{AFM Order}$$

The nature of magnetic order is strongly lattice dependent. For the non-bipartite $2d$ triangular lattice, what maximizes the bond-singlet amplitude is a $120^o$ structure with zero point fluctuations. The case of P-doped Si in the insulating state is described by a $3d$ random-lattice Heisenberg model. The nature of lattice constraint being very different, long-range magnetic order is believed to be absent. This results in a kind of singlet-bond glass state.

The role of the long-range AFM order should not be overemphasized in this context. As shown by Liang, Doucot and Anderson [36], the energy difference between a disordered spin liquid state with only short-range AFM correlations and the the best variational state with long-range AFM order is as small as one or two percent of the total energy:

$$\frac{\langle b_{ij}^{\dagger} b_{ij} \rangle_{SL} - \langle b_{ij}^{\dagger} b_{ij} \rangle_{G}}{\langle b_{ij}^{\dagger} b_{ij} \rangle_{G}} \sim 0.01 \tag{8.15}$$

At the level of variational wave functions, a small change in the long distance behavior of the pair function $\phi_{ij}$ takes us between a disordered and an ordered ground state. Hsu [21] also shows that the AFM order in $2d$ is a spinon density wave in a robust spin liquid state. *Thus the development of long-range AFM order in 2d is a result of a small final adjustment of the VB amplitude in a spin-liquid state in the maximization procedure.*

Even after doping, the proliferation of the singlet correlations continues, leading to the observed high-temperature superconductivity.

d) **Meaning of induced local gauge symmetry**

In discussing spin-half Hamiltonians, it is customary to use the Pauli spin operators to represent the spin degrees of freedom. While this representation is very useful for understanding ordered antiferromagnets, a spin liquid or RVB state cannot be discussed easily in this representation. A natural excitation in a spin liquid state is a spinon - an isolated upaired spin in the background of resonating singlets. Such excitations can not be easily discussed in terms of the magnetic moment (spin raising and lowering ladder) operators.

It is in this context the U(1) gauge theory of quantum antiferromagnets was developed. It described the Mott insulator in terms of electron variables rather than spin-half magnetic moment operators. An electron variable with a 'U(1) string' attached to it can represent a spinon excitation [37].

In terms of the electron variables, the superexchange processes conserve the local particle number, even though the original Hubbard model has only a total particle number conservation. The 'particle number conservation' at every site is a consequence of the opening of the charge gap in the Mott insulator. Thus we have new local conservation laws leading to a proliferation of symmetries. In particular the exchange interaction term

$$(\mathbf{S}_i \cdot \mathbf{S}_j - \frac{1}{4} n_i n_j) \equiv b_{ij}^\dagger b_{ij} \tag{8.16}$$

is invariant when we replace the electron operators $c_{i\sigma}^\dagger \to e^{i\theta_i} c_{i\sigma}^\dagger$, where $\theta_i$ is an arbitrary phase that is dependent on the site $i$. Thus the symmetry that we have is a local U(1) symmetry.

Once we have a U(1) symmetry it suggests the possibility of dynamically generated U(1) fields, a new type of excitation of the spin liquid state. The effective theory in terms of the dynamically generated gauge fields may look like a complicated U(1) gauge theory, studied in Lattice gauge theory literature. The notion of 'deconfinement' known in the gauge theory easily gets translated into the observability of low energy spin-half spinon excitatins. So the deconfinement of the quark, for example, is related to the deconfinement of spinons. Another way to think about this is as quantum number fractionization of a spin-one excitation in a spin liquid state. As mentioned earlier, an electron with a U(1) string attached to it represents a spinon:

$$\zeta_{i\sigma}^\dagger = c_{i\sigma}^\dagger b_{ii_1} b_{i_1 i_2}^\dagger b_{i_2 i_3} \cdots . \tag{8.17}$$

The above is the analogue of the operator $\psi^\dagger(\mathbf{r}) e^{\int_{\mathbf{r}}^{\infty} \mathbf{A} \cdot d\mathbf{l}}$ in gauge theories.

The local U(1) symmetry also brings in some useful ideas such as the area law or perimeter law characterization of confined and deconfined phases. Another useful idea is the Elitzur's therem. This theorem states that a local gauge symmetry, unlike a global symmetry can not be broken spontaneously. This theorem motivated and enabled one to find ways of making meaning out of the RVB mean field theories, where this theorem is violated.

In the original paper [5] in addition to the U(1) symmetry the persence of a particle-hole symmetry was also recongnized. But this additional symmetry was not made use of. Anderson soon pointed out that this particle-hole symmetry could enhance the local U(1) symmetry to an SU(2) symmetry [2, 41] by the following beautiful argument. In our limited Hilbert space of single particle occupation, the presence of an up-spin electron is the same as absence of down-spin electron. Thus an SU(2) transformation

$$c_{i\uparrow} \rightarrow u_i c_{i\uparrow} + v_i c_{i\downarrow}^{\dagger} \tag{8.18}$$

with $|u_i|^2 + |v_i|^2 = 1$ preserves the form of the Heisenberg Hamiltonian (a similar transformation applies for the down-spin electron).

In spite of the proliferation of the local symmetry in the Mott insulator, it was recognized in 1988 [38, 39], and independently by Senthil and Fisher more recently [40] that an irreducible minimum local symmetry is actually a $Z_2$ symmetry.

## ACKNOWLEDGMENTS

I wish to thank V. N. Muthukumar for many helpful comments, and K. Srinivasa Rao and S. Swaminathan for technical assistance.

## BIBLIOGRAPHY

[1] A. Bednorz and A. Muller, Z. Phys. B **64** 189 (1986).

[2] P. W. Anderson, *The theory of high temperature superconductivity.* (Princeton University Press, NY, 1996).

[3] P. W. Anderson, Science **235** 1196 (1987).

[4] G. Baskaran, Z. Zou and P. W. Anderson, Sol. State Commm. **63** 973 (1987).

[5] G. Baskaran and P. W. Anderson, Phys. Rev. B **37** 580 (1988).

[6] P. W. Anderson, G. Baskaran, T. Hsu and Z. Zou, Phys. Rev. Lett. **58** 2790 (1987).

[7] P. W. Anderson, in *Lecture Notes in Physics* (Springer), **337**, 1989.

[8] T. M. Rice, in *The physics and chemistry of oxide superconductors*, eds. Y. Iye and H. Yosuoka (Springer Verlag,1992) p.313; A.G. Loeser *et al.* Science **273** 325 (1996); H. Ding *et al.* Nature **382** 51 (1996).

[9] V. Emery, S. A. Kivelson and O. Zachar, Phys. Rev. B **56** 6120 (1997).

[10] I. Affleck and J. B. Marston, Phys. Rev. **B 37** 3774 (88); G. Kotliar, Phys. Rev. B **37** 3664 (1988).

[11] S. Elitzur, Phys. Rev. D **12** 3978 (1975).

[12] I. Dzhyaloshinskii, A. Polyakov and P. Wiegman, Phys. Lett. A **127** 112 (1988).

[13] Z. Zou, Phys. Lett. A **131** 197 (1988).

[14] M. Drzazga *et al.*, Z. Phys. B **74** 67 (1989).

[15] A. Nakamura and T. Matsui, Phys. Rev. B **37** 7940 (1988).

[16] L. Balents, M. P. A. Fisher and C. Nayak, cond-mat/9806164.

[17] S. Kivelson, J. Sethna and D. Rokshar, Phys. Rev. B **38** 8865 (1987).

[18] D. C. Johnston *et al.*, Mater. Res. Bull. **8** 777 (1973).

[19] A. W. Sleight *et al.*, Sol. St. Commn. **17** 27 (1975).

[20] F. Ronning *et al.*, Science, **282** 2067 (1998).

[21] T. Hsu, Phys. Rev. B **41** 11379 (1990).

[22] F. C. Zhang and T. M. Rice, Phys. Rev. B **37** 3759 (1988); M. A. Schluter and M. S. Hyberston, Physica. C **162** 583 (1989).

[23] C. T. Shih, Y. C. Chen and T. K. Lee, cond-mat/9910107; E. Dagotto, J. Phys. Chem. Solids **59** 1699 (1998).

[24] M. Calandra, S. Sorella, cond-mat/9911478.

[25] Th. Maier *et al.*, cond-mat/0002352.

[26] H. Fukuyama, Prog. Theor. Phys. (Supplement) **108** 287 (1992).

[27] Z. Zou and P. W. Anderson, Phys. Rev. B **37** 627 (1988).

[28] G. Baskaran, Physica Scripta, T **27** 53 (1989); P. B. Wiegman, Physica Scripta T **27** 160 (1989).

[29] P. B. Wiegman, Phys. Rev. Lett. **60** 821 (1988); Z. G. Wen and A. Zee, Phys. Rev. Lett. **62** 2873 (1989); Z. G. Wen, F. Wilczek and A. Zee, Phys. Rev. B **39** 11413 (89); L. Ioffe and V. I. Larkin, Phys. Rev. B **38** 8988 (1989); P. A. Lee and N. Nagaosa, Phys. Rev. B **45** 966 (1992); X.-G. Wen and P. A. Lee, Phys. Rev. Lett. **80** 2193 (1998); P. A. Lee *et al.* Phys. Rev. **57** 6003 (1998).

[30] D. H. Lee, cond-mat/9902287.

[31] P. W. Anderson and Z. Zou, Phys. Rev. Let., **60** 132 (1988).

[32] J. Wheatley, T.C. Hsu and P.W. Anderson, Phys. Rev. **37** 5897 (1988); Nature, **333** 121 (1988); P.W. Anderson, Physica, **185 - 189** 11 (1991).

[33] K.A. Moler *et al.*, Science **279** 1193 (1998); C. Panagopolos *et al.*, Phys. Rev. Lett. **79** 2320 (1997).

[34] S. Uchida *et al.* Phys. Rev. B **53** 14558 (1996); For recent measurements, see D. N. Basov *et al.* Science **283** 49 (1999).

[35] G. Baskaran, Cond-mat/0007137

[36] S. Liang, B. Doucot and P.W. Anderson, Phys. Rev. Lett., **61** 365 (88)

[37] G. Baskaran, Cond-mat/0008324

[38] G. Baskaran, Yu Lu and E. Tosatti, Int. J. Mod. Phys., **B1** 555 (88)

[39] B. Marston, Phys. Rev. Lett. **61** 1914 (88)

[40] T. Senthil and M. P. A. Fisher, Cond-mat/9910224.

[41] I. Affleck *et al.*, Phys. Rev. Lett., **38** 745 (88)

## CHAPTER 9

# QUANTUM CONFINEMENT AND CUPRATE CRITICALITY

T. SENTHIL AND MATTHEW P. A. FISHER
Institute for Theoretical Physics
University of California, Santa Barbara, CA 93106-4030

### ABSTRACT

Theoretical attempts to explain the origin of high-temperature superconductivity are challenged by the complexity of the normal state, which exhibits three regimes with increasing hole doping: a pseudo-gap regime when underdoped, strange power laws near optimal doping and more conventional metallic behavior when heavily overdoped. We suggest that the origin of this behavior is linked to a zero-temperature quantum phase transition separating the overdoped Fermi liquid from a spin-charge separated underdoped phase. Central to our analysis is a new $Z_2$ gauge theory formulation, which supports topological vortex excitations - dubbed visons. The visons are gapped in the underdoped phase, splitting the electron's charge and Fermi statistics into two separate excitations. Superconductivity occurs when the resulting charge $e$ boson condenses. The visons are condensed in the overdoped phase, thereby confining the charge and statistics of the electron leading to a Fermi liquid phase. Right at the quantum confinement transition the visons are in a critical state, leading to power-law behavior for both charge and spin.

## 9.1 INTRODUCTION

Despite the remarkable progress [1] in the experimental characterization of the cuprate high-$T_c$ materials, a theoretical consensus on the important underlying physics remains elusive. Experiments have revealed a rich phase diagram as the temperature and chemical doping are varied, with low temperature spin and charge ordering in addition to superconductivity. The normal phase at elevated temperatures is equally varied, exhibiting a pseudo-gap in the underdoped regime and strange power laws at optimal doping. In this paper, we propose a theoretical picture that provides a description of the basic aspects of all parts of the cuprate phase diagram.

It gives us great pleasure dedicating this paper to Phil Anderson in honor of his innumerable insights into condensed matter physics and - in particular - his profound clairvoyance in discerning the mysteries of high-temperature superconductivity. The approach we expound below shares many qualitative features with Phil's early RVB ideas. Towards the end of the paper we specifically address some of the commonalities and distinctions with RVB theory.

## 9.2 EXPERIMENTS

We begin with a brief discussion of experiment. In the last few years, angle-resolved photoemission spectroscopy (ARPES) has emerged as an important experimental probe [2] of the cuprates. The ARPES spectra provide a direct experimental measurement of the electron spectral function. In *any* conventional phase (such as a Fermi liquid, band insulator, spin density wave or superconductor), a sharp quasiparticle peak is expected as a function of frequency $\omega$ at some momentum $\vec{k}$ in the Brillouin zone. The experimental results in the underdoped [3] (*and* undoped [4]) cuprates in the non-superconducting state are in striking contrast to these expectations: the electron spectral function is highly smeared with no trace of a sharp quasiparticle peak. A sharp peak does appear, however, upon cooling *into* the superconducting state [5, 6]. With increasing doping the normal-state ARPES spectra sharpen somewhat, but even near optimal doping, the observed peak is far too broad to be consistent with a conventional quasiparticle description. Some representative data may be found in Figs. 9.1 and 9.2.

We take the absence of a quasiparticle peak in the ARPES data to be strong evidence that the electron *decays* into other exotic excitations in the underdoped cuprates. Further evidence for this comes from transport measurements. The $c$-axis d.c. resistivity shows "insulating" behavior increasing rapidly upon cooling, whereas the in-plane resistivity is typically "metallic" and much smaller in magnitude. Moreover, in a.c. transport a Drude peak is observed in the $ab$ plane, but not along the $c$−axis. This strangely anisotropic behavior, difficult to understand within a conventional framework, follows naturally if the electron decays into exotic excitations which reside primarily in the $ab$ plane. Transport along the $c$-axis requires hopping of *electrons* from layer to layer which is strongly suppressed at low energies.

## 9.3 NOVEL EXCITATIONS

If the electron indeed decays into other excitations, what is their character? There are two distinct possibilities: (a) The electron may decay into two or more other exotic particles, each of which carries some fraction of the quantum numbers of the electron (for instance, into separate spin and charge carrying excitations), or (b) The exotic excitations may admit no "particle" description at all - this is known to happen generically at quantum critical points. We hypothesize that (a) is realized in the underdoped cuprates. There are two reasons for doing so. First, the experiments strongly suggest that the electron decays throughout the underdoped region - fine-tuning to a critical point as in possibility (b) appears unnecessary. Second, as detailed below, the emergence of a sharp ARPES peak in the superconducting state points to the electron decaying into separate spin- and charge-carrying particle excitations.

## 9.4 THE $Z_2$ GAUGE THEORY

In a recent paper [7] we introduced a new theoretical formulation of strongly interacting electrons based on a $Z_2$ gauge theory, that enabled us to reliably demonstrate the possibility of electron "fractionalization" in two spatial dimensions. The theory is closely linked to an earlier "vortex field theory" approach by Balents et. al. [8], but is formulated in terms of particle excitations - a charge $e$, spin 0 boson (a chargon) and a charge 0, spin $1/2$ fermion (a spinon), which are minimally coupled to a fluctuating $Z_2$ gauge field. Of particular importance to issues of fractionalization are point-like vortex excitations in this $Z_2$ gauge field, called "visons". Fractionalization is obtained whenever the visons are gapped. When the visons *condense* the chargons and spinons are *confined*, effectively "glued" together to form an electron. This results in a conventional phase where the excitations are electrons (or electron composites such as a magnon or a Cooper pair).

The apparent decay of the electron in the underdoped cuprates, strongly suggests that the visons are gapped in this part of the phase diagram. On the other hand, in the heavily overdoped region Fermi liquid behavior is expected, implying a condensation of visons. Together, this implies that with increasing doping there must be a zero-temperature phase transition where the visons first condense. The existence of such a novel "quantum confinement transition" is essentially implied by the experimental data - the transition interpolates between the deconfined underdoped region (with no quasiparticle peak) and the heavily overdoped Fermi liquid regime.

## 9.5 PHASE DIAGRAM

A schematic zero temperature phase diagram paying attention only to the gross feature of whether the visons are gapped or condensed is shown in Fig. 9.3. Of particular interest is the quantum critical point associated with the confinement phase transition. It is clear that, at finite temperature, the crossover from underdoped to overdoped regions will be determined by the properties of the quantum

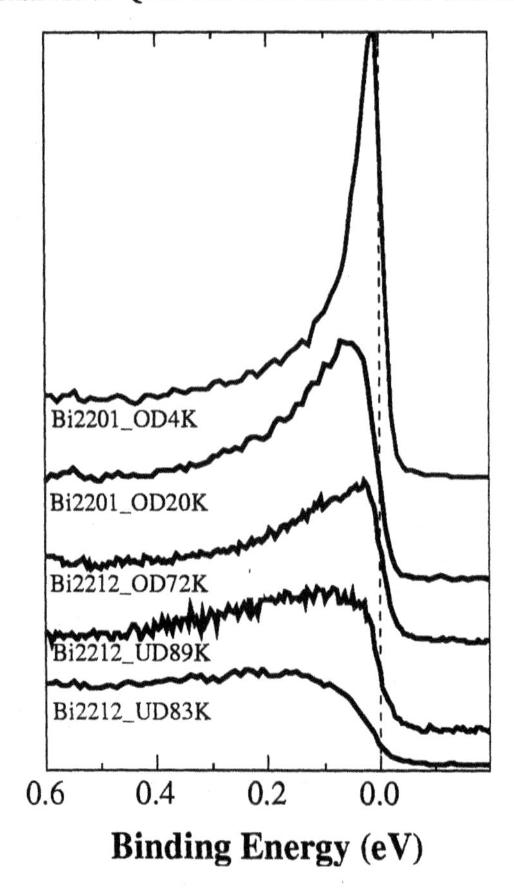


Figure 9.1: Evolution of normal state ARPES lineshape with doping at momentum $(\pi, 0)$. The two lower curves are for underdoped samples while the three upper curves are for overdoped samples. The data are from the group of J.C. Campuzano.

critical region associated with this quantum phase transition. In Fig. 9.4 we sketch the expected finite-temperature crossovers in the vicinity of this phase transition.

The existence of a quantum confinement critical point controlling the crossover from the underdoped to heavily overdoped regimes is in qualitative agreement with a number of experiments. It is well-known that this region is characterized experimentally by power-law temperature or frequency dependences of various physical quantities, as expected at a quantum critical point. But more specifically, the sharpening of the ARPES spectra on moving from the underdoped to the overdoped region strongly suggests that the critical point is associated with a *confinement* transition.

We now discuss the character of the two phases on either side of the confinement transition. For $x > x_c$, and below the finite temperature crossover

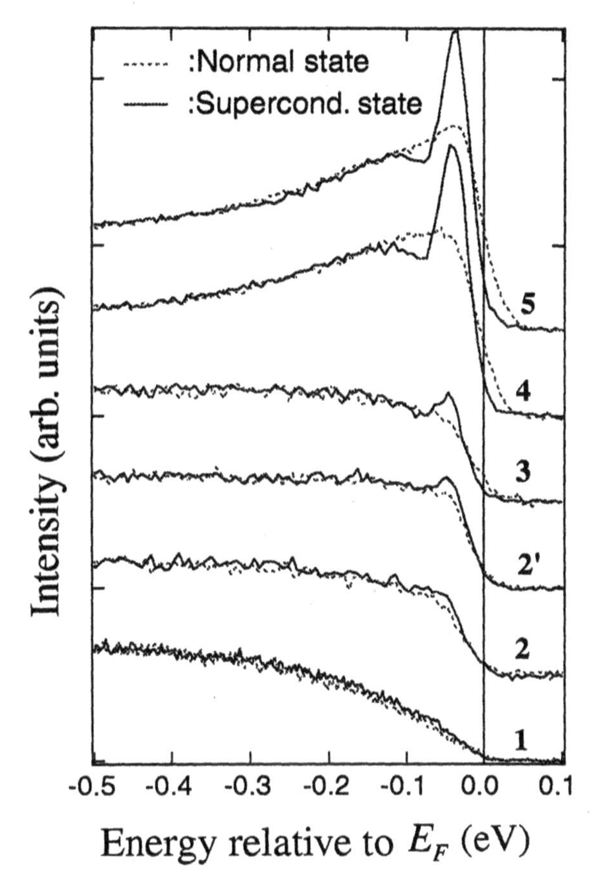


Figure 9.2: ARPES spectra of BSCCO-2212 at momentum $(\pi, 0)$. The data is from the group of Z.X. Shen. The dashed lines are spectra in the normal state, and the solid lines are in the same sample in the superconducting state. The highest curve corresponds to optimal doping with $T_c$ = 90 K. The lower curves correspond to underdoped samples with each successive curve corresponding to a lower value of $T_c$.

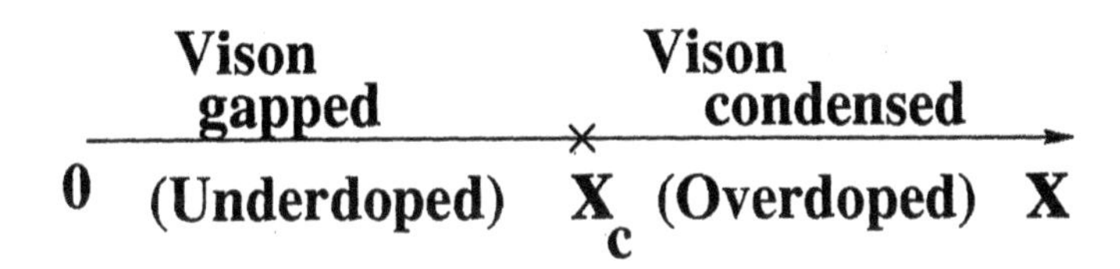


Figure 9.3: Schematic zero temperature phase diagram as a function of doping $x$

line, the system is presumably well-decribed by Landau Fermi liquid theory. In this theory the low-energy quasiparticle excitations near the Fermi surface are essentially electrons - they carry the electron quantum numbers, spin 1/2 and charge $e$ - perhaps with a renormalized effective mass. As such, the electron spectral function should exhibit sharp quasiparticle peaks for all momenta lying on the Fermi surface, which sharpen into delta functions at zero temperature. Unfortunately, samples are difficult to grow in this heavily overdoped regime, so that experimental data is rather limited.

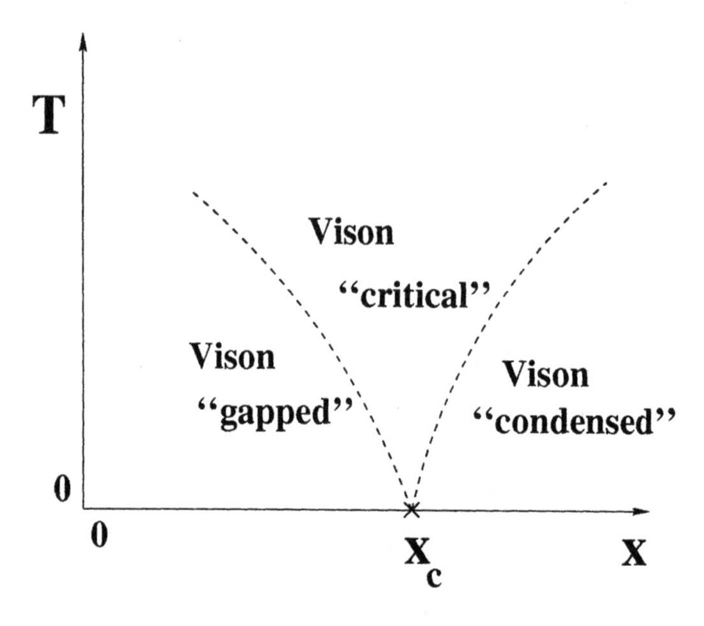


Figure 9.4: Finite temperature crossovers in the vicinity of the quantum confinement transition at $x = x_c$ and $T = 0$. The dashed lines denote crossovers, rather than finite temperature transitions.

But what is the character of the phase for $x < x_c$ where the "vison" excitations are gapped out? This follows readily by inspecting the effective action [7] for the $Z_2$ gauge theory:

$$S = S_c + S_s + S_K + S_B, \tag{9.1}$$

$$S_c = -t_c \sum_{\langle ij \rangle} \sigma_{ij}(b_i^* b_j + c.c.), \tag{9.2}$$

$$S_s = -\sum_{\langle ij \rangle} \sigma_{ij}(t_{ij}^s \bar{f}_{i\alpha} f_{j\alpha} + t_{ij}^{\Delta} f_{i\uparrow} f_{j\downarrow} + c.c) - \sum_i \bar{f}_{i\alpha} f_{i\alpha} \tag{9.3}$$

$$S_K = -K\sum_{\square}\prod_{\square}\sigma_{ij}. \tag{9.4}$$

Here, $b_i^\dagger$ creates a spinless, charge $e$ bosonic excitation - the chargon - and $f_i^\dagger$ creates the spinon, a fermion carrying spin $1/2$ but no charge. When created together, these two excitations comprise the electron. The field $\sigma_{ij}$ is a gauge field that lives on the links of the space-time lattice, and takes on two possible values: $\sigma_{ij} = \pm 1$. The kinetic term for the gauge field, $S_K$, is expressed in terms of plaquette products. Here, $S_B$ is a Berry's phase [7] term which depends on the doping $x$. The vison excitations are vortices in the $Z_2$ gauge field. Specifically, consider the product of the gauge field $\sigma$ around an elementary plaquette, which can take on two values, plus or minus one. When this product is negative, a vison excitation is present on that plaquette. Thus, when the visons are gapped and absent in the ground state, all the plaquette products equal plus one, and one can therefore put $\sigma_{ij} = 1$ on every link. In this case the chargon and spinon can propagate *independently*, and the electron is *fractionalized.* Once the electron is thus splintered, the character of the low temperature phase will depend sensitively on the doping. Based on knowledge of bosonic systems, one expects that the chargons will condense into a superconducting phase upon cooling, with $\langle b^\dagger \rangle \neq 0$. But this condensation can be easily impeded by commensurability effects from the underlying Copper-Oxygen lattice acting in concert with the long-ranged Coulomb interaction. Specifically, in the undoped limit with $x = 0$, there is one charge $e$ chargon per unit cell, and the chargons are expected to lock into a Mott insulating phase, rather than condensing. For very small $x$ with the doped holes well separated, the chargon motion will still be greatly impeded by the near commensurability, and the long-ranged Coulomb interactions should drive charge ordering into an insulating state. Thus, one only expects the chargons to condense into a superconducting phase for $x$ just less than $x_c$, as depicted schematically in Fig. 9.5, and consistent with experiment.

## 9.6 CHARGON CONDENSATION AND SUPERCONDUCTIVITY

What is the nature of the chargon condensed superconducting phase? In a conventional BCS description of superconductivity, two electrons near the Fermi surface pair, and the resulting charge $2e$ Cooper pairs condense. Within this charge $2e$ boson condensed superconductor, the flux quantum is halved, given by $\phi_0 = (1/2)(hc/e) = hc/2e$. This is the value of the observed flux quantum in the Cuprate superconductors, suggesting that the superconducting phase itself is of the BCS variety. But the chargons carry the electron charge $e$, so one might have thought that the BCS superconductor would be equivalent to a chargon-pair condensate - with $\langle b^2 \rangle \neq 0$ yet $\langle b \rangle = 0$ - rather than a single chargon condensate, with $\langle b \rangle \neq 0$. But, quite remarkably, this is not the case. As detailed in Ref. [7], it is the condensation of single charge $e$ chargons that corresponds to the

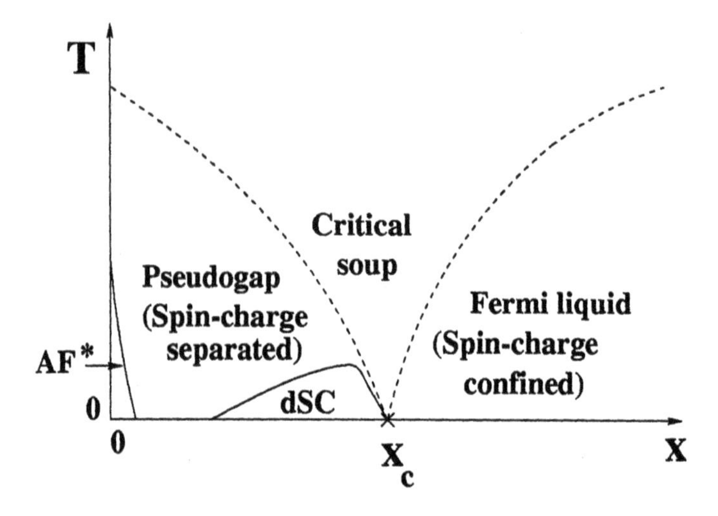


Figure 9.5: Schematic finite temperature phase diagram as a function of doping $x$, with the quantum confinement transition at $x = x_c$. For $x$ just less than $x_c$ and at low temperature, the superconducting state arises as an *inevitable consequence* of the fractionalization of the electron. At small $x$ and low temperature, the system orders antiferromagnetically. The resulting phase (denoted $AF^*$) is nevertheless spin-charge separated - see discussion in the text.

conventional BCS superconductor, whereas the chargon pair condensate describes an exotic non-BCS superconducting phase.

This remarkable fact indicates a new route to superconductivity, very different from a Cooper pairing of electrons. Instead, via a fractionalization process the electron charge is liberated from it's Fermi statistics - resulting in bosonic charge e particles. A direct condensation of these chargons gives the conventional BCS superconducting phase. Since fractionalization is tantamount to a gapping of the vison excitations, this occurs below the crossover line depicted in Fig. 9.5. Thus, below this crossover line one has "preformed" superconductivity, with liberated chargons poised to condense. The electron spin is carried by fermionic spinons in this regime, which are presumed to be gapped throughout the Brillouin zone, except for four gapless nodal points. This leads naturally to a gapping of spin excitations upon fractionalization. Thus, the non-superconducting vison "gapped" regime can account for the observed "pseudo-gap" phase in the underdoped cuprates.

## 9.7 QUANTUM CONFINEMENT CRITICAL POINT

Finally, we discuss the regime intervening between the pseudo-gap and Fermi-liquid phases, centered around $x = x_c$. In this regime the visons are neither gapped nor condensed, but in a critical state. The chargons and spinons which are separated in the vison gapped regime, and confined into the electron when the visons have condensed, are in a state of limbo near $x = x_c$. They cannot move as independent free excitaitons since they are both strongly coupled to the critical fluctutations of the visons, but they also cannot move together as a confined electron. The precise behavior in this critical regime will be controlled by the nature of the zero temperature quantum phase transition, at $x = x_c$ in the Fig. 9.3.

To our knowledge, the possibility and implications of a direct quantum phase transition between a d-wave superconductor and a Fermi liquid phase has not been discussed previously. Within conventional BCS theory there is *no* quantum phase transition separating the Fermi liquid and superconducting phases. Rather, the Fermi liquid phase in the presence of arbitrarily weak phonon mediated attraction between the electrons is *unstable* to the formation of Cooper pairs which then condense leading to superconductivity. Within a modern renormalization group framework, one would say that the fixed point describing the Fermi liquid phase is unstable and crosses over to the superconducting fixed point, as depicted schematically in Fig. 9.6. We are suggesting an alternate possibility interconnecting these two phases. As depicted in Fig. 9.7, we imagine the existence of an unstable fixed point, denoted QCCP (quantum confinement critical point), which controls the nature of a strong coupling zero temperature phase transition between the Fermi liquid and superconducting phases. The existence of such a fixed point is strongly implied by our $Z_2$ gauge theory formulation. To see this, imagine initially decoupling the chargons and spinons in Eqn. 9.1, by setting $t_c = t_s = t_\Delta = 0$ and putting $S_B = 0$. The remaining theory describes a pure $Z_2$ gauge theory, which has two phases [9] - a phase with gapped visons for $K > K_c$, and a vison condensed phase when $K < K_c$. Now recouple the chargon and spinon fields. When the visons are gapped, the chargons and spinons can propagate independently, forming a Bose and Fermi fluid, respectively. Presuming one is not too close to a strongly commensurate filling, the fluid of bosonic chargons should condense at low temperatures giving superconductivity. On the other hand, when the *visons* condense, they *confine* the spinons and chargons, giving fermionic charge $e$ carriers - the electron. Forming a fermionic fluid, these electrons of course cannot condense. Rather, one expects that away from commensurate fillings they will form a conventional metallic Fermi liquid. Finally, right at $K = K_c$ the visons will be in a critical state - described [9] by the classical 3d Ising model when the spinon and chargon coupling is ignored. Here, one expects the spinons and chargons to be strongly scattering off these critical fluctuations, forming a strongly interacting "soup".

Figure 9.6: A two dimensional section of the renormalization group flow diagram showing the instability of a Fermi liquid in the presence of arbitrary weak attractive interactions. The resulting Cooper pairs condense, leading to superconductivity.

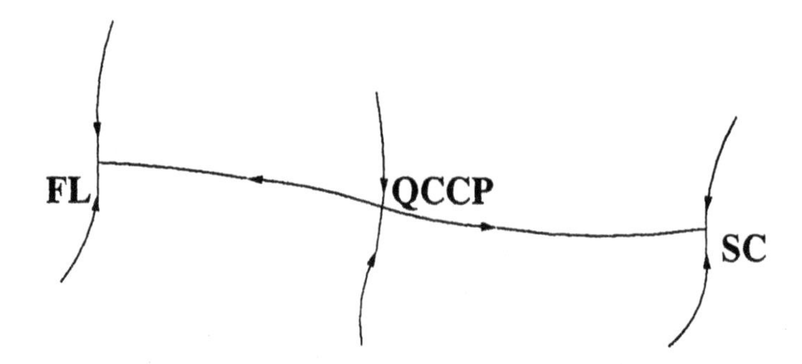


Figure 9.7: A two dimensional section of the renormalization group flow diagram illustrating the different route to superconductivity envisaged in this paper. Central to the proposal is the existence of an unstable fixed point (QCCP) controlling a quantum phase transition at the point of instability of a Fermi liquid toward fractionalization. The resulting spinless charge $e$ bosons condense, leading to superconductivity.

## 9.8 EXPERIMENTAL IMPLICATIONS

We now turn briefly to a few experimental implications of the above scenario, focussing initially on the vison gapped regime for $x < x_c$. Here, the electron is fractionalized - an electron added to the system will decay into a spinon and chargon. This has direct implications for electron photoemission experiments. Since the electron decays, one does not expect a sharp spectral feature in photoemission. More formally, in this regime the electron propogator, $G(r, \tau)$, can be roughly expressed as a product of the chargon and spinon propogators, $G_c$ and $G_s$:

$$G(r, \tau) \approx G_c(r, \tau) G_s(r, \tau). \tag{9.5}$$

The spectral functions for the spinons and chargons ($A(k, \omega) = -\frac{1}{\pi} ImG(k, \omega)$) will have sharp spectral features since these particles can propagate coherently when the visons are gapped, but the *electron* spectral function is a convolution of these two and will hence not exhibit any sharp spectral features. This is exactly as seen in the normal state ARPES spectra in the underdoped samples. Now consider cooling the system into the superconducting state. As explained above, this requires condensation of the chargons so that

$$G_c(r,\tau) \approx |\langle b\rangle|^2 . \tag{9.6}$$

Then, from Eqn. 9.5, the electron Green's function just reduces to

$$G(r,\tau) \approx |\langle b\rangle|^2 G_f(r,\tau), \tag{9.7}$$

and is simply proportional to the spinon Green's function inside the superconductor. Since the spinons propagate coherently, a sharp quasiparticle peak is expected - exactly as seen in the experiments [10]. Moreover, since the *amplitude* of the peak is proportional to $|\langle b\rangle|^2$, it should become smaller as the superconductivity weakens, for instance, by reducing the doping. This is also borne out by the photoemission data [6] - see Fig. 9.2. Thus, the qualitative trends in the underdoped photoemission experiments can be well-explained by assuming the electron decays into a chargon and a spinon.

For $x > x_c$, the low temperature properties of the system should be those of a Fermi liquid. This is commonly believed to be true. It would, however, be useful to have more detailed experimental support.

Now consider the "quantum critical" regime with $x \approx x_c$. As is usual near critical points, power-law temperature dependences are expected for various physical quantities. It is well-known that this is seen in a variety of experiments near optimal doping. In particular, the resistivity in the $ab$ plane exhibits a striking linear temperature dependance. In our scenario, the scattering of the chargons off the critical visons is expected to give a power law resistivity $\rho(T) \sim T^{\alpha}$ with an exponent $\alpha$ that is at present unknown. Calculation of this and other universal properties of this quantum confinement transition is an important challenge that we leave for future work.

Thus far we have primarily focussed on the doping regimes near the superconducting phase. We now turn to the *highly* underdoped and undoped materials. As discussed previously, upon approaching half-filling the condensation of the chargons is expected to be inhibited by commensurability effects together with the long-range Coulomb interactions. Instead, the chargons will localize. Away from half-filing, the charge localization will break the translational symmetry of the lattice. This is qualitatively consistent with the several experiments that observe stripe formation in this region at low temperatures. It is important to stress, however, that in our scenario charge localization and translational symmetry breaking *coexist* with electron fractionalization.

What is the fate of the gapped visons in the undoped material? It is very well established that the undoped insulator has antiferromagnetic long-ranged order. But magnetic order, just like charge order, is conceptually independent of whether or not the electron is fractionalized, in other words, whether the vison is gapped or not. One can therefore contemplate two possibilities - (a) the visons are gapped in the undoped antiferromagnetic insulator, denoted $AF^*$, or (b) the visons are condensed leading to a conventional antiferromagnetic insulator, denoted $AF$, with the electron in the spectrum. Note that the excellent description of the *low energy* spin physics by the quantum Heisenberg spin model is not sufficient to dispose of this question. In fact, the two alternatives are distinguished by the nature of the *gapped* excitations. Experimental evidence for possibility (a) follows from recent photoemission experiments [4] on undoped cuprates, which do not exhibit a sharp quasiparticle peak at *any* momentum in the Brillouin zone. Following Balents et. al. [8], we thus suggest that that the electron decays even in the undoped material, and that the visons are gapped. Further qualitative support is provided by mid-infrared optical absorption [11] and Raman [12] measurements in the undoped material which exhibit broad spectral features out to rather high energies, not expected for the simple Heisenberg model.

## 9.9 COMPARISON WITH ANDERSON'S RVB STATE

Since the original discovery of high temperature superconductivity, literally thousands of theories have been put forward to explain the phenomena. The scenario we describe above has some overlap with many earlier approaches, but is perhaps closest in spirit to the original Resonating Valence Bond (RVB) theory of Anderson [1]. Here, we briefly mention the key similarities and differences with the RVB theory. In the original RVB theory, spin-charge separation is intimately connected with the presence of a "spin-liquid" Mott insulating state, which was argued to support neutral spin one-half spinon excitations. It was soon established, however, that the undoped parent compounds are not spin-liquids but rather antiferromagnetically ordered. It then appeared that the RVB state, if present at all, required the presence of doped holes. In sharp contrast, within our $Z_2$ gauge theory approach spin-charge separation - or more generally electron fractionalization - is *not* directly linked to magnetic ordering. Rather, electron fractionalization occurs whenever the visons are gapped. This is possible even in the presence of long-range magnetic order, in which case the gapless magnons co-exist with gapped spinon excitations. We believe that this is a likely situation in the undoped cuprates.

The original motivation for the RVB approach was based on an analogy with the physics of spinons in one-dimension. But our approach demonstrates that spin-charge separation in two-dimensions requires the existence of a deconfined phase of the underlying $Z_2$ gauge theory. In one-dimension this gauge theory always confines, and spin-charge separation occurs via a different solitonic mechanism. Apparently, this solitonic mechanism of spin-charge separation encapsulated within the RVB approach [14], is not generally operative in higher dimensions. In more formal terms, one can attempt [15] to implement RVB theory directly in two-dimensions with a U(1) or SU(2) gauge theory. But despite the apparent similarity with our $Z_2$ gauge theory, these continuous gauge theories do not have a deconfined phase, and are thus apparently incapable of describing spin-charge separation.

Within the RVB framework, it is common to describe the valence bonds as being "Cooper pairs" pre-formed in the insulator [1], which become mobile upon doping and condense into the superconducting phase. In contrast, in our $Z_2$ gauge theory the picture underlying the superconductivity is the liberation of the electron's charge from it's Fermi statistics, to form bosonic charge $e$ particles - the chargons. Upon doping the chargons become mobile and can condense giving rise to superconductivity.

An entirely new aspect of our approach is the suggestion of a quantum confinement transition separating the spin-charge separated pseudo-gap regime from the heavily overdoped Fermi liquid phase. At this transition the visons are neither gapped nor condensed, but rather in a gapless critical state. Similarly, the spinons and chargons can neither propagate coherently as independent excitations nor as a confined electron. We believe that this quantum confinement transition might well account for much of the novel behavior observed near optimal doping in the cuprates. Developing a theoretical approach to access the properties of such confinement transitions remains as an important yet challenging task.

## ACKNOWLEDGMENTS

We are grateful to L. Balents, G. Baskaran, C. Lannert, P.A. Lee, T.V. Ramakrishnan, G. Sawatzky, R.R.P. Singh, and Doug Scalapino for illuminating conversations. We would especially like to thank Chetan Nayak for emphasizing to us the importance of the crossover to the heavily overdoped portion of the cuprate phase diagram, and J.C. Campuzano and Z.X. Shen for permission to reproduce their experimental data. This research was generously supported by the NSF under Grants DMR-97-04005, DMR95-28578 and PHY94-07194.

## BIBLIOGRAPHY

[1] For a recent review of some of the phenomena, see T. Timusk and B. Stratt, Rep. Prog. Phys. **62**, 61 (1999).

[2] For a review, see for instance M. Randeria and J. -C. Campuzano, *Varenna Lectures*, cond-mat/9709107.

[3] A.G. Loeser et. al., Science **273**, 325 (1996); D.S. Marshall et.al., Phys. Rev. Lett., **76**, 4841 (1996); H. Ding et. al., Nature **382**, 51 (1996).

[4] F. Ronning et. al., Science, **282**, 2067 (1998).

[5] J.C. Campuzano et. al. Phys. Rev. Lett., **83**, 3709 (1999); A. Kaminski et. al., cond-mat/9904390.

[6] Z.-X. Shen, private communication.

[7] T. Senthil and Matthew P.A. Fisher, cond-mat/9910224.

[8] L. Balents, M.P.A. Fisher, and C. Nayak, Phys. Rev. B **60** , 1654 (1999); L. Balents, M.P.A. Fisher, and C. Nayak, cond-mat/9903294.

[9] F. Wegner, J. Math. Phys. **12**, 2259 (1971).

[10] This is particularly clear in the data near momentum $(\pi, 0)$. It is less clear, however, along the "nodal directions" ($k_x = \pm k_y$) at optimal doping - see the very recent work by T. Valla et. al., Science **285**, 2110 (1999). Future experiments will hopefully clarify the issue of the nature of the nodal excitations.

[11] J.D. Perkins et. al., Phys. Rev. B **58**, 9390 (1998) and references therein; M. Gruninger, Ph.D Thesis, Univ. of Groningen (1999); J. Lorenzana and G.A. Sawatzky, Phys. Rev. Lett. **74**, 1867 (1995); J. Lorenzana, J. Eroles, and S. Sorella, cond-mat/9911037.

[12] Some representative papers are G. Blumberg et.al., Phys. Rev. B **53**, R11930 (1996); R.R.P. Singh, P.A. Fleury, K.B. Lyons, and P.E. Sulewski, Phys. Rev. Lett. **62**, 2736 (1989).

[13] P.W. Anderson, Science, **235**, 1196 (1987).

[14] S. Kivelson, D.S. Rokhsar, and J. Sethna, Phys. Rev. B **35**, 8865 (1987).

[15] G. Baskaran, Z. Zou, and P.W. Anderson, Solid State Commun. **63**, 873 (1987); G. Baskaran and P.W. Anderson, Phys. Rev. B **37**, 580 (1988); I. Affleck and J.B. Marston, Phys. Rev. B **37**, 3774 (1988); L. Ioffe and A. Larkin, Phys. Rev. B **39**, 8988 (1989); P.A. Lee and N. Nagaosa, Phys. Rev. B **45**, 966 (1992); P.A. Lee, N. Nagaosa, T.-K. Ng, and X.G. Wen, Phys. Rev. B **57**, 6003 (1998).

# CHAPTER 10

# SPIN-TRIPLET SUPERCONDUCTIVITY OF $SR_2RUO_4$

YOSHITERU MAENO

Department of Physics, Kyoto University
Kyoto 606-8502, Japan

ABSTRACT

We discuss the unconventional superconducting properties of $Sr_2RuO_4$, which has the same layered perovskite structure as high-temperature cuprate superconductors. There are now several key experiments that strongly indicate that $Sr_2RuO_4$ is a spin-triplet, $p$-wave superconductor. We review current experimental progress in determining the details of the gap function.

## 10.1 INTRODUCTION

Discoveries of unconventional superconductivity in heavy Fermion compounds, organic compounds, and copper-based oxides (cuprates) in the last two decades have expanded our knowledge of superconductivity in rich and unexpected ways. It is now widely recognized that strong correlation between the electrons plays an essential role in determining their physical properties, including the unconventional nature of the superconductivity.

The discovery of the high-$T_c$ superconductivity in cuprates [1] has had an enormous impact both experiemtally and theoretically on the study of superconductivity. While searching for superconductivity among the compounds in the same layered perovskite structure (Fig. 10.1), we found superconductivity in the ruthenate $Sr_2RuO_4$ [2]. With the improvement of the quality of single crystals, it soon became clear that its superconductivity is unconventional [3, 4, 5].

The symmetry of conventional superconductivity is spin-singlet, $s$-wave pairing with $\ell = 0$ ($\ell$ is the angular momentum of the Cooper pair). In contrast, unconventional superconductivity with non-$s$-wave pairing may occur among strongly-correlated systems. In these systems, strong Coulomb repulsion between quasi-

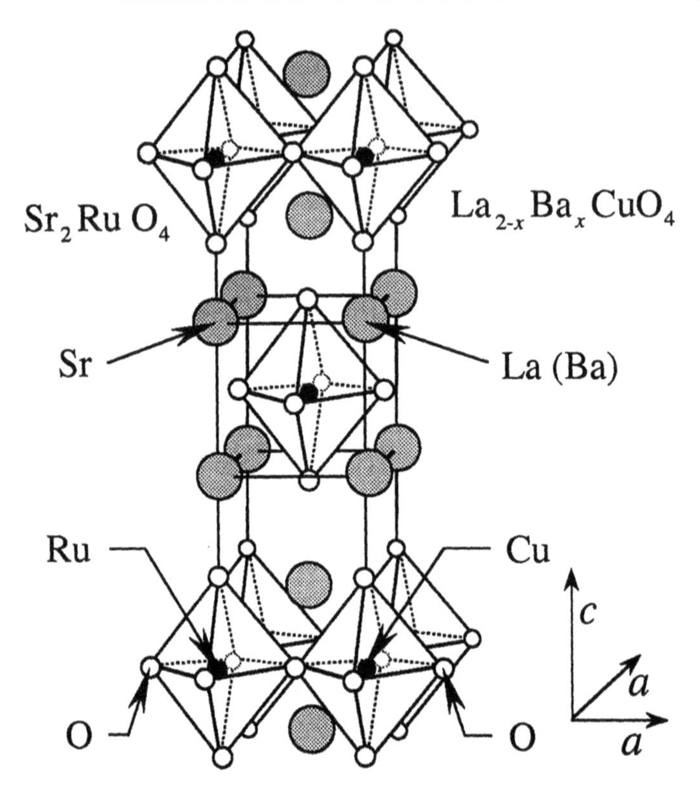


Figure 10.1: Layered crystal structure commom to the ruthenate and cuprate superconductors. The $RuO_2$ and $CuO_2$ planes play essential roles in the superconducting instability.

particles disfavors $s$-wave pairing, since the latter implies a large amplitude for the wave function at close separation; the result is pairing with $\ell \neq 0$. In the high-$T_c$ cuprates, spin-singlet, $d$-wave pairing ($\ell = 2$) has been established [6]. There are now several examples among the heavy Fermion compounds [7] and organic compounds [8] which are believed to be $d$-wave superconductors.

In addition to spin-singlet pairing, spin-triplet pairing with odd $\ell$ is also possible. The heavy Fermion UPt3 [9] and the layered oxide $Sr_2RuO_4$ [2] have recently been shown to be triplet superconductors [10, 11]. In this article, we review the superconducting properties of $Sr_2RuO_4$ with emphasis on the determination of its pairing symmetry. It is remarkable that the possibility of $p$-wave pairing in $Sr_2RuO_4$ was pointed out on theoretical grounds soon after the discovery of its superconductivity [12, 13]. The normal state properties, including the details of the Fermi surface parameters, have been well characterized (section 2). It has been shown that the spin state of the Cooper pair wave function is a triplet (section 3). Some of the unconventional superconducting properties deviate strongly from the expectations based on the simplest triplet gap function. Theories to explain them, based on features in the actual Fermi Surface have been proposed (section 4). There is now evidence for multiple superconducting phases, which is understood as a manifestation of the two-component order parameter (section 5).

## 10.2 QUASI-2D FERMI LIQUID PROPERTIES

Before describing the superconducting properties in detail, let us summarize its normal-state properties. $Sr_2RuO_4$ (with transition temperature $T_c$ = 1.50 K) has the same layered perovskite structure as one of the best studied high-$T_c$ cuprates, $La_{2-x}Sr_xCuO_4$, but the differences in electronic configuration make their physical properties quite different. The states near the Fermi level are derived from four $4d$ electrons in $t_{2g}$ orbitals of $Ru^{4+}$ ions, which hybridize with oxygen $2p$ electrons to form anti-bonding $\pi^*$ states. The Fermi Surface (FS) is quasi-two-dimensional (Q2D) and consists of three nearly cylindrical sheets, $\alpha$, $\beta$ and $\gamma$ [14]. The sheets $\alpha$ and $\beta$ are mainly derived from hybridization of quasi-one-dimensional networks of FS sheets based on the $d_{yz}$ and $d_{zx}$ orbitals, whereas $\gamma$ has the character of a 2D-network of $d_{xy}$ orbitals. Measurements of the de Haas-van Alphen effects have clarified the parameters of all three FS sheets, including the thermodynamic masses [15] and the symmetries and amplitudes of warping of the FS cylinders [16]. Moreover, cyclotron resonance [17] and angular-dependent magnetoresistance oscillation (AMRO) measurements [18] have provided additional information on the quasi-particle mass and the shape of the Fermi surfaces, respectively.

Reflecting the Q2D nature of the FS, the normal state of $Sr_2RuO_4$ exhibits strongly anisotropic transport. Although coherent transport is established in all three directions at low temperatures, the electronic behavior is characteristic of a 2D Fermi liquid: various universality relations for Fermi liquids are satisfied only with the parameters associated with the in-plane conduction [19]. Moreover, above 130 K, metallic conductivity in the inter-layer direction is lost. We also note that it is now possible to grow $Sr_2RuO_4$ crystals with residual resistivities as low as 0.05 $\mu\Omega$cm, corresponding to a quasi-particle mean free path of 2 $\mu$m.

The thermodynamic masses of the quasiparticles are enhanced from the band masses by a factor of 3 to 5, corresponding to a Sommerfeld coefficient in the specific heat of $\gamma_N = 39\text{mJ}/\text{K}^2\text{mol}$. This value is in quantitative agreement with that evaluated from the three thermodynamic masses on the assumption of a 2D FS. The heaviest mass is on the $\gamma$ FS, and equals 14.6 times the bare electron mass. Such a large mass enhancement over the band mass is a clear indication of strong correlation. In fact, a metal-insulator transition occurs in $Ca_{2-x}Sr_xRuO_4$ at $x$ = 0.2, and $Ca_2RuO_4$ is identified as a correlation-driven Mott insulator [20, 21].

Although triplet pairing has been experimentally demonstrated in $Sr_2RuO_4$, the compound is not in the immediate vicinity of a ferromagnetically ordered phase, as might be naively expected from 'parallel spin' pairing. Moreover, in contrast to the case in $UPt_3$, antiferromagnetic magnetic order is not observed in $Sr_2RuO_4$. The uniform susceptibility is only weakly temperature dependent. Since the Wilson ratio is only 1.6, the susceptibility enhancement is mainly attributable to mass enhancement, rather than Stoner enhancement. Consequently, ferromagnetic spin fluctuation at the wave number $q = 0$ is probably not strong. Recently, Sidis *et al.* [22] have observed incommensurate spin fluctuation at $\mathrm{q}_0 = (\pm 0.6\pi/\mathrm{a}, \pm 0.6\pi/\mathrm{a}, 0)$, which is identified as the predicted nesting wave vector between $\alpha$ and $\beta$–FS sheets [23]. However, it is not known at present whether this mode of spin fluctuations is favorable or unfavorable for pairing.

The characteristics of spin fluctuations in $Sr_2RuO_4$ have also been studied with NMR (nuclear magnetic resonance) [24, 25]. Consistent with the results of the neutron scattering, there is no indication of significant spin fluctuation at small $q$. Instead, the data indicates that the susceptibility $\chi(\mathbf{q}, \omega)$ is enhanced over a wide range of $q$'s. This behavior is ascribable to exchange-enhanced ferromagnetic coupling at short distance combined with quasi-two dimensionality [24].

## 10.3 SPIN-TRIPLET SUPERCONDUCTIVITY

A prominent characteristic of the superconducting state in $Sr_2RuO_4$ is the extreme sensitivity of $T_c$ to non-magnetic impurities and lattice defects [26, 27]. Reflecting the unconventional pairing, superconductivity is completely suppressed when the quasiparticle mean free path becomes comparable to the in-plane coherence length, $\xi_{ab}(0) = 66$ nm [28] (an impurity concentration of a few hundred ppm is sufficient to suppress $T_c$ completely). The relation between $T_c$ and the residual resistivity $\rho_{ab,0}$ is well fitted with the Abrikosov-Gor'kov function modified to describe a non-$s$-wave superconductor with non-magnetic impurities.

Superconductor-normal metal-superconductor (SNS) tunneling has also been employed by Jin *et al.* [29] to probe $Sr_2RuO_4$. It was found that the critical current of a Pb/$Sr_2RuO_4$/Pb junction, induced by the proximity effect, is abruptly suppressed on cooling through the $T_c$ of $Sr_2RuO_4$. This phenomenon may be explained if one assumes interference between the pair-tunneling paths of singlet and triplet pairs [30].

As further evidence for unconventional pairing, we have reported that reduction in $T_c$ results in a substantial increase in the residual electronic coefficient of the specific heat, $C_e/T$ [31]. This behavior arises from an increase in the DOS (density of states) at low energies coming from impurity states, as expected in unconventional superconductors. Therefore, single crystals of very high purity are needed to study the intrinsic superconducting properties.

From the experimental results described above, it is clear that the superconductivity in $Sr_2RuO_4$ is unconventional. To determine the spin state of the Cooper pairs, however, a more direct microscopic probe is required. Measurement of the spin susceptibility by the shift of the NMR resonance peaks (the Knight shift) is probably the most reliable means for this purpose.

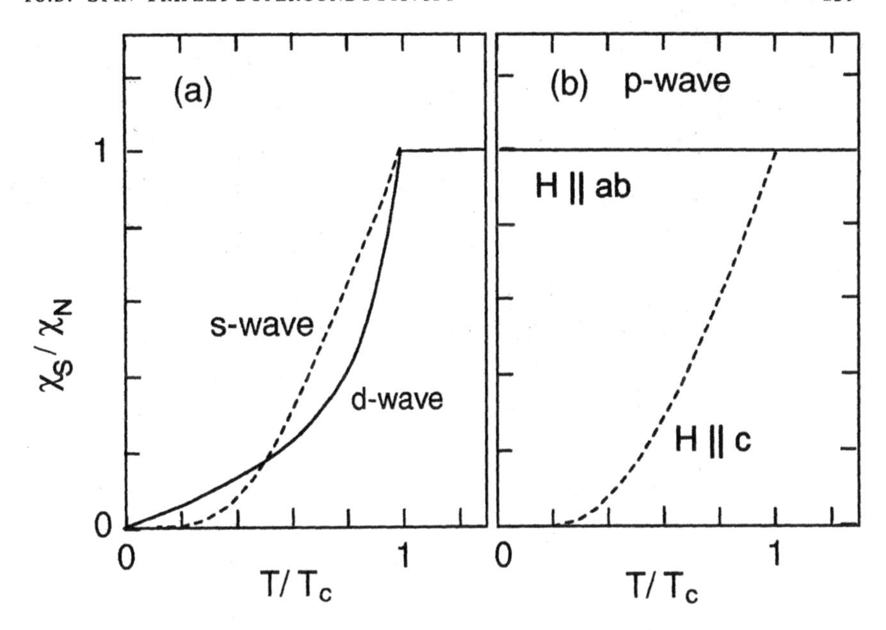


Figure 10.2: Ratio of the spin susceptibilities in the superconducting state to the normal state, $\chi_S/\chi_N$. (a) Spin-singlet $s$-wave and $d$-wave states; (b) Spin-triplet state $\mathbf{d}(\mathbf{k}) = \hat{\mathbf{z}} f(\mathbf{k})$.

In a spin-singlet state, the spin susceptibility $\chi$ decreases as $T \to 0$ regardless of the field direction, as shown in Fig. 10.2a. Because of the differences in their gap structures, the $s$ and $d$-wave states display different temperature dependences. In high-$T_c$ cuprates, a clear reduction in the spin susceptibility is observed [32].

In a spin-triplet state, the gap function is expressed in terms of a vector order parameter called a $\mathbf{d}$-vector [4, 12], which points along the direction in which the component of the spin vanishes (hence, the parallel spins lie in the plane normal to $\mathbf{d}$). Consequently, $\chi$ remains unchanged if the field is normal to the $\mathbf{d}$-vector (Fig. 10.2b). For fields parallel to $\mathbf{d}$, $\chi$ decreases just as in the singlet state.

Unlike the situation in $f$-electron systems, we do not encounter the problem of extracting the spin susceptibility from the NMR Knight shift in $d$-electron systems such as $Sr_2RuO_4$. First, the spin-orbit coupling is weaker, and the pairing is well-described as occurring in spin space. This makes the interpretation of the Knight shift simpler. Secondly, the contribution of the orbital susceptibility is rather small, and it is relatively straightforward to extract the spin part of the Knight shift. In $Sr_2RuO_4$, the Knight shifts at both the in-plane oxygen sites are, in fact, dominated by the spin susceptibility.

Measurements of the spin susceptibility by the $^{17}$O NMR Knight shift were carried out by Ishida *et al.* [11] ($^{17}$O was substituted for $^{16}$O by annealing of the crystals). For fields applied || [100], the Knight shift does not change across $T_c$,

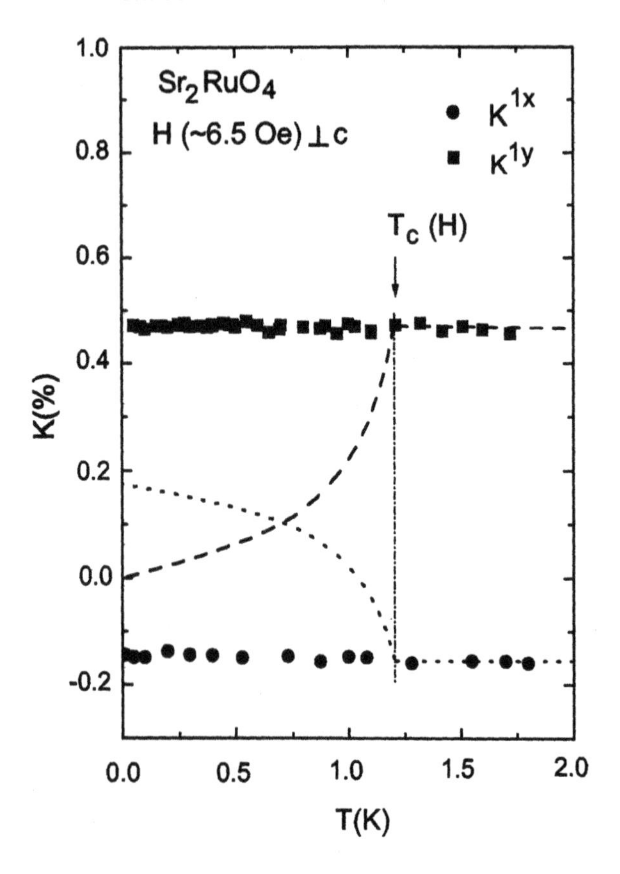


Figure 10.3: Knight shift of the $^{17}O$ NMR peaks in $Sr_2RuO_4$ [11].

as shown in Fig. 10.3. This invariance shows that the spin state of the Cooper pair is a triplet. Furthermore, it indicates that the d-vector is perpendicular to [100]. Since the Knight shift is invariant also for $\mathbf{H} \parallel [110]$ in a more recent experiment, we conclude that the d-vector is most probably along [001], viz. $\mathbf{d}(\mathbf{k}) = \hat{\mathbf{z}} f(\mathbf{k})$.

Another piece of important information on the gap comes from the observation of a spontaneous internal magnetic field by muon spin relaxation ($\mu$SR) [33], as shown in Fig. 10.4. The observation strongly suggests that time-reversal symmetry (TRS) is broken in the superconducting state [33]. Since $\hat{\mathbf{z}}$ does not break the TRS, we may conclude that it is the orbital part that breaks the TRS. Among the allowed superconducting symmetries of $Sr_2RuO_4$, the one that satisfies this constraint is $\mathbf{d}(\mathbf{k}) = \hat{\mathbf{z}}\Delta_0(\hat{k}_x + i\hat{k}_y)$, where $\hat{\mathbf{k}}$ is a unit vector specifying a direction in orbital space [12, 34]. The wave function expressed by this d-vector is schematically shown in Fig. 10.5. In the presence of a weak spin-orbit

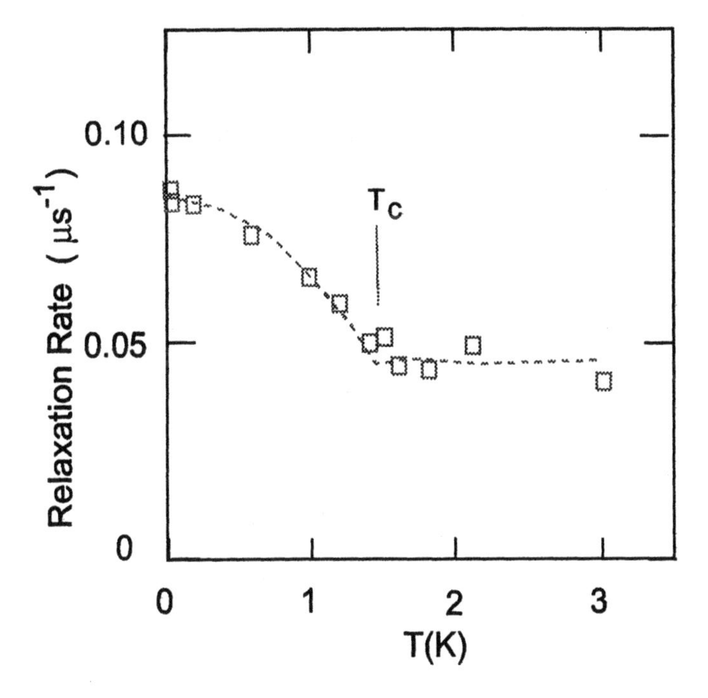


Figure 10.4: Muon relaxation rate $R$ in $Sr_2RuO_4$ upon zero-field-cooling. The increase in $R$ below $T_c$ is identified with the spontaneous emergence of an internal field [33].

coupling, which should be the case in $Sr_2RuO_4$, it is expected that the direction of **d** is pinned along the interlayer direction **c**.

The spin part, $\hat{\mathbf{z}} = |S_z = 0\rangle$, is shown by the thin arrows as the superposition of the up-up and down-down spin pairs (equal spin pairs) with the quantization axis along any direction in the $ab$-plane. The orbital part, $\hat{k}_x + i\hat{k}_y$, refers to the $p$-wave state with an unquenched angular momentum $L_z = +1$, as indicated by the thick open arrows.

For this **d**-vector, the magnitude of the gap given by $|\ \Delta(\mathbf{k})\ | = \Delta_0\sqrt{(\hat{k}_x^2 + \hat{k}_y^2)}$ is isotropic on the 2D FS; the DOS in the superconducting state is the same as that for the $s$-wave state. Therefore, one expects the specific heat to exhibit the familiar BCS behavior. However, the results in $Sr_2RuO_4$ are quite different. We discuss below how the specific heat and the nuclear relaxation rate may be understood in terms of a modified gap function.

## 10.4 ANISOTROPY OF THE SUPERCONDUCTING GAP

Figure 10.6 shows the temperature ($T$) dependence of $C_e/T$, the electronic specific heat divided by $T$ [35]. The transition midpoint $T_c$ = 1.48 K corresponds to

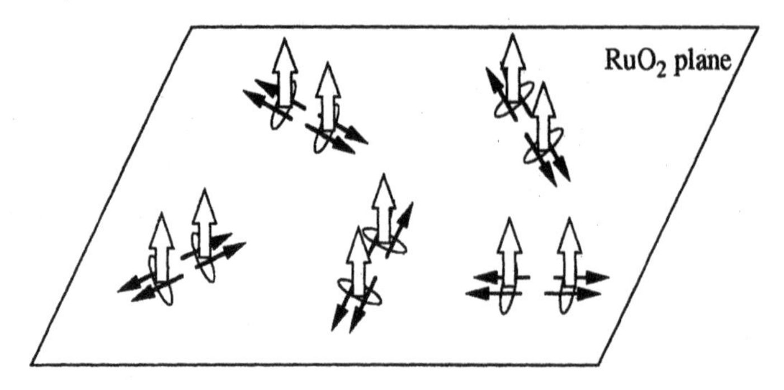


Figure 10.5: Schematic representation of the spin-triplet, $p$-wave state with the $\mathbf{d}$-vector $\mathbf{d}(\mathbf{k}) = \hat{\mathbf{z}}\Delta_0(\hat{k}_x + i\hat{k}_y)$. The thin arrows indicate spin-wave states of equal spin pairing. Open arrows indicate the orbital function with $L_z = +1$.

$T_c/T_{c0} = 0.99$, for which $T_{c0} = 1.50$ K has been deduced from the impurity effect. We first note that the normalized jump at $T_c$ is $\Delta C_e/\gamma_N T_{c0} = 0.74 \pm 0.02$, where $\gamma_N$ is the value of $C_e/T$ in the normal state.

This is substantially less than the value 1.43 expected for an isotropic gap in 2D, but is comparable to that expected for state with line nodes (0.96). Secondly, $C_e/T$ linearly extrapolates to $\gamma_0 = 3 \pm 3$ m/K$^2$mol as $T \to 0$ K. It is likely that $\gamma_0$ is intrinsically zero. The linear $T$-dependence is consistent with a line-node gap, but the experimental results show even stronger entropy increase.

The results of nuclear relaxation rate studied by Ru NMR [36, 37], shown in Fig. 10.7, are consistent with the behavior of the specific heat. First, $1/T_1$ drops sharply below $T_c$ with no Hebel-Slichter coherence peak. Secondly, the relaxation rate $1/T_1$ decays as $T^3$ at low temperatures, in accordance with the behavior for a line-node. The value of $1/T_1$ at 0.1 K indicates that the DOS at the Fermi energy $\epsilon_F$ is less than 10% of the DOS in the normal state.

While we cannot derive a definitive conclusion on the nodal structure of the gap simply from the observed $T$ dependence, we may at least conclude from the results of $C_e/T$ and $1/T_1$ that the DOS is similar to that of a superconducting gap with line nodes. The possibility of the line-node state as the ground state [38] was discarded in the original treatment by Rice and Sigrist [12]. In the weak-coupling limit, a fully-gapped state always provides a lower energy, unless a special mechanism is operative in the pairing interaction. In fact, the $\mu$SR results, combined with the NMR Knight shift results, require a nodeless gap as we described in section 3.

A coherent explanation may be given if the orbital wave function $f(\mathbf{k})$ is assumed to be strongly anisotropic, yet still nodeless. The physical origin of the

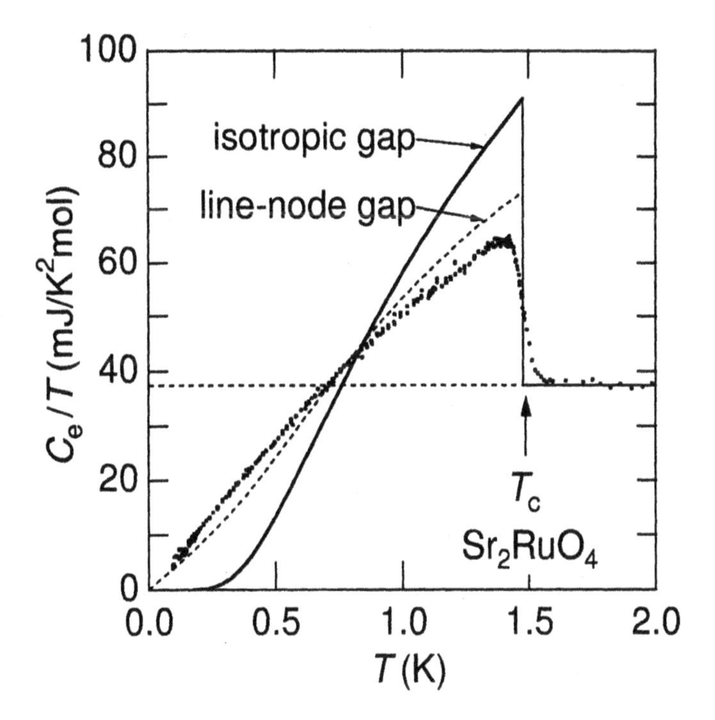


Figure 10.6: The electronic specific heat divided by temperature, $C_e/T$, plotted against $T$ in $Sr_2RuO_4$.

gap anisotropy may arise from anisotropy in the DOS and the Fermi velocity, or from anisotropy in the pairing interaction [39, 40]. For the latter, Miyake and Narikiyo have considered short-range ferromagnetic interaction between the Ru spins, and derived a strongly anisotropic gap structure with a tiny gap for spin-triplet superconductivity, $\mathbf{d} = \hat{\mathbf{z}}\Delta_1[\sin(\pi R\hat{k}_x) + i\sin(\pi R\hat{k}_y)]$, where $R \simeq 0.9$ is a fitting parameter determined by experiment [40]. The magnitude of the gap on the FS is given by $|\ \Delta(\mathbf{k})\ | = \Delta_1[\sin^2(\pi R\cos\phi) + \sin^2(\pi R\sin\phi)]^{1/2}$, where $\phi$ is the angle that $\hat{\mathbf{k}}$ makes with the $x$ axis.

## 10.5 $H-T$ PHASE DIAGRAM

For $Sr_2RuO_4$, the most probable **d**-vector is that obtained from a two-component order parameter with nodeless gap. Thus, under a symmetry-breaking field that lifts the degeneracy of the two components, such as a field precisely aligned with the *ab*-plane, it is predicted that a second superconducting phase with a different symmetry becomes stable. We will discuss our search for such superconducting multiple phases of $Sr_2RuO_4$ in this section.

On the basis of the Ginzburg-Landau approximation, Agterberg suggested that

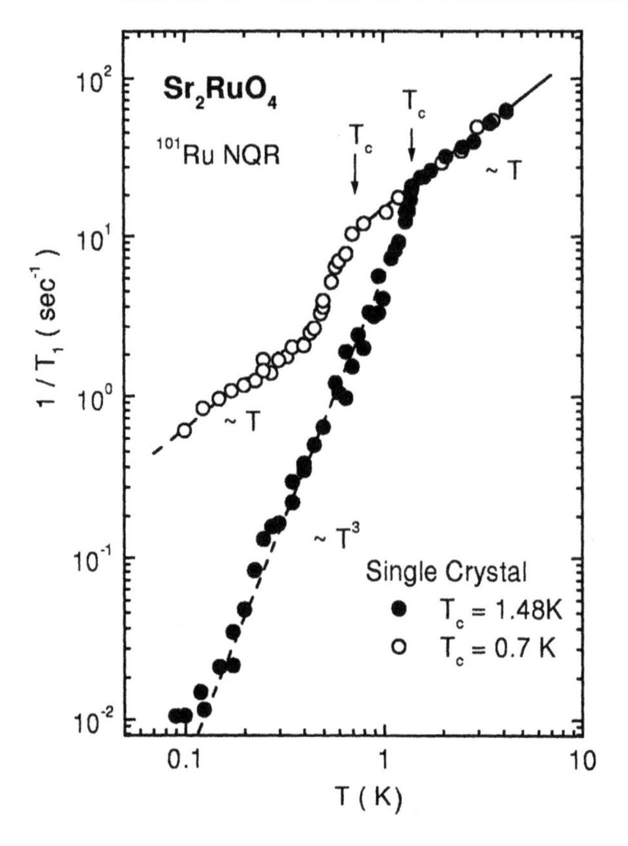


Figure 10.7: The nuclear spin relaxation rate $1/T_1$ of Ru NQR in $Sr_2RuO_4$ [36].

the state with a line-node gap with $\mathrm{d} = \hat{\mathbf{z}}\Delta_0\hat{k}_{x'}$, where $\hat{\mathbf{x}}'$ is along the applied field, becomes stable near $H_{c2}$ [41]. This state has a gap maximum along the external field **H**. Since the line nodes rotate with **H**, we expect that the in-plane anisotropy of $H_{c2}$ is enhanced beyond the value expected from the tetragonal FS anisotropy alone. Moreover, since the state expected in low fields and low temperatures has a different symmetry from the line-node state, we naturally anticipate a second-order phase transition between these two states. If the field is not exactly parallel to the basal plane, however, the line-node singularity formation is incomplete and we expect only a crossover behavior.

We have investigated the in-plane anisotropy of $H_{c2}$ and searched for the second superconducting transition by measuring both the AC susceptibility and the specific heat [42]. By rotating the sample in **H**, we were able to attain alignment of the *ab*-plane to within 0.02 deg. At low $T$, we observed an in-plane 4-fold symmetric anisotropy in $H_{c2}$ of about 3 %. The maximum in $H_{c2}$ is along the [110] direction, diagonal to the direction of the Ru-O bonding. The in-

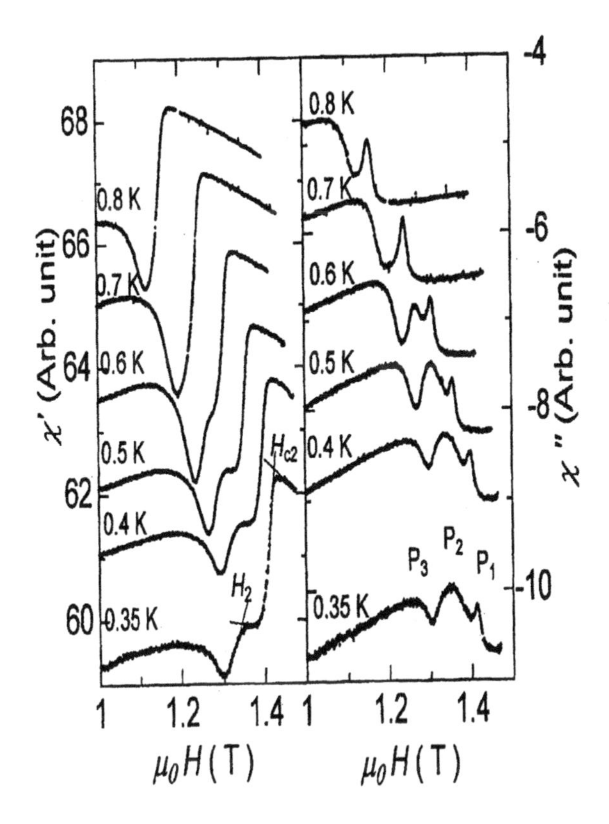


Figure 10.8: AC suseptibility of $Sr_2RuO_4$ in a field $\mathbf{H} \parallel [110]$. Both the in-phase, shielding component $\chi$' and the out-of-phase, energy-absorption component $\chi$" exhibit unusual features below ~0.7 K. P1 ($H_{c2}$) is the superconducting-normal state transition. P2 ($H_2$) corresponds to a second superconducting transition, and P3 corresponds to the vortex synchronization peak.

plane $H_{c2}$ anisotropy exhibits a non-monotonic $T$ dependence: it remains large only below 0.8 K, rapidly decreases at higher temperatures, and even reverses sign above 1.1 K. The anisotropy is suppressed also by slight field misalignments of less than 1deg. away from the plane.

In addition, we observed a peculiar absorption peak in the AC susceptibility at the field $H_2$, slightly below $H_{c2}$ (see Fig. 10.8). The peak at $H_2$ becomes undetectable if $T$ is raised above 0.7 K, or if $\mathbf{H}$ is tilted out of the $ab$-plane by a mere 0.6 deg. $H_2$ appears to merge with $H_{c2}$ at 0.8 K and 1.2 T, in the $H - T$ phase diagram (Fig. 10.9).

Moreover, in the region of the phase diagram where the feature at $H_2$ is observed, we have observed substantial entropy release in the specific heat (Fig. 10.10). Thus the feature is ascribable to the presence of additional thermodynamic

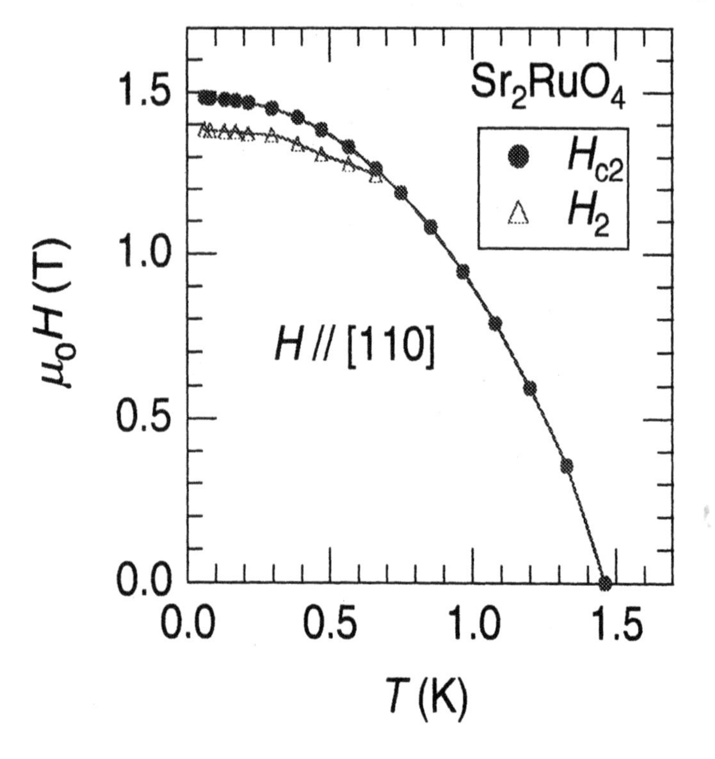


Figure 10.9: The $H - T$ phase diagram of $Sr_2RuO_4$. The field $H_{c2}$ ($H_2$) is indicated by circles (triangles).

transitions in the superconducting state. To our knowledge, the only plausible interpretation for the presence of the second transition for **H** $\parallel$ $ab$-plane is the emergence of the line-node state.

## 10.6 CONCLUDING REMARKS

We have described the superconducting properties of $Sr_2RuO_4$ with emphasis on its pairing symmetry. There is now convincing evidence that its spin state is a triplet. However, there exist important deviations from the simple $p$-wave state with isotropic energy gap. The $H - T$ phase diagram with multiple superconducting phases may provide an important clue towards the understanding of the symmetry of the wave functions.

Of all the known unconventional superconductors, $Sr_2RuO_4$ may be the simplest in many respects, and may well become the best characterized. There is a real hope that its superconductivity will be understood in unprecedented detail and accuracy. For this reason, we envisage that $Sr_2RuO_4$ will serve as an important

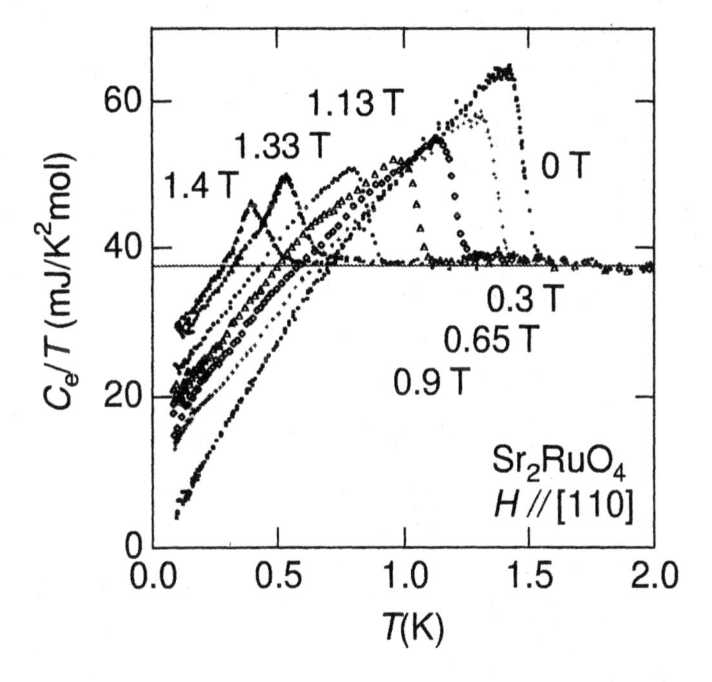


Figure 10.10: Specific heat of $Sr_2RuO_4$ in fields $\mathbf{H} \parallel [110]$. The peak in $C_e/T$ near the transition is anomalously narrowed in fields above 1.3 T.

model system, both experimentally and theoretically, to expand our knowledge of the physics of spin-triplet superconductivity.

Acknowledgements: The author is most grateful to T. Ishiguro for his support. He also wishes to thank his colleagues, especially T. Akima, S. NishiZaki, and Z. Q. Mao, for their contributions. He has benefited from active collaborations with many groups. In particular, he thanks K. Ishida, M. Sigrist, A. P. Mackenzie, T. Ohmi, D.F. Agterberg, T. Imai and K. Miyake for useful comments and suggestions. This work is supported by a grant from CREST, Japan Science and Technology Corporation. Finally, he sincerely thanks N. P. Ong for his help in preparing the manuscript.

## BIBLIOGRAPHY

[1] J.G. Bednorz and K.A. Müller, Z. Phys. B **64** 189 (1986).

[2] Y. Maeno, H. Hashimoto, K. Yoshida, S. Nishizaki, T. Fujita, J. G. Bednorz, and F. Lichtenberg, Nature **372** 532 (1994).

[3] A.P. Mackenzie and Y. Maeno, Physica B **280** 148 (2000); Y. Maeno, Z. Q. Mao, S. NishiZaki, and T. Akima, *ibid* 285 (2000).

[4] Y. Maeno, S. NishiZaki, and Z.Q. Mao, J. Supercon. **12** 535 (1999); A.P. Mackenzie, *ibid.*, 543.

[5] S.R. Julian, A. P. Mackenzie, G.G. Lonzarich, C. Bergemann, R.K.W. Haselwimmer, Y. Maeno, S. NishiZaki, A.W. Tyler, S. Ikeda, and T. Fujita, Physica B **259-261** 928 (1999).

[6] For example, C.C. Tsuei and J.R. Kirtley, Physica C **282-287** 4 (1997); D.J. Van Harlingen, *ibid.*, 128.

[7] For example, J-P. Brison, L. Glémot, H. Suderow, A. Huxley, S. Kambe, and J. Flouquet, Physica B 280 165 (2000), and references therein.

[8] For example, T. Ishiguro, K. Yamaji, and G. Saito, *Organic Superconductors*, 2nd ed. (Springer, Heidelberg, 1998), Ch. 5.

[9] G.R. Stewart, Z. Fisk, J.O. Willis, and J.L. Smith, Phys. Rev. Lett. **52** 679 (1984).

[10] H. Tou *et al.*, Phys. Rev. Lett. **77** 1374 (1996); H. Tou, Y. Kitaoka, K. Ishida, K. Asayama, N. Kimura, Y. Onuki, E. Yamamoto, Y. Haga, and K. Maezawa, Phys. Rev. Lett. **80** 3129 (1998).

[11] K. Ishida, H. Mukuda, Y. Kitaoka, K. Asayama, Z. Q. Mao, Y. Mori, and Y. Maeno, Nature **396** 658 (1998).

[12] T. M. Rice and M. Sigrist, J. Phys. Condens. Matt. **7** L643 (1995).

[13] G. Baskaran, Physica B **223-224** 490 (1996).

[14] T. Oguchi, Phys. Rev. B **51** 1385 (1995); D.J. Singh, Phys. Rev. B **52** 1358 (1995).

[15] A.P. Mackenzie, S.R. Julian, A. J. Diver, G.J. McMullan, M.P. Ray, G.G. Lonzarich, Y. Maeno, S. Nishizaki, and T. Fujita, Phys. Rev. Lett. **76** 3786 (1996).

[16] C. Bergemann, S. Julian, A.P. Mackenzie, S. NishiZaki, and Y. Maeno, Phys. Rev. Lett. **84** 2662 (2000).

[17] S. Hill, J. Brooks, Z.Q. Mao, and Y. Maeno, Phys. Rev. Lett. **84** 3374 (2000).

[18] E. Ohmichi, H. Adachi, Y. Maeno, and T. Ishiguro, Phys. Rev. B **59** 7263 (1999).

[19] Y. Maeno, H. Hashimoto, K. Yoshida, T. Fujita, A.P. Mackenzie, J. Phys. Soc. Jpn. **66** 1405 (1997).

[20] S. Nakatsuji and Y. Maeno, Phys. Rev. Lett. **84** 2666 (2000).

[21] H. Fukazawa, S. Nakatsuji, and Y. Maeno, Physica B 281-282 613 (2000).

[22] Y. Sidis, M. Braden, P. Bourges, B. Hennion, Y. Mori, and Y. Maeno, Phys. Rev. Lett. **83** 3320 (1999).

[23] I.I. Mazin and D.J. Singh, Phys. Rev. Lett. **82** 4324 (1999).

[24] H. Mukuda, K. Ishida, Y. Kitaoka, K. Asayama, Z.Q. Mao, Y. Mori, and Y. Maeno, J. Phys. Soc. Jpn. **67** 3945 (1998).

[25] T. Imai, A.W. Hunt, K.R. Turber, and F.C. Chou, Phys. Rev. Lett. **81** 3006 (1998). Their statement of the observation of the ferromagnetic spin fluctuation should be taken with caution. Their data indicate the absence of strong commensurate AF fluctuations, but not necessarily the presence of strong spin fluctuations at $q = 0$.

[26] A. P. Mackenzie, A. W. Tyler, S. R. Julian, K. Haselwimmer, G. G. Lonzarich, Y. Maeno, S. Nishizaki, Phys. Rev. Lett. **80** 161 (1998); *ibid.* **80** 3890(E) (1998).

[27] Z.Q. Mao, Y. Mori, and Y. Maeno, Phys. Rev. B **60** 610 (1999).

[28] T. Akima, S. NishiZaki, and Y. Maeno, J. Phys. Soc. Jpn. **68** 694 (1999).

[29] R. Jin, Yu. Zadorozhny, Y. Liu, D.G. Schlom, Y. Mori, and Y. Maeno, Phys. Rev. B **59** 4433 (1999).

[30] C. Honerkamp and M. Sigrist, Prog. Theor. Phys. **100** 53 (1998).

[31] S. NishiZaki, Y. Maeno, and Z. Q. Mao, J. Low Temp. Phys. **117** 1581 (1999).

[32] See for e.g. S.E. Barrett *et al.*, Phys. Rev. B **41** 6283 (1990).

[33] G.M. Luke, Y. Fudamoto, K.M. Kojima, M.I. Larkin, J. Merrin, B. Nachumi, Y.J. Uemura, Y. Maeno, Z. Q. Mao, Y. Mori, H. Nakamura, and M. Sigrist, Nature **394** 558 (1998).

[34] M. Sigrist, D. Agterberg, A. Furusaki, C. Honerkamp, K.K. Ng, T.M. Rice, and M.E. Zhitomirsky, Physica C **317-318** 134 (1999).

[35] S. NishiZaki, Y. Maeno and Z.Q. Mao, J. Phys. Soc. Jpn **69** 572 (2000).

[36] K. Ishida, H. Mukuda, Y. Kitaoka, K. Asayama, Y. Maeno, Phys. Rev. B **56**, R505 (1997).

[37] K. Ishida, H. Mukuda, Y. Kitaoka, K. Asayama, Y. Maeno, Phys. Rev. Lett. **84** xxx (2000), *in press*.

[38] Y. Hasagawa, K. Machida, and M. Ozaki, J. Phys. Soc. Jpn. **69** 336 (2000).

[39] Y. Okuno and M. Sigrist, J. Superconductivity **12** 563 (1999).

[40] K. Miyake and O. Narikiyo, Phys. Rev. Lett. **83** 1423 (1999).

[41] D. F. Agterberg, Phys. Rev. Lett. **80** 5184 (1997).

[42] Z.Q. Mao, Y. Maeno, T. Akima, and T. Ishiguro, Phys. Rev. Lett. **84** 991 (2000).

# CHAPTER 11

# TRIPLET QUASI-ONE-DIMENSIONAL SUPERCONDUCTORS

S. E. BROWN[1], M. J. NAUGHTON[2], I. J. LEE[3]
E. I. CHASHECHKINA[3], AND P. M. CHAIKIN[3]

[1]Department of Physics, Univ. of California, Los Angeles, CA 90024
[2]Department of Physics, Boston College, Boston, MA 02167
[3]Department of Physics, Princeton University, Princeton, NJ 08544

ABSTRACT

The highly anisotropic, quasi-one-dimensional organic conductors $(TMTSF)_2X$, the Bechgaard salts, are correlated materials with many interesting properties and ground states. A magnetic field component between the layers serves to dephase the interlayer coupling and control the dimensionality. However, at certain "magic" angles, when the field is aligned along a crystallographic axis, interlayer tunneling is possible along the field lines. As suggested by Strong, Clarke and Anderson, the interlayer decoupling leads to different ground states as evidenced by a metallic conductivity for the magic angles and a non-metallic conductivity for non-magic angles. In the superconducting state, the interlayer decoupling or dimensional crossover produces a greatly enhanced orbital critical field and allows the evaluation of spin pair-breaking effects. A critical field more than four times the Clogston-Pauli limit for singlet superconductors strongly suggests that these materials are triplet superconductors. Recent $^{77}$Se NMR Knight shift measurements confirm the triplet state.

## 11.1 INTRODUCTION

The Bechgaard salts are among the most remarkable electronic materials ever discovered. A single crystal of $(TMTSF)_2PF_6$ is metallic at room temperature but upon cooling at different pressures enters a variety of different low temperature states including, antiferromagnetic, spin and charge density wave insulator and semiconductor, semimetal, metal and superconducting [1]. Application of a moderate magnetic field introduces some phases previously unknown in other materials, including a cascade of transitions to different spin density wave states coincident with quantum Hall plateaus. In these different states a single crystal exhibits all of the electronic transport mechanisms yet discovered including, single particle, sliding density wave, quantum Hall and superconducting. The subject of the present talk are two more remarkable phenomena, magnetic field-induced interlayer decoupling and triplet superconductivity. These materials are an appropriate contribution to a symposium in honor of Phil Anderson. The vast range of phenomena observed cover a large part of condensed matter physics and he has made fundamental contributions to all of them. However, there are two more direct contributions to these studies. When the field of low dimensional organic conductors was in its infancy, in the early 70's, Phil and his colleagues at Bell, P. A. Lee and T. M. Rice [3] pointed out many of the distinct properties later observed. They addressed the characteristic energies and interactions and particularly the role of fluctuations in low dimensions to both the phase transitions and the transport. His most direct contribution to our specific work came in the models he and his postdocs, Steve Strong and Dave Clarke [4] proposed to explain some of our early results on angular dependent magnetotransport. They have profoundly influenced the direction of this work and led to the observation of even more intriguing phenomena.

The (TMTSF) molecule is flat with relatively large Se orbitals sticking out from the plates. The molecules stack face-to-face in a zigzag pattern forming linear chains along the **a** direction with a bandwidth of $4t_a$ ~1eV and a room temperature conductivity $\sigma_a \sim (1\ \text{m}\Omega\text{-cm})^{-1}$ (about .001 of copper). Between the chains along the **b** direction the electron transfer is still sizable with a bandwidth of $4t_b \sim 0.1$eV and $\sigma_a/\sigma_b \simeq 100$. The conducting **a-b** layers are separated by the $PF_6$ anions which greatly reduce the transfer. In the interlayer direction the bandwidth is $4t_c \sim 0.003$ eV and $\sigma_a/\sigma_c \simeq 10^5$ [2]. The charge transfer to the anions is precisely one electron per 2 TMTSF's leaving a three quarter filled band. Since there is a slight dimerization at room temperature, there are two bands, and the upper band is half-filled. If the system were strictly one-dimensional, the one half-filled band would certainly lead to an insulating state. If the system were strictly one-dimensional, the Fermi surface would consist of two parallel sheets at $\pm k_F$. The bandwidths perpendicular to the chains slightly warp the Fermi surface and the lack of Fermi surface nesting reduces the one-dimensional instabilities. At ambient pressure the warping is insufficient to prevent a Peierls-like spin density wave (SDW) transition at 12 K. Application of ~ 6 kbar pressure increases the transverse bandwidths sufficiently to suppress the SDW transition, and the material goes superconducting at 1.2 K.

## 11.2 EARLY RESULTS ON P-WAVE PAIRING

$(TMTSF)_2PF_6$ was the first organic superconductor discovered [5]. Since there had been extensive theoretical work in the 60's and 70's anticipating one dimensional (1-D) conductors, the ground states for interacting 1-D systems had been extensively investigated. In the '$g$-ology' picture, the superconducting state which bordered the SDW phase was a triplet [6]. So the first guess for the symmetry of the order parameter was $p$-wave triplet. One of the first tests of the order parameter symmetry was the effect of small amounts of non-magnetic impurities. Using radiation damage, it was found that even 100 ppm of damage was enough to kill the superconductivity [7]. Experiments with demonstrably non-magnetic impurities [8] found similar results. The conclusion of these studies was that the superconductivity did not involve $s$-wave pairs, a point in favor of triplet pairing.

The next bit of experimental evidence came from the anisotropic critical field [9]. According to Ginzburg-Landau theory the critical field anisotropy should be $H_{c2i}/H_{c2j} \propto t_i a_i/t_j a_j$ where $i = x, y, z$ and $a_i$ is the lattice constant in the $i$ direction. While the ratio of critical fields between **c** and **b** was reasonable, the **b**-**a** ratio was too small. Analysis indicated that the **a** axis critical field was limited by spin pair-breaking and $H_{c2a}$ was consistent with the Clogston limit [10] for singlet superconductivity (although the measurements were only performed down to $0.66T_c$). Clogston and Chandrasekhar argued that in the absence of orbital effects the normal metal would be restored. The difference between the magnetic spin energy of the normal and superconducting states $-(\chi_{metal} - \chi_{super})H^2/2$ became lower than the condensation energy of the superconductor $-N(0)\Delta^2/2$. Since $\chi_{metal} = \chi_{Pauli} = 2N(0)(g\mu_B)^2$ and $\chi_{super,singlet} = 0$, the Clogston limit is

$$H_P \equiv \frac{\Delta}{\sqrt{2}g\mu_B} \approx 1.8\, T_c \text{ Tesla.K}$$

assuming the BCS relation $2\Delta = 3.5k_BT_c$. This was considered strong evidence for singlet superconductivity, and when the experiment was repeated, a similar result was found [12].

NMR $T_1$ (nuclear spin lattice relaxation) measurements were performed and showed no Hebel-Slichter peak [11], indicating that the transition was not BCS-like. At low temperature the relaxation rate followed a power law of $\approx T^3$ instead of the BCS exponential behavior [13]. The conclusion from the NMR was that the superconductivity was unusual and that the Fermi surface was not completely gapped. The $T^3$ suggests a line of nodes for the order parameter. The experiment was repeated with the same conclusion by Takahashi [14] and Brown [15]. On the other hand, thermodynamic measurements such as specific heat [16] and thermal conductivity [17] indicate a conventional BCS-like mean-field transition with a complete gap and an exponential behavior at low temperature.

The evidence during the mid 80's - 90's was therefore evenly divided: thermodynamic and critical field measurements indicated conventional BCS singlet superconductivity and full gapping of he Fermi surface, whereas impurity-doping effects and NMR suggested nodes in the wavefunction and non time-reversed pairing. Interest then turned to some of the other remarkable features of these materials, the striking and unusual effects of magnetic fields aligned in different directions, and these eventually led to new ways to investigate the superconducting state.

## 11.3 DIMENSIONALITY REDUCTION IN A MAGNETIC FIELD

The Bechgaard salts are very clean, with mean-free-paths at low temperatures that are microns long and scattering rates $\hbar/\tau$ that correspond to an energy of 100 mK or less. The non-intersecting Fermi surface sheets with no closed orbits (for fields perpendicular to the highly conducting chains, Fig. 11.1a) would suggest that magnetotransport effects would be innocuous and Schubnikov de Haas (SdH) oscillations could not appear.

It was therefore very surprising when early experiments indicated the SdH effects and even the hint of the quantum Hall effect [18]. So some effort was put into understanding the effects of a magnetic field on these quasi-1-D open orbit materials. The earliest theoretical treatment [19] suggested an interesting phenomenon, *the magnetic field reduced the dimensionality of the system.* Consider first the most conducting layers. There is a two-dimensional dispersion relation of the form

$$\epsilon(k_x, k_y) = \hbar v_F(|k_x| - |k_F|) - 2t_b cos(k_y b).$$

In the presence of a magnetic field along $\mathbf{z}$, the dispersion relation becomes a function of $k_x$ only. $\epsilon(k_x)$ is a solution of

$$\hbar v_F(\partial_x - |k_F|)\psi + 2t_b cos(2\pi x/\lambda)\psi = \epsilon(k_x)\psi.$$

This results from a Landau-Peierls substitution ($\vec{k} \to i\vec{\nabla} - e\vec{A}/\hbar c$) with the choice of a Landau gauge $\vec{A} = (0, Hx)$. Thus the dimensionality is reduced by one. (This contrasts with the closed orbit case where the dimensionality is reduced by two, the Landau levels are completely degenerate and have no dispersion in $x$ and $y$.) Physically, we can see this one dimensionalization from the quasi-classical electron trajectories. In Fig. 11.1a the electron sweeps across the Brilluion Zone, Umklapps back, and repeats its path with a frequency $\omega_b = eHv_F b/\hbar c$. Since it traverses all values of $k_y$ its energy cannot depend on $k_y$. The $k$-space trajectory is shown in a repeated zone scheme in Fig. 11.1b, and the real space trajectory in 11.1c. The real space trajectory is just the $k$-space path rotated by 90 degrees since $\hbar\partial_t\vec{k} = (e/c)\partial_t\vec{r} \times \vec{H}$. The effect of the magnetic field is to limit the electron motion along $\hat{\mathbf{y}}$ (from chain to chain) to $(4t_b/\hbar\omega_b)b$ ($b$ is the interchain lattice constant) and to introduce a new periodicity $\lambda$ which is the one-dimensional magnetic length, $\lambda bH = \phi_0 = hc/e$. The dispersion relation is effectively one-dimensional as soon as the characteristic frequency for traversing the Brillouin

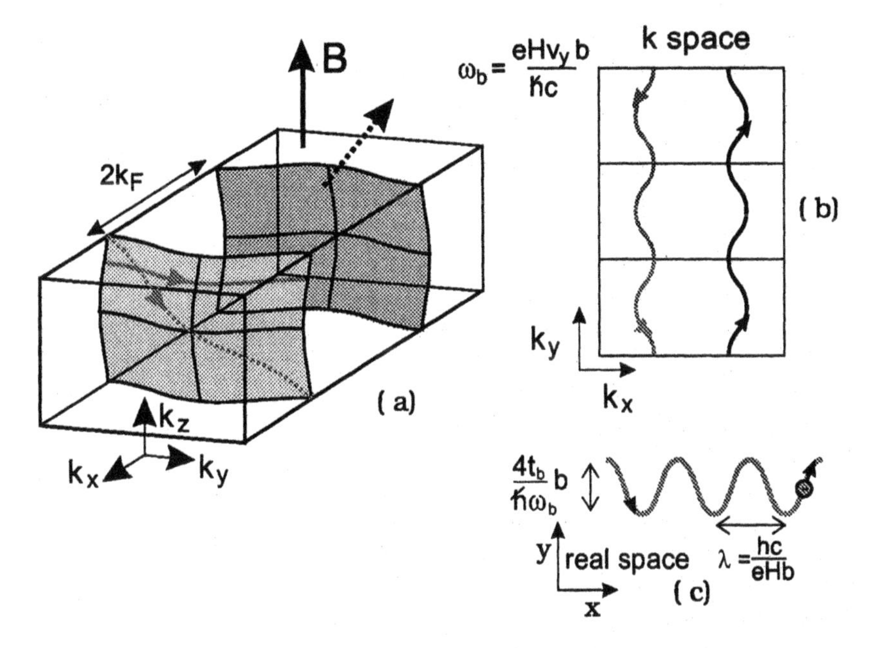


Figure 11.1: (a)The quasi-one-dimensional Fermi surface consists of two non-intersecting warped sheets. For magnetic fields oriented in the **b**-**c** plane all of the electron orbits are open. Two repeating orbits are shown. (b) An electron trajectory for **H** || **c** is shown in the repeated zone scheme. The electron crosses the Brillouin Zone with a frequency $\omega_b = eHv_Fb/\hbar c$. (c) The real-space electron trajectory for the orbit in (b) is just rotated by 90 degrees. There is a new periodicity along **x** given by the magnetic length $\lambda = hc/eHb$. The extent of the quasi-classical motion in the **y** direction (between the chains) is $(4t_b/\hbar\omega_b)b$. When $\hbar\omega_b > 4t_b/$, the electron wavefunction shrinks to less than the distance between chains $b$.

Zone is greater than the temperature or the scattering rate, and the electron motion is effectively restricted to a single chain when this frequency is greater than the interchain hopping rate.

Some of the physics associated with this field-induced one dimensionalization was quickly realized by Gor'kov and Lebed [20] in formulating a model for the unusual magnetotransport in terms of a cascade of Field-Induced Spin Density Wave (FISDW) transitions. They suggested that the lowered dimensionality of the dispersion relation in a magnetic field led to the Peierls-like SDW transition. Further development of this model led to the explanation of the previously seen SdH-like oscillations [21] and the presence of quantum Hall states [22]. The first observation of the QHE in a bulk crystal, in these materials, was shortly presented [23]. Soon after the flurry of theoretical and experimental work on the FISDW's, Lebed considered the effect of a magnetic field oriented at different angles in the plane perpendicular to the highly conducting **a**-axis [24]. He suggested that the

FISDW's would have a different character when the magnetic field was applied at "magic angles" where the frequencies for traversing the Brillouin Zone in the **b** and **c** direction was commensurate, i.e. when $\omega_c/\omega_b = p/q$ where $p$ and $q$ are integers. At these special angles an electron retraces its path across the Fermi surface upon repeated Umklapp scattering. For other angles, the electron path completely covers the Fermi surface without retracing. At the special angles, the dimensionality of the system is reduced by one dimension ($\perp$ to the applied field and the **a** axis). At other angles, it is reduced in both directions. Later Lebed and Bak [25] predicted that these magic angles would show up as peaks in the magnetoresistance.

The first observations of magic angle effects were presented as small dips in the magnetoresistance of the Bechgaard salt $(TMTSF)_2ClO_4$ [26]. Much more dramatic effects were observed in $(TMTSF)_2PF_6$ by Woowon Kang [27] during his thesis work at Princeton. He found strikingly sharp dips at the Lebed magic angles. Since the features were dips rather than the peaks predicted by Lebed and Bak, there was need for a new model [28]. The one presented by Woowon at his thesis defense was a classical "hot spot" model introduced by his advisor, P. Chaikin [29]. This explanation envisioned electrons swept across the Fermi surface to regions where the scattering was particularly strong. At non-magic angles, all electrons would eventually be swept to the hot spots. For magic angles, some of the retracing trajectories would lead to a fraction of the electrons which missed the hot spots. The resistance at the magic angles would be lower. Phil Anderson was present at Woowon's thesis defense and although he thought the experiments interesting and allowed his doctorate, he thought that the explanation was completely unreasonable and missed the physics of these systems completely. How could one treat a clearly strongly-correlated, reduced-dimensionality material as if conventional one-electron theory would apply? Clearly one could not.

## 11.4 INTERLAYER DECOUPLING

Phil began to work on the problem with his postdocs Steve Strong and Dave Clarke, and the basic idea of their work appeared several months later [4]. What Strong, Clarke and Anderson (SCA) suggested was not a magnetoresistance effect, but effectively a change in the nature of the quantum state of the system as the field was rotated. Consider first the isolated layers of TMTSF coupled chains. According to SCA the large Coulomb interaction would render this two-dimensional system a non-Fermi liquid with power-law correlations. Any orbital effects of the magnetic field would involve only the component of the field perpendicular to the layers and one would expect a power law. Now introduce an interlayer hopping. According to SCA the system remains a decoupled non-Fermi liquid in the presence of a small interlayer hopping, but as the interlayer tunneling integral is increased the decoupled non-Fermi liquid gives way to a three-dimensional Fermi liquid.

Now for the magic angles. As it turns out, the Lebed condition of commensurate orbits in $k$ space is precisely the same as the condition that the magnetic field is aligned along the real space translation vectors. (This works for any crystal structure, even the triclinic structure of the Bechgaard salts. It results from **k**-space being perpendicular to real-space and electron orbits being perpendicular to the magnetic field.) Thus there can be a real-space interpretation of the magic angle effects. Start with the system as a 3-D Fermi liquid. Suppose we turn on a field that has a component along the **b** axis. It can dephase the interlayer hopping and restore the decoupled non-Fermi liquid. However, if the field is along a lattice translation vector, then an electron can tunnel from a molecule in one layer to a molecule in the next layer along the field line without ever being dephased. The layers are again coupled. The scenario is then: magic angle – 3D Fermi liquid, conventional transport, non magic angle – 2D non Fermi liquid, $(\delta\rho(H)/\rho) \propto (H_\perp)^\alpha$, where $(H_\perp)$ is the field perpendicular to the layers and $\alpha$ is an exponent that depends on which direction the resistance is being measured.

One of the main flaws of this model was that, although it wasvery hard to prove, if it was wrong it would be very easy to disprove. Thus an effort was made to disprove it. The resistance should vary as $(\delta\rho(H)/\rho) \propto (H_\perp)^\alpha$ except at magic angles. It does, along the most and least conducting directions where it has been tested. (This is an especially powerful result when resistance is measured in the least conducting **c** direction. Here SCA predicted the magnetoresistance is largest when the current and field are almost parallel and minimum when they are perpendicular, exactly opposite to what is expected from conventional Lorentz force arguments. From the experiments [30], what is important is the orientation of the field relative to the crystal axis rather than to the current - the SCA prediction.) The Danner-Kang-Chaikin [31] resonances are well accepted as Fermi surface probes that measure the transverse bandwidths using coherent interlayer transport and a well-defined Fermi Surface. If we had a three dimensional Fermi liquid at all field orientations, they would persist. If we developed a two-dimensional non-Fermi liquid when the field was tilted to give a component along **b**, they would disappear. Experimentally, application of a small decoupling field parallel to the layers made them disappear [32], providing more support for SCA. There is a threshold field that separates a classical magnetotransport regime from one where decoupling appears to be important.

If there is really a difference in the quantum state of the system when the field is oriented in different directions, we should expect some spectacular differences, and they should show up in a more pronounced fashion as we lower the temperature.

Since the interesting state of the $PF_6$ salt occurs under pressure, the easiest experiment to perform is resistance. We therefore performed a series of resistance measurements as a function of temperature for transport along the three principal axes [33, 34, 35]. The results are shown in Figs. 11.2, 11.3 and 11.4. The transfer integral in the interlayer direction $t_c \approx 0.0007eV$ corresponds to a temperature

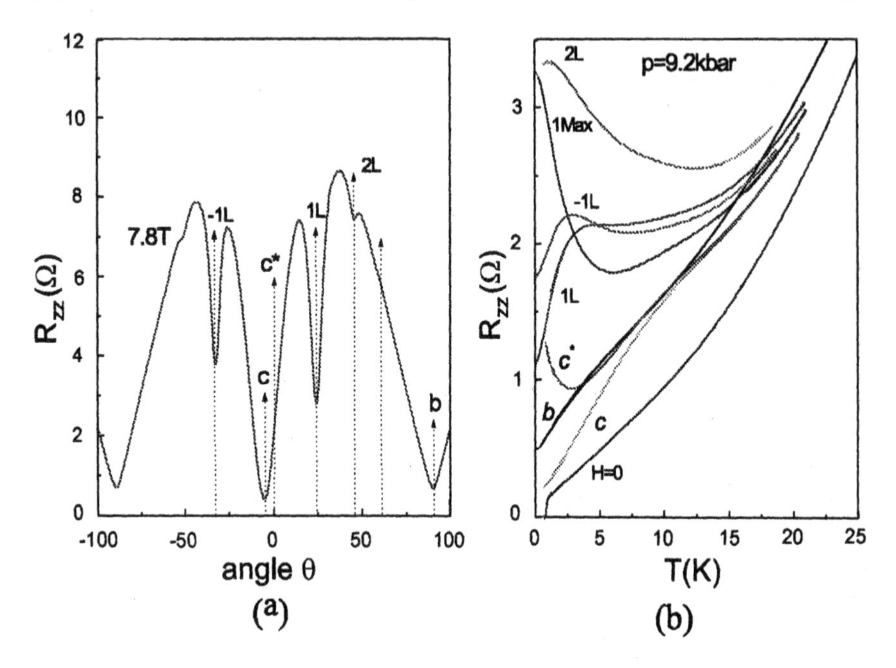


Figure 11.2: (a) The angular dependence of the **c** axis resistance for $(TMTSF)_2PF_6$ at 9 kbar, 4 T, 0.5 K for rotations in the **b-c** plane. Note the sharp dips at the "magic" angles where the field is aligned along the lattice translation vectors. Note also the behavior away from the magic angles where the resistance increases as the field is rotated toward the current direction (along **c**). Classically the magnetoresistance is largest when field and current are perpendicular - here for field along **b**, the opposite of what we have here. (b) Temperature dependence of the resistivity in the least conducting direction for $(TMTSF)_2PF_6$ at several of the angles noted in the rotation diagram above. Note that at all magic angles the low temperature resistance decreases with temperature while at other angles the resistance increases with decreasing temperature.

of $\approx 8K$. We therefore ignore what is happening at higher temperatures and see what happens at temperatures from 8 K down. Here we do see a most striking behavior. Independent of the current direction and for a wide range of fields we find that field alignment at magic angles produces a metallic resistance ($dR/dT <$ 0) whereas for any non-magic angle the resistance is non-metallic ($dR/dT < 0$). In particular, look at Fig. 11.2b. For field aligned along the $c$-axis (a magic angle) the resistance is metallic. For field perpendicular to the **a-b** plane the resistance is non-metallic. For field along the (1,1) -1L magic angle the resistance is metallic whereas 6 degrees away, where the angular trace (Fig. 11.2a) shows a maximum, the resistance is non-metallic (1Max). Figure 11.5 shows the dependence of the $c$-axis resistance on temperature for different angles and field strengths. Here we see that the basic distinction between the magic and non-magic angles shows up for all fields. As mentioned briefly above, the classical behavior for magnetoresistance

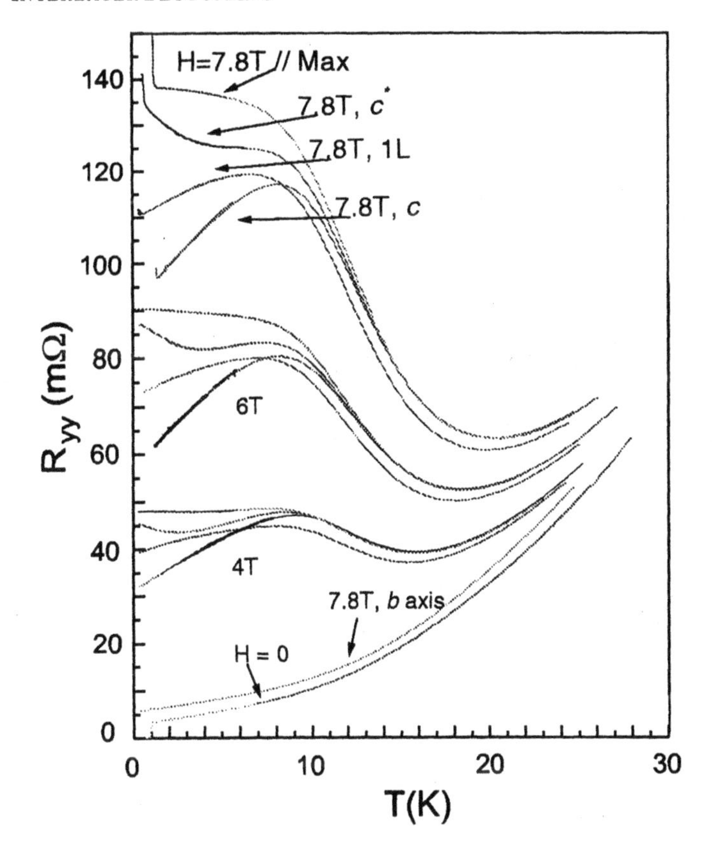


Figure 11.3: Temperature dependence of the resistivity in the intermediate conducting direction **b** for $(TMTSF)_2PF_6$ at 11 kbar at several angles and at several field values. Note again that below 8 K at all magic angles $dR/dT > 0$ (metallic), while at other angles $dR/dT < 0$.

when the current is along the **c** axis is that it should be largest when the component of **H** perpendicular to the current is largest. The behavior suggested by SCA says that what is important is the component parallel to **c** (perpendicular to the decoupled sheets) except at magic angles.

In Fig. 11.5, we see that at high temperature the classical behavior is obeyed (e.g. compare the data at 10 K, 7.8 Tesla for 1L and 1Max, the magnetoresistance for 1Max is lower than for 1L because the field is closer to being parallel with **c**). However, the data cross over to the SCA behavior as temperature is lowered below 5 K. The crossover temperature between these regimes is seen to increase as the field increases.

It appears from our data that the magic angles are a discrete set of angles for which the behavior is metallic to low temperatures; any slight orientational devi-

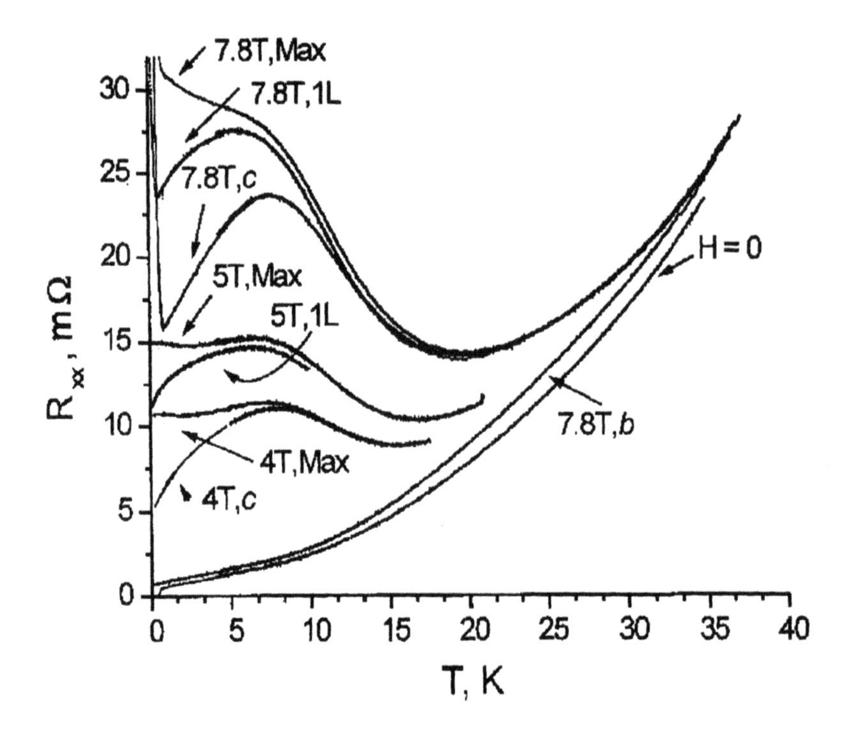


Figure 11.4: Temperature dependence of the resistivity in the most conducting direction **a** for $(TMTSF)_2PF_6$ at 9kbar at several of angles and for several fields. Note again that below 8 K at all magic angles $dR/dT > 0$ - metallic while at other angles $dR/dT < 0$. (Exceptions are the sharp rise for all angles at 7.8 T and low temperature where the samples enter the FISDW insulating phase. The fact that the resistance shows the same metallic - non metallic behavior at magic - non magic angles indicates that this results from a change is state, rather than a magnetoresistance effect.

ation from the magic angles will lead to a resistance increase a low temperature. The smaller the angular deviation, the lower you have to go in temperature to begin to observe an upturn in resistance. The metallic state seems to exist at any rational field (an infinite set of measure zero) at zero temperature. This would imply a dense set of quantum critical points (of measure zero) at rational fields ($\omega_b/\omega_c = p/q$, with $p$ and $q$ integers).

The big question that remains in this area is the nature of the ground state when the system is not at one of the magic angles. It has also been argued that the decoupling effect can simply be a result of the quasi-classical one dimensionalization suggested in the non-interacting model. What argues against this interpretation is the small value of field component between the sheets that causes the decoupling. From the quasi- classical model one would expect such effects

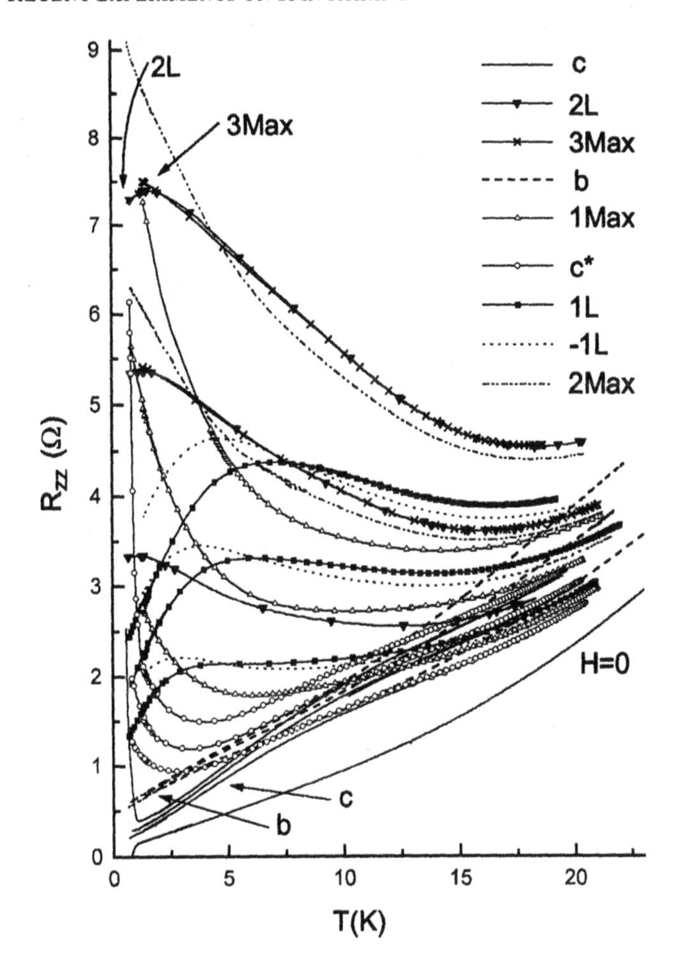


Figure 11.5: Here the **c** axis resistance is shown for several fields, and angles as a function of temperature, illustrating the robustness of the low temperature slope change distinguishing magic angle and non magic angle behavior.

at a field such that $\hbar\omega_c \approx t_c$ ( if not $4t_c$) which would correspond to a field of $\approx$ 8 Tesla (30 Tesla). The actual cross over field is less than 1 Tesla, which suggests that another mechanism (such as the competition between Fermi-liquid and correlated non-Fermi liquid as in SCA) is aiding the decoupling.

## 11.5 RECENT EXPERIMENTS ON SPIN PAIRING

We now return to the superconductivity. By the end of the 80's the idea that a magnetic field might decouple the layers and reduce the dimensionality of the system was established both theoretically and experimentally. Lebed suggested an interesting way to use this decoupling to find the nature of the superconducting state [36]. For a magnetic field aligned precisely along the **b** direction the superconducting screening currents flow long the chains (**a** direction) and between the layers (**c** direction). The kinetic energy associated with these screening currents is what lowers the condensation energy of the superconducting state and gives rise to the orbital critical field. If the layers were decoupled, then there would be no screening currents and the condensation energy and critical temperature would be unaffected by the magnetic field - the critical field would be infinite. In our case the presence of sufficient field along the **b** direction decouples the planes. Thus for weak fields, we should get the usual depression of $T_c$ (a linear critical field curve). As the field is increased, the coupling between the planes decreases and the orbital pair-breaking decreases. There should be strong deviations from the usual critical field plot with an upward curvature - higher critical fields than extrapolated from near $T_c$. When the field is strong enough to decouple the layers, the transition temperature should start to increase toward its zero-field value. All of the above discussion just concerns the orbital part of the pair-breaking. If the superconductivity were singlet, then the critical field would still be limited to the Clogston value. Only if there were triplet superconductivity would the critical field continue to increase, and possibly become reëntrant. In any event, the decoupling mechanism would allow a substantial increase in the orbital critical field up to the level where the spin effects should be dominant. These effects are also presented in several more recent theories [37].

The first experimental indication of these effects was the observation of the upward curvature of the critical field of $(TMTSF)_2ClO_4$ for field along **b** [38]. A more thorough study was done a few years later on $(TMTSF)_2PF_6$ [41]. The critical field along the principal directions is shown in Fig. 11.6. There are several interesting aspects to this data. As found in previous data, the **c**-axis critical field is very low. It is for this reason that we need precise field alignment along the other axes to see the intrinsic behavior; a small component of field along **c** will kill the superconductivity. (The decoupling effect would be present as long as there is a component of the field along **b**.) For the field along **b**, we see the striking upward curvature of $H_{c2b}(T)$ starting from $T_c$. This is consistent qualitatively with the interlayer decoupling - field-induced dimensional crossover model. In addition, with the precisely aligned field, we were able to increase the critical field well into the region where it should be dominated by spin effects. The Clogston limit for the measured $T_c$ is $H_P \approx 2.1$ Tesla at $T = 0$. We see that $H_{c2b}(T = 0.1K) \approx$ 6 Tesla is already about three times the limit for a singlet superconductor. One way around the Clogston limit without invoking triplet superconductivity is strong spin-orbit scattering. The spin-orbit scattering rate depends on scattering from heavy (large Z atoms) and can be estimated from the transport scattering rate and the spin-orbit coupling of the constituent atoms. Assuming the worst-case scenario (all of the resistance coming from scattering from the Selenium atoms -

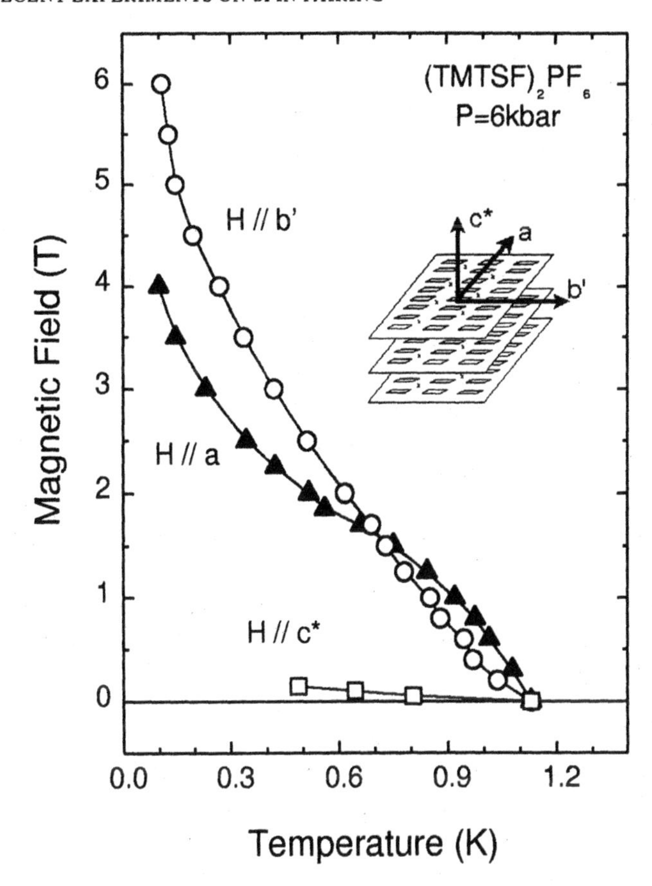


Figure 11.6: The critical field for fields precisely aligned along the **a, b** and **c** directions for a sample at 6 Tesla, and for **c** alignment, for a sample at 5.5 kbar. The critical field value is constructed by the intersection of the extrapolated high temperature resistance and the extrapolated maximum slope line from the superconducting transition. The Pauli limit for the critical field is 2.1 Tesla..

the heaviest in these materials) we find that the spin-orbit scattering is four orders of magnitude too small to give such a large enhancement of the spin critical field $H_P$. This is because the materials are very clean with long mean-free-paths, and Se is relatively light.

The **a**-axis critical field is also unusual. Near $T_c$, $H_{c2a}$ is larger than $H_{c2b}(T)$, as would be expected since the **a** axis bandwidth $t_a$ is larger than $t_b$. However, it is not as large as it should be from the ratio $t_a a/t_b b$. It is concave down until $\sim 0.5K$ and then turns upward. At low temperature, it is also significantly larger than $H_P$. Note that for field along **a** we do not expect dimensional crossover, or decoupling. For this configuration (and range of fields $< 10T$) the electron velocities perpendicular to the field are not large enough to constrain the electrons or dephase an interchain jump. Some of our present understanding of $H_{c2a}(T)$ comes from recent calculations of Lebed [39]. He suggests that the triplet state is equal-spin-paired (only up-up and down-down ) as in the Anderson Brinkman Morel [40] state of $He^3$ but with an anisotropy field that aligns the $\vec{d}$ vector (perpendicular to the spins) along the crystallographic **a** axis. A magnetic field along **b** would then change the spin populations of the up and down pairs and give the same susceptibility as in the normal state - there would be no Clogston limit. A magnetic field along **a** would not reorient the spins. The susceptibility would be zero as for a singlet superconductor, and the critical field would be Clogston limited. In fact, Lebed has well fitted $H_{c2a}$ (0.5 K $< T <$ 1.2 K) to an expression with only spin limiting, and with no orbital effects. (From the slope of $H_{c2b}(T)$ near $T_c$ and the ratio $t_a a/t_b b$ we would expect orbital pair-breaking to be important at low temperatures at fields of ~15 Tesla.) Thus the conclusion from studies in the early 80's of paramagnetic limiting are probably correct, but they do not legislate against triplet. They point to a particular orientation of the triplet pairs.

In Fig. 11.7, we show data from critical field studies of a number of other samples for different pressures close to the boundary between the SDW and superconducting phases. This is where $T_c$ and the critical fields are highest. Note that in all cases, there is the strong upward curvature, and the critical fields are raised here to more than 4 times $H_P$, the singlet limit.

The classic signature of singlet superconductivity is the vanishing of the NMR Knight shift upon cooling into the superconducting state. Conversely, the absence of a change in the Knight shift is evidence that the spin susceptibility remains the same as in the normal metal [42]. In free space, the magnetic moment of a nucleus would precess in an applied field **H** at a frequency $\omega_N = \mu_N H/\hbar$. The nuclear spin in the solid precesses in a field which is the applied field plus an internal field $H_{int}$, viz.

$$\omega_N = \mu_N(H + H_{int})/\hbar.$$

Part of the internal field comes from chemical and other shifts and part from the polarization of the conduction electrons by the applied field. This polarization acts on the nuclear spin through the contact hyperfine interaction. Thus, $H_{int} = (a + b\chi_{spin})H$ and the Knight shift $K = \Delta\omega_N \propto (a' + b'\chi_{spin})H$ have a component related to the spin susceptibility. The proportionality constant can be obtained by comparing the temperature dependence of the Knight shift to the temperature dependence of the spin susceptibility in a $K - \chi$ plot. From such a plot we can extrapolate to $\chi = 0$ to see what the Knight shift should be like if the electron susceptibility goes to zero, as it should for a singlet superconductor.

In Fig. 11.8 we show the low-$T$ Se77 NMR lineshape for $(TMTSF)_2PF_6$ at a pressure of 0.64 GPa from above $T_c$ to well into the superconducting state [43]. The upper two traces were taken with a spin-echo technique while the lower

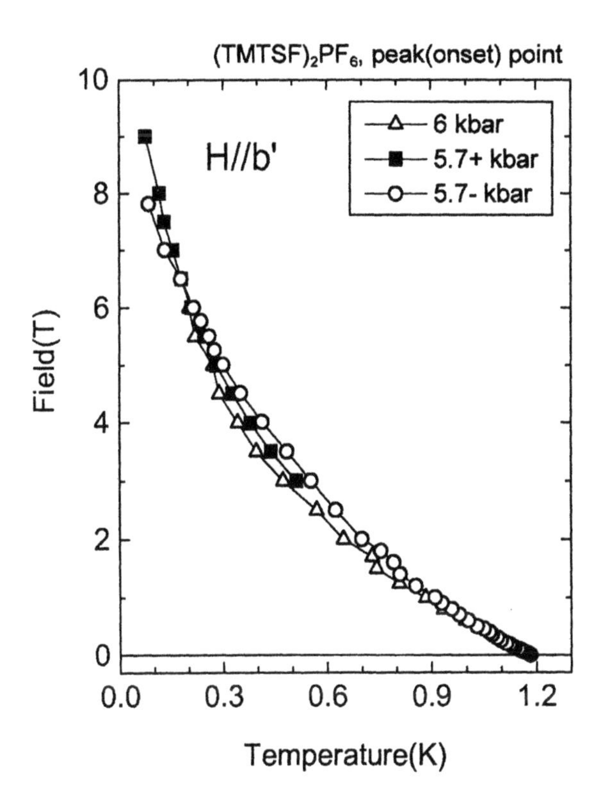


Figure 11.7: The critical field for fields precisely aligned along the **b** direction for three samples at ~6 kbar. $H_{c2}$ at low temperature is a factor of 4 larger than the Pauli limit.

traces are from the free induction decay (FID). The fact that there is no significant change on cooling through $T_c \sim 0.6K$ at 2.4 T along **b** shows that the Knight shift does not change on entering the superconducting state.

The Se77 nucleus was chosen for this study since it sits in a region of the molecule with reasonably large overlap with the conduction electron wavefunctions. It should therefore have a Knight shift sensitive to the spin susceptibility. In the normal state from room temperature to 20 K there are four distinct peaks in the NMR lineshape corresponding to the four inequivalent positions for Se atoms in the crystal structure. An extrapolation of the Knight shift for the centroid of these lines to zero spin susceptibility gives the shaded region in Fig. 11.8. Since any possible shift is much less than the distance from the centroid of the

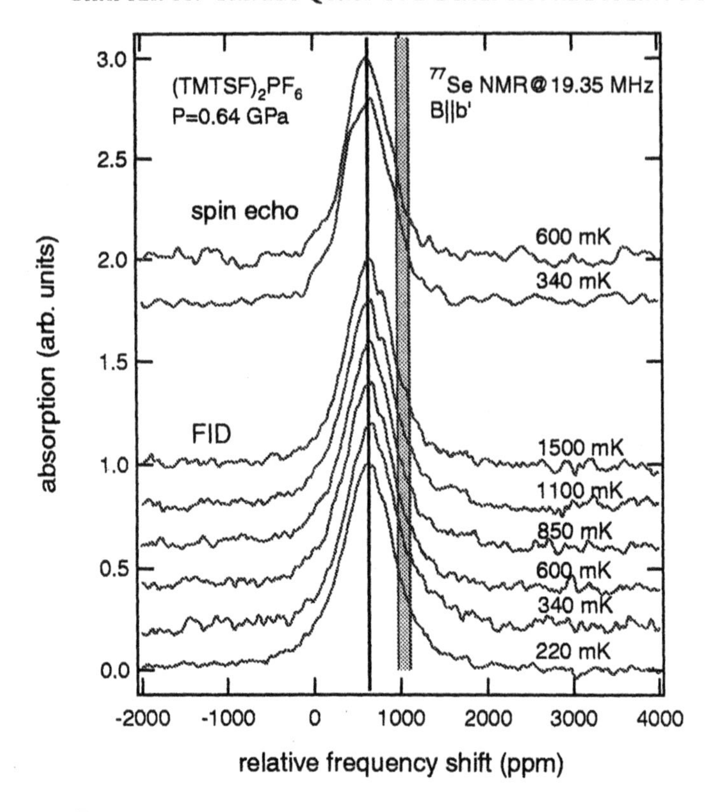


Figure 11.8: $^{77}$Se spectra from spin-echo and free induction decay measurements at temperatures below and above $T_c$ for the magnetic field $B$ = 2.38 T oriented parallel to the molecular layers to within 0.1 degree. The solid line marks the measured first moment, and the hashed region marks the expected first moment for a singlet ground state.

data to this extrapolated value we conclude that the material is not a singlet superconductor and in fact retains its susceptibility in a way consistent with triplet superconductivity.

The results shown in Fig. 11.8 are correct and convincing, but we had similar results during the first week of the experiment which tend to show the real difficulty of this experiment. First, it should be realized that the experiment sounds very difficult. The sample is under pressure, in a dilution fridge at low temperature, in a precisely aligned magnetic field along the **b** axis, and simultaneous measurements of pulsed NMR and transport are made on the same sample. The transport measurement on the sample is dictated by the condition that sample alignment to better than 0.1 degree is needed and is most readily obtained by the angular dependence of the magnetoresistance. It is most unusual to have transport leads on the sample at the same time as pulsed NMR is performed, but here we were very lucky that they were present.

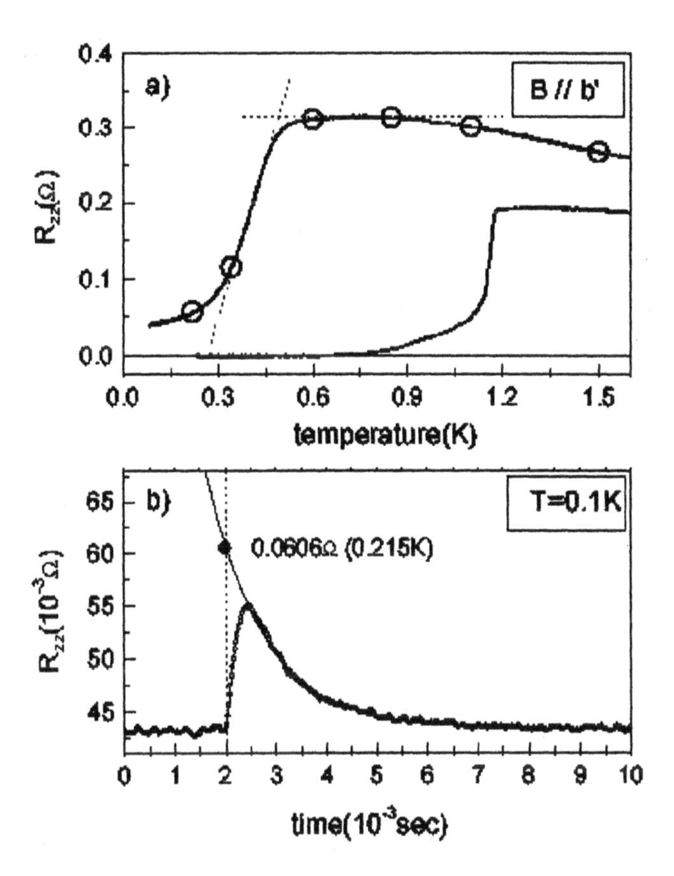


Figure 11.9: (a) Interlayer resistance $R_{zz}$ vs. temperature at zero applied magnetic field, at 1 Tesla and at the $^{77}$Se measuring field of $B = 2.38$ T. In (b) we show the time-synchronous resistance measurements recorded, triggered simultaneous with the rf-pulses of the NMR measurements. (The rf pulse and its trigger occur at 2 msec in this plot.)

So what was really difficult? Normally we measure the resistance of the sample with an AC technique using a lock-in amplifier. We can observe the superconducting transition with or without the magnetic field to assure that the sample is superconducting during the NMR measurement. The **c** axis resistance in different fields is shown in Fig. 11.9. We can also do nuclear spin relaxation rate, $1/T_1$, measurements on the sample to see the superconducting transition. The $1/T_1$ data are shown in Fig. 11.10. Note the cusp in $T_1T$ at $\sim 0.7$ K indicating the superconducting transition at the applied field of 2.4 Tesla. When we performed the spin-echo measurement of the Knight shift we were excited at the fact that there was no difference between the line shapes above and below $T_c$.

However, from what we know now, we were probably not in the supercon-

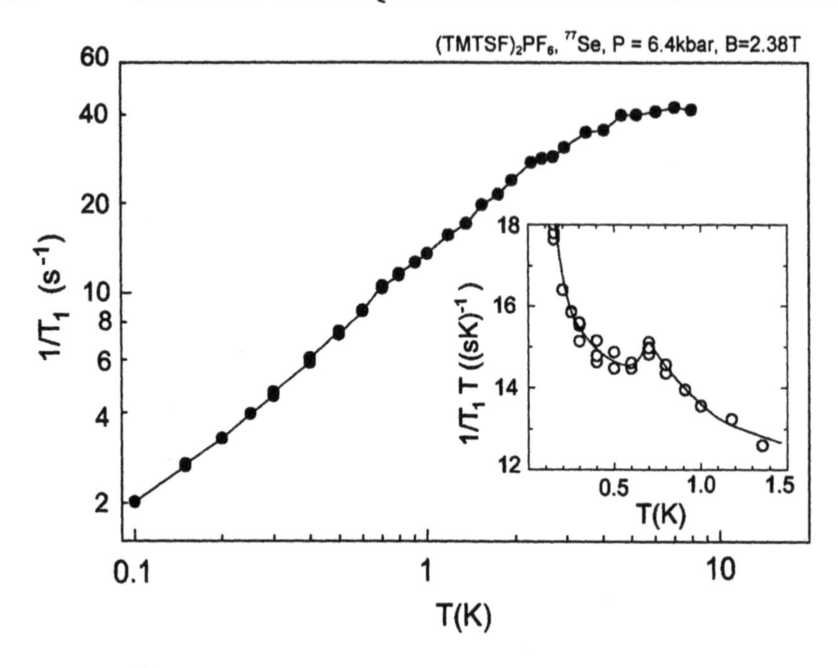


Figure 11.10: $^{77}$Se $1/T_1$ vs. temperature for $B = 2.38$ T || **b**. In the inset at the bottom are $1/T_1T$ data carefully taken and repeated over several runs to establish the anomaly at the superconducting transition.

ducting phase for these initial measurements. Nature had conspired against us. The pulsed spin-echo measurement takes place on a time scale of less than 100 $\mu$s, and is repeated once a second. The thermal relaxation time of the sample is on the scale of a millisecond. $T_1$ is on the scale of a second and the thermometry and resistance were measured with a time constant on the lock-in of 1 second. During an RF pulse the sample was heating significantly, probably to above $T_c$, by either direct induction or from the Joule heating of the NMR coil (in the pressure cell surrounding the sample). The electron temperature goes up, the sample goes normal and the Knight shift is that of the normal metal during the time of the spin-echo measurement $\sim$ 100$\mu$s. The sample then cools down to low temperature in 1 msec and remains there for 99.9 percent of the time until the next pulse. The thermometry and the sample resistance measured on a second time scale see the time average and therefore register the sample as remaining cold. The $1/T_1$ measurement also averages over a second and sees the sample as superconducting. We can also determine a temperature by integrating the spin-echo line to obtain the total nuclear spin alignment in the field (naively a Curie law). This measurement is precisely on the time scale of the spin-echo measurement itself. Unfortunately, the temperature it measures is the nuclear spin temperature, not the electron spin temperature of interest for the Knight shift. Since $T_1$ is about a second, the nuclei are temperature decoupled from the electrons on shorter time scales. Thus, everything would look as if we were observing the superconducting state while the Knight shift was measuring the normal metallic state.

We discovered what was actually happening by turning the frequency of the lock-in to 30kHz and the time constant to 0.3 msec for the resistance measurement of the sample. We then triggered the resistance measurement with the NMR RF pulse and averaged the transients on a digital scope. The results are shown in Fig. 11.9b. Here, we see the resistance increase from heating and then decrease with a thermal relaxation time of about 1 millisecond. The delay in heating after the pulse (at 2 second offset on this plot) probably indicates that most of the heating is from the NMR coil itself. The heat then conducts in the pressure medium to the sample and the temperature then exponentially decays, as heat is conducted away. It is possible that very little heating occurs at the sample during the 100 microseconds for the free induction decay or spin-echo measurement, that the heat has not yet reached the sample. But to be on the conservative side we have extrapolated the data back to the instant of the pulse and this gives us an upper limit to the temperature of the sample. The data shown in Figs. 11.8 and 11.9 all refer to this maximum possible temperature. Our initial pulse sequences were much hotter than shown in Fig. 11.8 but once we could measure the heating we could control it. So the actual data shown were laboriously taken with very weak pulses and long averaging periods to avoid heating.

## 11.6 CONCLUSIONS

The combination of critical fields above the Pauli limit along **b** with the absence of a change in the Knight shift strongly suggest that $(TMTSF)_2PF_6$ is a triplet superconductor. The problem with this interpretation for the paramagnetic limiting previously seen for $H_{c2a}$ is now understood as being the result of an anisotropy field. The problem with the thermodynamic behavior showing a BCS-like transition and a fully gapped Fermi surface is now naturally understood as the result of a $p$-wave order parameter which has its nodes where the Fermi surface doesn't exist. This is the most favorable configuration in these quasi-one-dimensional materials with their Fermi surface of disconnected sheets. We can have one sheet + and the other -. What is left to understand is why the original and subsequent $1/T_1$ measurement all tend to indicate non-BCS like behavior and in particular the presence of zeroes of the gap on the Fermi surface.

So there are still many things to explain and discover about these unique materials in both the metallic and superconducting states. They are excellent examples of some of the ideas on unusual and fascinating phenomena suggested and/or explained by Phil Anderson.

BIBLIOGRAPHY

[1] D. Jerome and H.J. Schultz, Adv. Phys. **31**, 299 (1982); R.L. Greene and P.M. Chaikin, Physica **126B**, 431 (1984), Articles by Heritier, Montambaux, Chaikin, Ribault in *Low Dimensional Conductors and Superconductors*, edited by D. Jerome and L. G. Caron, NATO Advanced Study Institute, Series B, Vol. **155** ( Plenum, New York, 1987), T. Ishiguro, K. Yamaji, and G. Saito, *Organic Superconductors*, 2nd ed., *Springer Series in Solid* State Science 88 (Springer-Verlag, Berlin, 1998) p. 214-257, Articles by Montambaux, Chaikin, Jerome, Gor'kov, Yakovenko in J. Physique 1, **6**, (1996).

[2] K. Bechgaard et al., Solid State Comm., **33**, 1119 (1980), C. S. Jacobsen, D. B. Tanner, K. Bechgaard, Phys. Rev. Lett. **28**, 7019 (1983), P. M. Grant, J. Phys. (Paris), Colloq. **44**, C3-847 (1983), L. Ducasse et al., J. Phys. C **18**, L947 (1985)P. M. Grant, J. Physique **44**, C3-847 (1983).

[3] P. A. Lee, T. M.Rice, and P. W. Anderson, Phys. Rev. Lett. **31**, 462 (1973), P. A. Lee, T. M.Rice, and P. W. Anderson, P. A. Lee, T. M.Rice, and P. W. Anderson, Solid State Commun. **14**, 703, (1974).

[4] S. P. Strong, D. G. Clarke, and P. W. Anderson, Phys. Rev. Lett. **73**, 1007 (1994).

[5] D. Jerome, A. Mazaud, M. Ribault, and K. Bechgaard, J. Physique **41**, L-95 (1980).

[6] J. Solyom, Adv. Phys. **28**, 201 (1979); H. J. Schulz, *Proceedings of Les Houches Summer School LXI*, ed. E. Akkermans, G. Montambaux, J. Richard, J. Zinn-Justin (Elsevier, Amsterdam, 1995), p.533; H. J. Schulz, Int. J. Mod. Phys. **B5**, 57 (1991).

[7] P. M. Chaikin, M. Y Choi, R. L. Greene, Jnl. Magn. Magnetic Mater., **31-34**, 1268-1272 (1983), M. Y. Choi, P. M. Chaikin, S. Z. Huang, P. Haen, E. M. Engler, and R. L Greene, Phys. Rev. B **25**, 6208-17, (1982)

[8] S. Bouffard, M. Ribault, R. Brusetti, D. Jerome, K. Bechgaard, J. Phys. C: Solid State Phys., **15**, 2951-2964 (1982).

[9] P.M. Chaikin, M. Choi, and R.L. Greene, J. Mag. Mag. Mater. **31-34**, p. 1268 (1983).

[10] A. M. Clogston, Phys. Rev. Lett. **9**, 266 (1962); B. Chandrasekhar, Appl. Phys. Lett. **1**, 7 (1962).

[11] L.C. Hebel, P. C. Slichter, Phys. Rev. **113**,1504, (1959).

[12] K. Murata, M. Tokumoto, H.Anzai, K. Kajimura, T. Ishiguro, Jpn. J. Appl. Phys. **26** Suppl. 26-3, 1367 (1987).

[13] M. Takigawa, H. Yasuoka, G. Saito, J. Phys. Soc. Jpn., **56**, 873-876 (1987).

[14] T. Takahashi, H. Kawamura, T. Ohyama, Y. Maniwa, K. Murata, G. Saito, G., J. Phys. Soc. Jpn., **58**, 703-709, (1989).

[15] D. Chou, S. E. Brown, *to be published.*

[16] F. Pesty, P. Garoche, and K. Bechgaard, Phys. Rev. Lett.**55**, 2495 (1985); N.A. Fortune, J.S. Brooks, M. J. Graf, G. Montambaux, L. Y. Chiang, A. A. J. Jos Perenboom, and D. Althof, Phys. Rev. Lett. **64**, 2054 (1990).

[17] S. Belin, K. Behnia, Phys. Rev. Lett., **79**, 2125-2128 (1997).

[18] J. F. Kwak, J. E. Schirber, R. L. Greene, and E.M. Engler, Phys. Rev. Lett. **46**, 1296 (1980); M. Ribault et al., J. Phys. Lett. (Paris), **44**, 953 (1983); P. Chaikin et al., Phys. Rev. Lett. **51**, 233 (1983); M. J. Naughton et al., Phys. Rev. Lett. **55** (1985) 969.

[19] P. M. Chaikin, T. Holstein and M. Ya. Azbel, Philo. Mag. **48**, 457 (1983), P.M. Chaikin, Phys. Rev. B **31**, 4770 (1985).

[20] L. P. Gor'kov and A. G. Lebed, J. Physique Lett. **45**, L433 (1984), M. Héritier, G. Montambaux and P. Lederer, J. Physique Lett. **45**, L 943 (1984); K. Yamaji, J. Phys. Soc. Jap. **54**, 1034 (1985).

[21] G. Montambaux, et al., J. Phys. Lett. **45**, L-533 (1984), M. Héritier, et al., J. Physique Lett. **45**, L-943 (1984); M. Ya Azbel, et al., Phys. Lett. **A117**, 92 (1986); K. Maki, Phys. Rev. **B33**, 4826, (1986); D. Poilblanc et. al., Phys. Rev. Lett. **58**, 270 (1987); K. Machida et. al., Phys. Rev. B **50**, 921 (1994).

[22] K. Yamaji, Synth. Metals, **13** ,29 (1986); D. Poilblanc, G. Montambaux, M. Héritier, and P. Lederer, Phys. Rev. Lett. **58**, 270 (1987), V. M. Yakovenko, Phys. Rev. **B43**, 11353 (1991).

[23] S. T. Hannahs, J. S. Brooks, W. Kang, L. Y. Chiang, and P. M. Chaikin, Phys. Rev. Lett. **63** (1989) 1988; J. R. Cooper, W. Kang, P. Auban, G. Montambaux, D. Jérome, and K. Bechgaard, Phys. Rev. Lett. **63** (1989) 1984.

[24] A.G. Lebed, JETP Lett. **43**, 174 (1986).

[25] A. G. Lebed and P. Bak, Phys. Rev. Lett. **63**, 1315 (1989).

[26] T. Osada, A. Kawasumi, S. Kagoshima, N. Miura and G. Saito, Phys. Rev. Lett. , **66**, 1525 (1991).

[27] W. Kang, S. T. Hannahs, P. M. Chaikin, Phys. Rev. Lett. **70**, 3091 (1993); W. Kang, *Ph.D. thesis*, Princeton University (1993).

[28] T. Osada, et al., Phys. Rev. **B46**, 1812 (1992); K. Maki, Phys. Rev. B **45**, 5111 (1992); V. M. Yakovenko, Phys. Rev. Lett. 68, 3607 (1992); K. Maki, Phys. Rev. B **45**, 5111 (1992).

[29] P. M. Chaikin, Phys. Rev. Lett. **69**, 2831 (1992).

[30] W. Kang , S. T. Hannahs and P. M. Chaikin, Phys. Rev. Lett. **70**, 3091 (1993); W. Kang , S. T. Hannahs and P. M. Chaikin, Phys. Rev. Lett. **69**, 2827 (1992).

[31] G. Danner, W. Kang, and P. M. Chaikin, Physica **B201**, 442 (1994).

[32] G. M. Danner and P. M. Chaikin, Phys. Rev. Lett., **75**, 4690 (1995).

[33] E. I. Chashechkina, *Ph.D. thesis*, Princeton University, 1998.

[34] E. I. Chashechkina, and P. M. Chaikin, Phys. Rev. Lett. **80**, 2181 (1998).

[35] D. G. Clarke, S. P. Strong, P. M. Chaikin, E. I. Chashechkina, Science **279**, 2071 (1998).

[36] A. G. Lebed, JETP Lett. **44**, 114 (1986).

[37] L. I. Burlachkov, L. P. Gor'kov and A. G. Lebed, Physica **148B** 500 (1987); N. Dupuis and G. Montambaux, Phys. Rev. Lett. **68**, 357 (1992); N. Dupuis, G. Montambaux and C. A. R. Sa de Melo, Phys. Rev. Lett. **70**, 2613 (1993).

[38] I. J. Lee, A. P. Hope, M. J. Leone, M. J. Naughton, Synth. Metals **70**, 747-750, (1995).

[39] A. G. Lebed, Phys. Rev. Lett., *submitted.*

[40] see chapter by P. W. Anderson and W.F. Brinkman in P. W. Anderson "Basic Notions of Condensed Matter Physics", (The Benjamin/Cummings Publishing Co. Inc., Menlow Park, 1984), P. W. Anderson and P. Morel, Phys. Rev. **123**, 1911 (1961).

[41] I. J. Lee, M. J. Naughton, G. M. Danner and P. M. Chaikin, Phys. Rev. Lett. **78**, 3555, (1997).

[42] K. Ishida, H. Mukuda, Y. Kitaoka, K. Asayama, Z. Q. Mao, Y. Mori, Y., Maeno, Nature **396**, 658-660, (1998).

[43] I.J. Lee, D.S. Chow, W.G. Clark, Strouse J., M.J. Naughton, P.M. Chaikin, S.E. Brown, *to be published.*

# Chapter 12

# Magnetic Moments in Metals

H. R. Ott

Laboratorium Für Festkörperphysik, ETH-Hönggerberg
8093 Zürich, Switzerland

Abstract

What started with a seemingly simple problem - understanding the effects after placing a single magnetic moment into a metallic environment - developed into a major field of research during the last 50 years of condensed matter physics. An attempt is made to briefly review or simply mention some of the outstanding achievements in these activities. The present state of affairs indicates that this field touches upon very fundamental questions regarding many-body physics and the possible ground states of condensed matter.

## 12.1 Introduction

One of the most fascinating chapters in condensed matter physics during the last 50 years has been related with the question: How do magnetic moments behave in a metallic environment? Incompletely filled electron shells of atoms of the $d$- and $f$-transition metal elements may be the source of localized magnetic moments. Considering the available time and space, it seems quite impossible to cover the whole story. Nevertheless, an attempt is made to describe the main events that launched extended research activities in the beginning, and also to mention some of the new developments that still emerge periodically.

In early work, the behavior of magnetic moments in, and their influence on a metallic environment, was treated by considering a single local moment as an impurity by Friedel [1], Blandin [2] and others. Subsequent experimental work, mainly by Matthias, Clogston and co-workers [3, 4], which indicated more complications than previously suspected, led Anderson to extend the existing model by creating what is now known as the Anderson Hamiltonian [5]. The model was intended to explain the experimental fact that Fe ions retain a localized moment in some metals and alloys, but not in others of the 4*d* transition series, as shown schematically in Fig. 12.1. Two important parameters in this model are the on-site Coulomb repulsion $U$ between electrons, and the hybridization strength $V$ between the wave functions of localized and itinerant electrons. Depending on the ratio of these two parameters, the persistence or suppression of the localized moment may be expected.

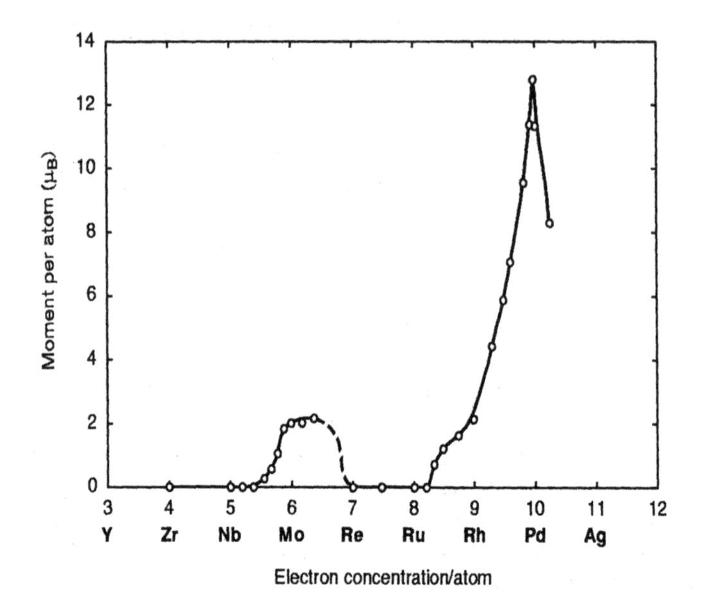


Figure 12.1: Magnetic moment of Fe atoms in various metals (after Ref. [4].)

At about the same time, it was also recognized experimentally, again by Matthias and coworkers [6], that the introduction of elements with localized magnetic moments as impurities into superconductors has a detrimental effect on the latter's critical temperature $T_c$. The explanation for this feature was given by Abrikosov and Gor'kov [7], who introduced the concept of pair breaking, and by Anderson [8], who emphasized the time-reversal, symmetry-breaking character of magnetic moments and fields, which perturbs the wave functions of Cooper pairs formed by electronic states that are related by a time-reversal operation. In the course of these investigations, it was found that in a narrow range of parameter space, gapless superconductivity is also possible.

Only shortly after these developments, a long-standing problem in the physics of metals, the experimental observation of a resistance minimum in some metals at low temperatures [9] was explained theoretically by the work of Kondo [10]. He pointed out that a single, localized magnetic moment, embedded in a metallic matrix, may be shielded by the accumulation of opposite moments due to the spin of the conduction electrons around the impurity moment. This shielding inevitably leads to a partial rearrangement of the electronic excitation spectrum and to an increase of the electrical resisitivity with decreasing temperature according to

$$\rho \sim C[1 + D(E_F) J \ln(k_B T/W) + ...] , \tag{12.1}$$

where $C$ is a constant, $D(E_F)$ the density of electronic states at the Fermi energy, $J$ the exchange coupling between localized and itinerant states, $k_B$ the Boltzmann constant and $W$ the band width of the itinerant electron states. The energy scale for this interaction is fixed by the Kondo temperature

$$k_B T_K = W \exp[1/D(E_F) J]. \tag{12.2}$$

Typical experimental results [11] reflecting this behavior are shown in Fig. 12.2, where the excess resistivity of Eq.12.1 leads to the characteristic Kondo minimum for a simple metal with a small amount of magnetic impurities. The quenching of the magnetic moment, probed by measurements of the magnetic susceptibility, seems to occur gradually if the temperature of the material is lowered below $T_K$. The so-called Kondo problem, i.e., the impact of a single magnetic moment on its metallic environment and vice versa, was later solved in more rigorous ways by Wilson [12], by Andrei [13] and by Wiegmann [14]. Another development in relation with dilute magnetic impurities in metals led to the concept of spin glasses.

As time went on, the main interest shifted away from magnetic impurities towards systems in which magnetic ions occupy regular lattice sites of an intermetallic compound. This situation is often encountered when elements of the $4f$ and $5f$ transition series serve as cations in intermetallic compounds, the so-called rare-earth and actinide compounds which, in many cases, order magnetically at low temperatures. The necessary interaction is provided by the conduction electrons in the form of the Ruderman-Kittel-Kasuya-Yoshida (RKKY) interaction [15, 16]. Its strength is measured by

$$T_{RKKY} \sim NS(S+1)J^2 , \tag{12.3}$$

where $N$ is the number of moments and $S$ the quantum number fixing the moments. In principle, the situation may also be modeled by assuming an array or a lattice of interacting Kondo centers, a problem that has proven to be exceptionally difficult to solve.

Again, two directions of experimental investigations emerged almost simultaneously. The first was triggered by the observation that potentially magnetic intermetallic compounds did not order magnetically, even at very low temperatures. This was particularly intriguing, because the Curie-Weiss type behavior of the magnetic susceptibility at elevated temperatures indicated the presence

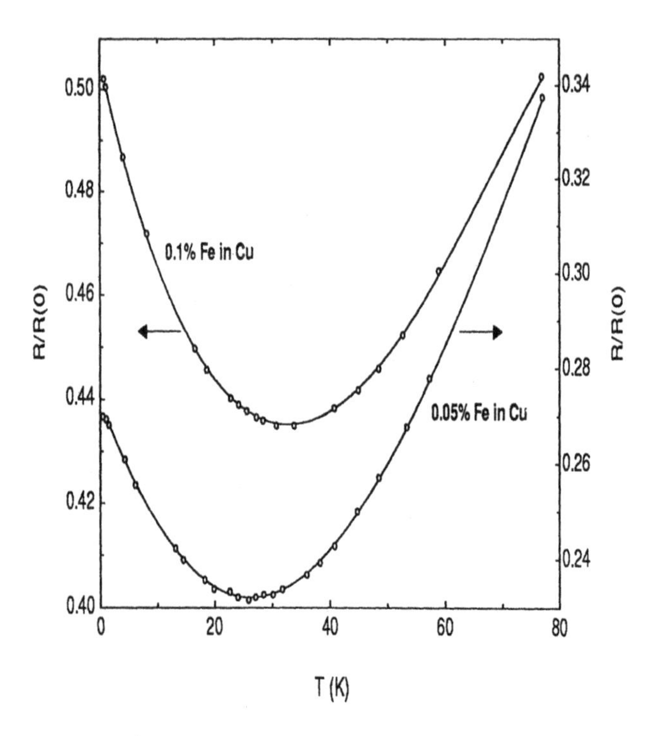


Figure 12.2: Low temperature electrical resistivity of Cu doped with Fe impurities (after Ref. [11].)

of localized and well-defined magnetic moments. The obvious instability of magnetic moments was thought to arise from a fairly strong hybridization between localized $f$-electron and itinerant or conduction electron states. As a consequence, the concept of intermediate valence for the magnetic ions was introduced and the quenching of the magnetic moments at low temperatures was ascribed to rapid fluctuations between two ionic configurations of the $f$-electron elements, preventing the formation of well-defined magnetic moments, and hence the onset of magnetic order [17]. This type of phenomenon seems to occur mainly in compounds containing Ce, Tm, Yb and U, i.e. elements at the beginning and at the end of the $f$-transition series [18].

The second direction of studies took off with the discovery of superconductivity in materials with a sizeable concentration of magnetic ions on regular lattice sites, the rare-earth rhodium borides [19] and the rare-earth Chevrel phases [20]. Here, the main surprises were the observations that the pair-breaking of the magnetic ions was not sufficient to quench the superconducting state in these materials, and that the superconducting state may coexist with antiferromagnetic order.

In the following sections, some recent developments in condensed matter physics, that have their roots in the investigation of magnetic moments in metallic environments, are briefly discussed.

## 12.2 HEAVY (SLOW) ELECTRONS

Meanwhile, another many-body phenomenon, the formation of itinerant electrons with exceptionally large effective masses $m^*$, synonymous with very low Fermi velocities $v_F$, was recognized to occur in $f$-electron metals [21]. A comparison of basic physical properties, such as the electrical resistivity $\rho$, the magnetic susceptibility $\chi$, and the electronic specific heat $C^{el}$ between common metals and heavy-electron materials indicates the striking differences between ordinary metals and heavy-electron systems. This is schematically shown in Fig. 12.3. First, we note that the quantities $\rho$, $\chi$ and $C^{el}$ are larger in heavy-electron metals than in common metals by factors between 100 and 1000. The temperature dependences are also quite distinctive. We summarize the general features in the heavy-electron metals. The resistivity at room temperature is usually of the order of 100 $\mu\Omega$cm, and more or less temperature independent. With decreasing temperature, $\partial\rho/\partial T$ may be positive or negative, depending on the material. A quite generic feature, however, is the distinct decrease of the resisitivity, sometimes by orders of magnitude, without a cooperative phase transition, and a subsequent $T^2$ variation as $T$ approaches zero. Instead of a rather small and temperature independent magnetic susceptibility, a Curie-Weiss type variation of $\chi(T)$, due to rather well-defined localized magnetic moments, is observed at elevated temperatures. At low temperatures, the behavior of $\chi(T)$ is not universal but often exhibits a tendency to a constant value that is greatly enhanced. Finally, we note that the ratio $C^{el}/T$ increases steadily with decreasing temperature, reaching values close to $T =$ 0 K that are two to three orders of magnitude larger than the corresponding values in ordinary metals. This increase of $C^{el}/T$, if related with heavy electron formation, reflects the low value of the Fermi energy $E_F$ of these electrons and is characteristic for these materials.

An early example (now archetypal) of a metal exhibiting these features is $CeAl_3$, a compound that also doesn't order magnetically down to very low temperatures [21]. In Figs. 12.4 and 12.5 we show $\rho(T)$ and $C^{el}/T$ of this compound in the relevant temperature regimes. Quite unusual are the values for $A = 35\mu\Omega\text{cm}/\text{K}^2$, the prefactor of the $T^2$ term in the low temperature resisitivity, represented as $\rho(T) = \rho(T) + AT^2$ below 0.3 K, and $C^{el}/T = 1600\text{mJ}/\text{molK}^2$ below 0.1 K. At first, it appeared that the conduction electrons in $CeAl_3$ form a Fermi liquid with very strongly renormalized parameters. Subsequent microscopic experiments [22, 23] indicate that this interpretation might be too naive. Nevertheless, it seems clear that the electronic degrees of freedom providing the large specific heat are of magnetic origin and must be related with the $4f^1$ configuration of the Ce ions.

A particularly strong impetus to the field of heavy electrons was provided

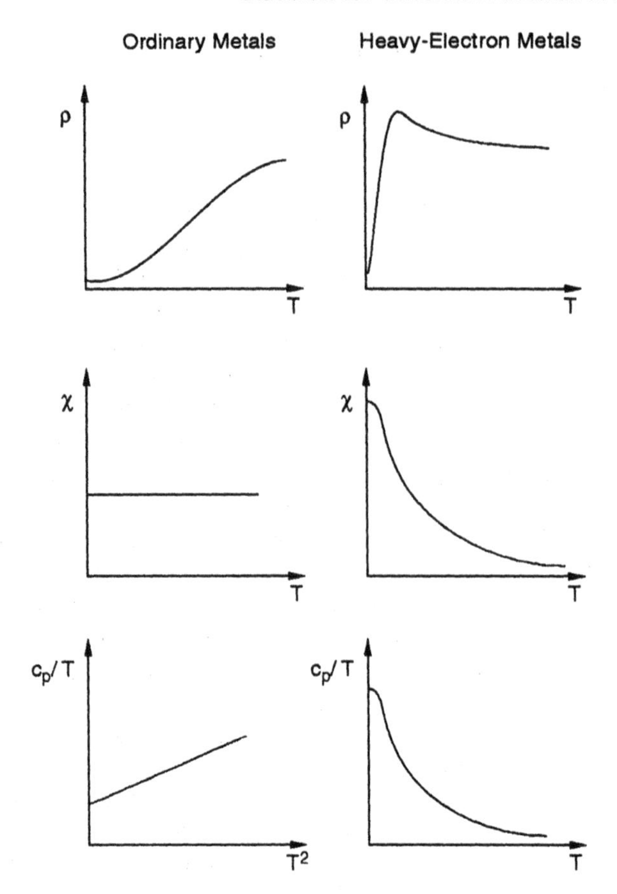


Figure 12.3: Comparison of the temperature dependences of important physical quantities in ordinary and heavy-electron metals (schematic).

by the discovery of superconductivity involving heavy-mass electrons [24]. The situation that electrons, which contribute to localized magnetic moments at elevated temperatures, also participate in the instability leading to a superconducting state at low temperatures seems most intriguing, at least if the occurrence of superconductivity is regarded from a conventional point of view. The proof that heavy electrons form the superconducting condensate is provided by the large anomaly of the specific heat at the critical temperature $T_c$, which matches the value of $C^{el}/T$ of the normal state at $T_c$. An example of this observation is shown in Fig. 12.6 for $UBe_{13}$.

It turns out that heavy-electron metals are at the border of magnetic order provided by the RKKY interaction mentioned above via the conduction electrons. In this sense, the question of the stability of magnetic moments in metals again

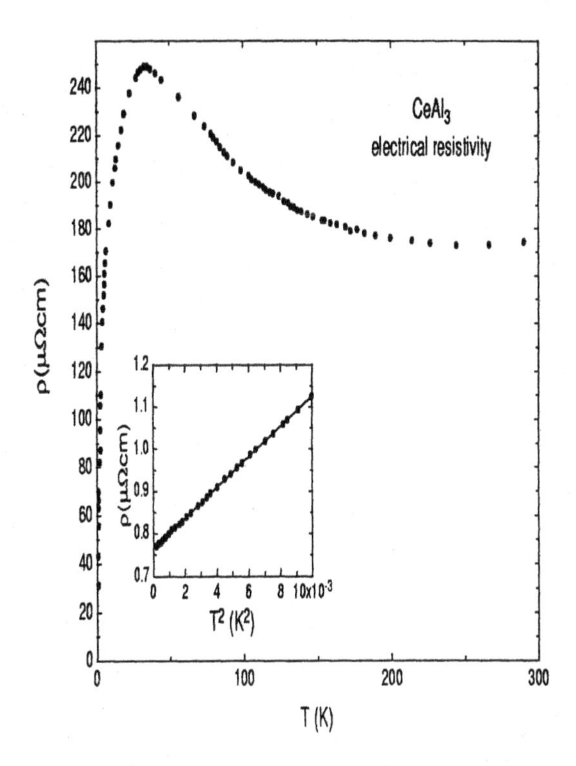


Figure 12.4: Low temperature electrical resistivity of $CeAl_3$. The inset emphasizes the $T^2$ variation at very low temperatures.

arises. The problem is now more complicated than in the dilute case, because the interaction between the magnetic moments cannot be neglected. Many different theoretical approaches have been made to address this issue. Doniach [25] has proposed a schematic way to interpret the trend from well-defined localized moments to their complete quenching by comparing the strength of the two most essential interactions, the Kondo interaction measured by $T_K$ and the RKKY interaction, measured by $T_{RKKY}$. The diagram based on this view is shown in Fig. 12.7, in which $T_K$ and $T_{RKKY}$ are plotted as a function of the exchange interaction $J$. For low values of $J$, an increase in $J$ leads to an enhancement of the Néel temperature $T_N$, defining the onset of antiferromagnetic order. Upon further increasing $J$, the competition between the two interactions becomes more severe, finally leading to a decrease of $T_N$ towards $T = 0$ K, into a so-called quantum-critical regime (QCR). This regime is of particular interest, because theoretical arguments suggest that the conduction electrons in a metal in this regime cannot be regarded as the physical realization of a Landau Fermi-liquid [26, 27]. Landau's Fermi liquid model [28] was long regarded as the master model for describing systems of interacting Fermions, i.e., for conduction electrons in

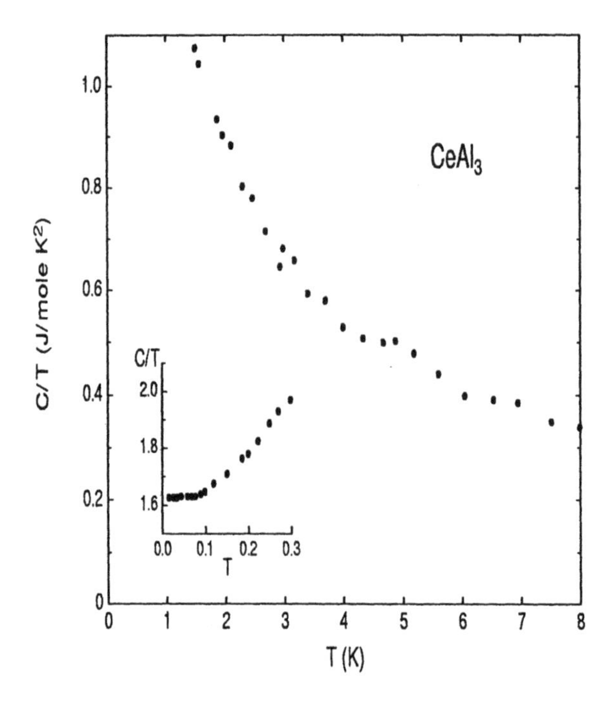


Figure 12.5: Low temperature electronic specific heat of $CeAl_3$ plotted as $C^{el}/T$ versus $T$. The inset emphasizes the behavior below 0.3 K.

metals in particular. Naturally, the suggestion of a possibly new kind of ground state in metals is appealing to many, especially because related issues arise in the investigation of high-$T_c$ superconductivity in cuprates (discussed elsewhere in this book). There are various ways to identify experimentally the so-called non-Fermi-liquid (NFL) effects. Before we address this aspect, we present a case in which the competition between the Kondo- and the RKKY interaction is such that a coexistence of magnetic order and heavy-electron formation is observed.

## 12.3 COEXISTENCE OF MAGNETIC ORDER AND HEAVY ELECTRONS

The coexistence of magnetic order and heavy-electron formation, suggestive of a strange inhomogeneity in $k$-space, was first noted in the compound $UCu_5$ [29]. Here, the characteristic increase of the ratio $C^{el}/T$ with decreasing temperature was observed only well-below $T_N = 15$K. Another particularly informative example is the compound $CePd_2In$, where the energies associated with $T_K$ and $T_{RKKY}$ are both small and similar in magnitude. $CePd_2In$ orders antiferromagnetically at 1.23 K [30]. From various experiments, $T_K$ is estimated to be less than 10 K. The ordered moment well below $T_N$ is only of the order of 0.1 $\mu_B$/Ce

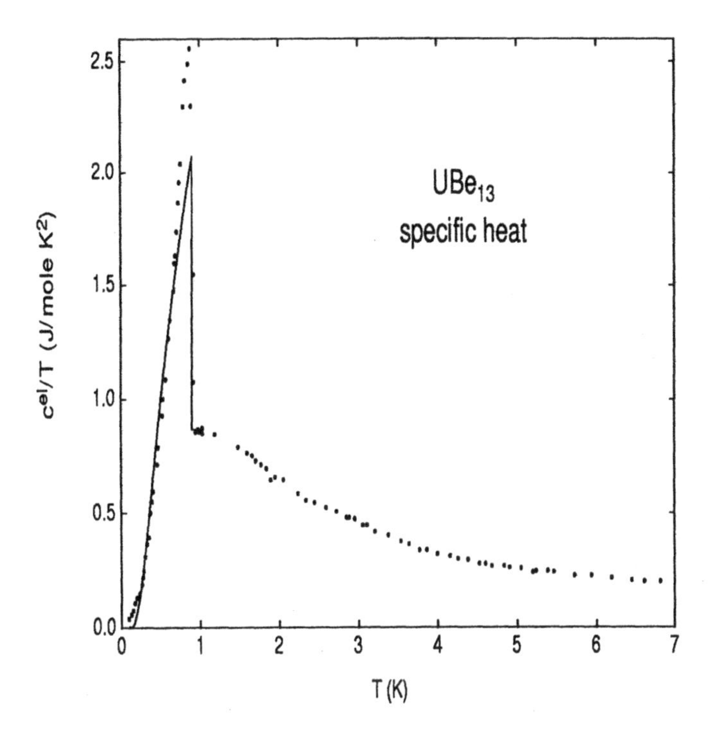


Figure 12.6: Low temperature electronic specific heat of $UBe_{13}$. The solid line represents the expectation for a weak coupling BCS superconductor.

[31], much less than the value 1.3 $\mu_B$/Ce expected from the most likely ground state of the $4f$ electron on the $Ce^{3+}$ ion, as inferred from measurements of $\chi(T)$. Concomitant with the onset of magnetic order, the ratio $C^{el}/T$ increases from 30 mJ/molK$^2$ far above $T_N$ to 137 mJ/molK$^2$ well below $T_N$. This reflects an enhancement of the conduction-electron effective mass which occurs in parallel with the onset of magnetic order among moments that are drastically reduced from their apparent values at higher $T$. The enhancement in $C^{el}$ is also reflected in the $^{115}$In NMR spin-lattice relaxation rate $T_1^{-1}$, which is greatly increased over its value in $LaPd_2In$ at these temperatures [32]. The two relaxation rates are compared in Fig. 12.8. It may be seen that the enhancement below 1 K may be quenched by applying an external magnetic field of moderate strength of the order of 3.5T which, of course, also suppresses the antiferromagnetically ordered state.

The coexistence of heavy electrons with an ordered state involving very small moments has meanwhile been observed in various substances. It is still not clear why the quenching of the moments is not complete in these cases.

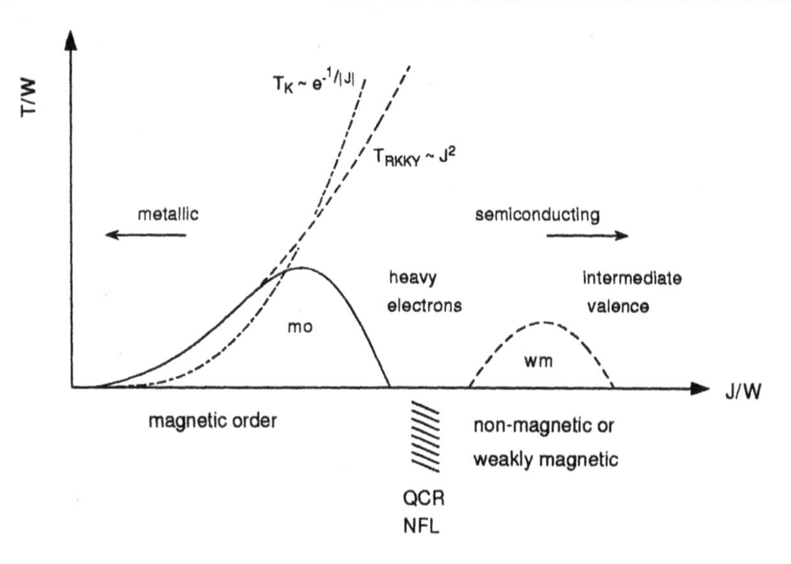


Figure 12.7: Low temperature features of metals experiencing the competition between the RKKY- and the Kondo interaction, as suggested by Doniach [25]. For explanations see text.

## 12.4 NON-FERMI-LIQUID FEATURES OF HEAVY-ELECTRON METALS

As shown in Fig. 12.9, the resistivity $\rho(T)$ of $CeCu_6$ varies linearly with $T$ below 1 K, except at the very lowest temperatures [33]. It has been found that this $T$-linear dependence persists to very low temperatures, if a small amount of Cu is replaced by Au [34]. At the same time, the ratio of $C^{el}/T$ of $CeCu_{5.9}Au_{0.1}$ continues to increase logarithmically with decreasing temperature down to at least 0.1 K. Both these features are clearly not compatible with the expectations of the Fermi-liquid model. Since a further increase of the Au content in these alloys leads to the onset of magnetic order [35], it seems quite likely that the situation of a quantum critical point, schematically depicted in Fig. 12.7, is realized in $CeCu_{5.9}Au_{0.1}$. It appears that Fermi-liquid type behavior of this particular alloy can be restored by applying external magnetic fields of moderate strengths [34].

There are other scenarios, such as multi-channel Kondo effects [36] and distributions of Kondo temperatures [37] that may lead to non-Fermi-liquid behavior in heavy-electron metals. Indeed, experimental evidence for this type of behavior have been found for an increasing number of compounds and alloys, mainly with Ce and U as the cationic constituents [38]. Experimentally, non-Fermi liquid behavior is, in most cases, indicated by unusual temperature dependences of physical properties at low temperatures, for which the Fermi liquid model provides rather rigorous predictions. These are the $T^2$ dependence of the electrical resisitivity $\rho$, the linear-$T$ dependence of the electronic specific heat $C^{el}$ or

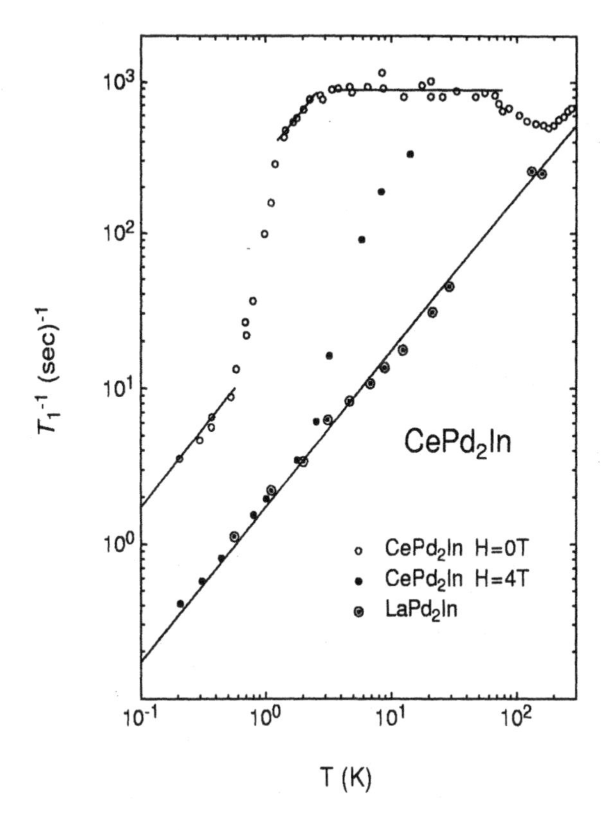


Figure 12.8: $^{115}$In NMR spin-lattice relaxation rates for $CePd_2In$ and $LaPd_2In$.

the NMR spin lattice relaxation rate $T_1^{-1}$, the temperature independence of the magnetic susceptibility $\chi$, the $\omega^2$ dependence of the optical resistivity at constant temperature, etc. For the anomalous cases, it is found that $\rho \sim 1 \pm aT^n$ with $n < 2$, $C^{el}/T \sim \ln T, \chi \sim 1 - bT^{1/2}$, etc.

## 12.5 SUPERCONDUCTIVITY OF HEAVY ELECTRON METALS

The first observation [24] of superconductivity in a heavy-electron metal, $CeCu_2Si_2$, was greeted with a lot of skepticism. Nevertheless, subsequent discoveries of superconductivity in other heavy electron materials [39, 40] confirmed that the phenomenon is real. By now, the list of heavy electron superconductors is quite long. Because of the strong magnetic background in heavy electron metals, the conventional expectation is that superconductivity ought to be quite unfavorable. This led some to suggest that the superconducting state is of an unconventional nature [41, 42]. These conjectures soon received support from experiments which

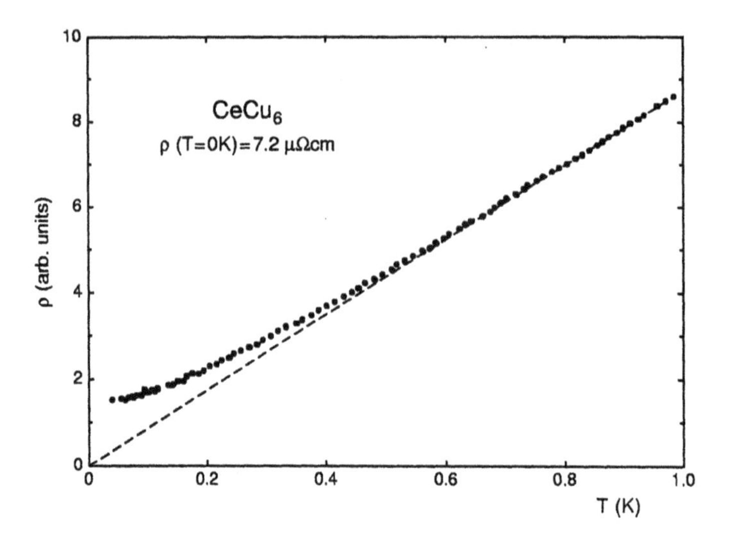


Figure 12.9: Electrical resistivity of $CeCu_6$ below 1 K.

obtained evidence for nodes in the gap function. The experiments primarily involved measuring the $T$ dependences, well below the critical temperature $T_c$, of physical quantities that probe the density-of-states of the quasiparticles $D(E)$ close to $E_F$ [43, 44, 45, 46]. If gap nodes exist, the $T$ dependences are power-laws, instead of the activated form in superconductors with nodeless gaps. The existence of nodes does not necessarily imply unconventional pairing, i.e. one that does not involve phonon-mediated pairing. A more persuasive evidence for unconventional pairing is the existence of different superconducting phases below $T_c$. This has indeed been observed for $U_{1-x}Th_xBe_{13}$ [47] and $UPt_3$ [48].

Applying the criteria discussed above, it is also quite clear that, in some of these materials, superconductivity appears, even though the electronic properties reveal non-Fermi-liquid behavior. This again provides strong arguments in favour of unconventional superconductivity.

If the case of $UBe_{13}$ is considered further, we note that, at $T_c$, the resistivity is of the order of 100 $\mu\Omega$cm [39] and far from exhibiting a $T^2$ dependence. As both the large mass enhancement and the dominant scattering mechanism are related with magnetic degrees of freedom and interactions, it seems quite conceivable that the superconducting state is also caused by the same type of interaction. This notion is supported by recent observations of superconductivity in rare-earth compounds under externally applied pressure [49] (see Fig. 12.10). The compound $CePd_2Si_2$ orders antiferromagnetically at ~10 K at ambient pressure. Under hydrostatic pressure, the Néel temperature decreases substantially, until, around 25 kbar, the ground state switches from being magnetically ordered to superconductivity [50]. A similar behavior has also been reported for other ambient-

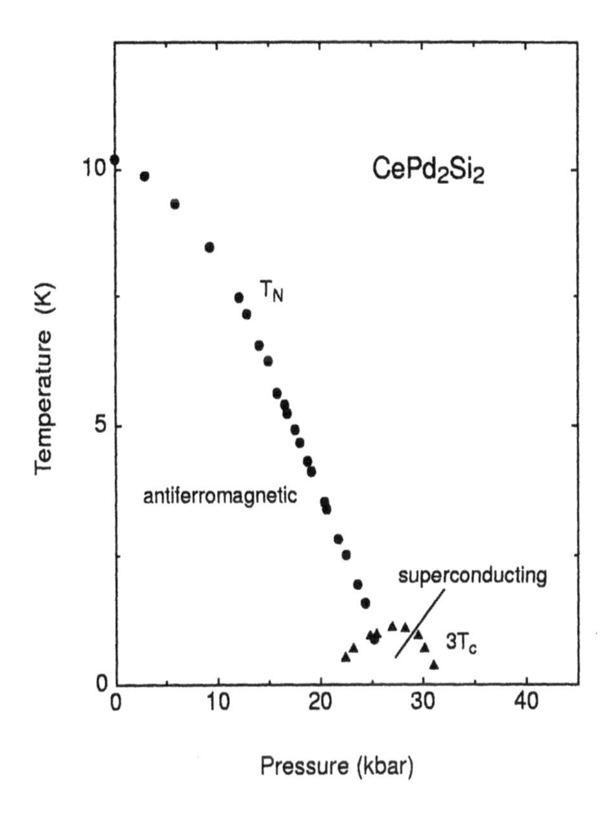


Figure 12.10: Low-temperature $[T, p]$ phase diagram for $CePd_2Si_2$ (after Ref. [49]).

pressure low-temperature antiferromagnets, such as $CeCu_2Ge_2$ [51], $CeRh_2Si_2$ [52], $CeCu_2$ [53] and $CeIn_3$ [49]. The available simple experimental facts suggest that external pressure drives the system out of magnetic order into the quantum critical region, as indicated schematically in Fig. 12.7. Additional evidence [49] supporting this view is the observation that above $T_c$, the $T$ dependence of the electrical resistivity adopts a power-law $T^n$ for which $n$ is distinctly less than 2. Not much is known yet about the microscopic aspects of the antiferromagnetic and the superconducting ground states in the boundary regime. Needless to say, experiments addressing these questions should have a high priority.

The schematic phase diagram shown in Fig. 12.11, which emerges from the few experiments mentioned [49], strikingly resembles, at least in part, the generic phase diagram established for cuprate high-$T_c$ superconductors, in which the antiferromagnetic order of the insulating parent compounds is gradually quenched by doping with itinerant charge carriers, giving way to a superconducting ground state beyond a certain critical value, but in a limited range, of doping. Further, in

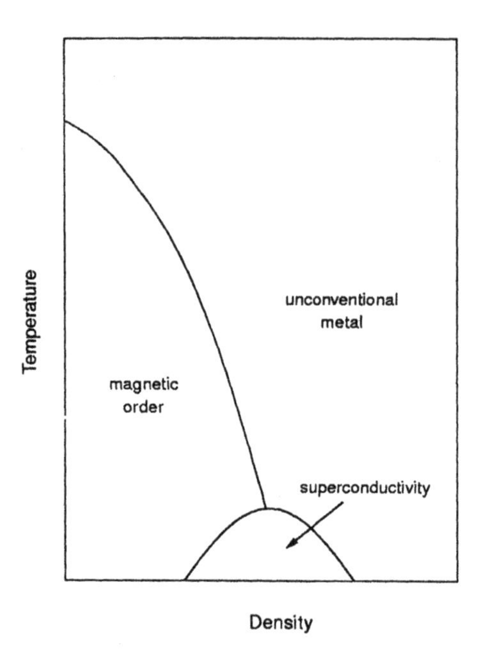


Figure 12.11: Schematic low temperature phase diagram for exotic metals (after Ref. [49]).

the cuprates, the normal-state above $T_c$ has for a long time been claimed to exhibit distinct deviations from Fermi-liquid behavior [54].

## 12.6 CONCLUSIONS

What started with the investigation of the seemingly simple problem of a single magnetic moment in an ensemble of itinerant electrons has turned into one of the most exciting and long-lived fields of research in condensed matter physics. Because of the many-body aspects of the problem, the theoretical treatment is notoriously difficult. The field encompasses many fundamental problems that address the possible ground states of condensed matter. These include metal-insulator transitions, magnetic order, superconductivity and, quite generally, a new theory of metals with strong interactions that goes beyond the standard Fermi-liquid model.

## BIBLIOGRAPHY

[1] J. Friedel, Can. J. Phys. **34** 1190 (1956).

[2] A. Blandin and J. Friedel, J. Phys. Radium **19** 573 (1958).

[3] B.T. Matthias, M. Peter, H.J. Williams, A.M. Clogston, E. Corenzwit, and R.C. Sherwood, Phys. Rev. Lett. **5** 542 (1960).

[4] A.M. Clogston, B.T. Matthias, M. Peter, H.J. Williams, E. Corenzwit, and R.C. Sherwood, Phys. Rev. **125** 541 (1962).

[5] P.W. Anderson, Phys. Rev. **124** 41 (1961).

[6] B.T. Matthias, H. Suhl, and E. Corenzwit, Phys. Rev. Lett. **1** 92 (1958).

[7] A.A. Abrikosov and L.P. Gor'kov, Zh. Eksp. Teor. Fiz. **39** 1781 (1960); Soviet Physics JETP **12** 1243 (1961).

[8] P.W. Anderson, J. Phys. Chem. Solids **11** 26 (1959).

[9] G.J. van den Berg, in *Progress in Low Temperature Physics*, Vol. IV, ed. G.J. Gorter, (North-Holland, Amsterdam 1964) p. 194; and references therein.

[10] J. Kondo, Progr. Theoret. Phys. (Kyoto) **32** 37 (1964).

[11] J.P. Franck, F.D. Manchester, and D.L. Martin, Proc. Roy. Soc. (London) **A263** 494 (1961).

[12] K.G. Wilson, Rev. Mod. Phys. **47** 773 (1976).

[13] N. Andrei, K. Furuya, and J.H. Lowenstein, Rev. Mod. Phys. **55** 331 (1983).

[14] P.B. Wiegman, Zh. Eksp. Teor. Fiz. Pis'ma Red. **32** 392 (1980) (Sov. Phys. JETP Lett. **31** 364 (1980)).

[15] T. Kasuya, Progr. Theoret. Phys. (Kyoto) **16** 58 (1956).

[16] K. Yosida, Phys. Rev. **107** 396 (1957).

[17] See, e.g., D.K. Wohlleben and B.R. Coles, in *Magnetism*, Vol. V, eds. G.T. Rado and H. Suhl (Academic Press, New York 1973).

[18] See, e.g., in *Valence Instabilities*, eds. P. Wachter and H. Boppart (North-Holland, Amsterdam 1982).

[19] B.T. Matthias, E. Corenzwit, J.M. Vandenberg, and H.E. Barz, Proc. Natl. Acad. Sci. USA, **74** 1334 (1977).

[20] O. Fischer, A. Treyvaud, R. Chevrel, and M. Sergent, Solid State Commun. **17** 721 (1975).

[21] K. Andres, J.E. Graebner, and H.R. Ott, Phys. Rev. Lett. **35** 1979 (1975).

[22] S. Barth, H.R. Ott, F.N. Gygax, B. Hitti, E. Lippelt, A. Schenck, and C. Baines, Phys. Rev. B **39** 11695 (1989).

[23] J.L. Gavilano, J. Hunziker, and H.R. Ott, Phys. Rev. B **52** R13106 (1995).

[24] F. Steglich, J. Aarts, C.D. Bredl, W. Lieke, D. Meschede, W. Franz, and H. Schäfer, Phys. Rev. Lett. **43** 1892 (1979).

[25] S. Doniach, Physica B **91** 231 (1977).

[26] J.A. Hertz, Phys. Rev. B **14** 1165 (1976).

[27] A.J. Millis, Phys. Rev. B **48** 7183 (1993); A.M. Tsvelik and M. Reizer, Phys. Rev. B **48** 9887 (1993); M.A. Continentino, Phys. Rep. **239** 179 (1994); L.B. Ioffe and A.J. Millis, Phys. Rev. B **51** 16151 (1995).

[28] L.D. Landau, Zh. Eksp. Teor. Fiz. **30** 1058 (1956) (Sov. Phys. JETP **3** 920 (1957)).

[29] H.R. Ott, H. Rudigier, E. Felder, Z. Fisk, and B. Batlogg, Phys. Rev. Lett. **55** 1595 (1985).

[30] A.D. Bianchi, E. Felder, A. Schilling, M.A. Chernikov, F. Hulliger, and H.R. Ott, Z. Phys. B **99** 69 (1995).

[31] J.L. Gavilano, P. Vonlanthen, B. Ambrosini, J. Hunziker, F. Hulliger, and H.R. Ott, Europhys. Lett. **32** 361 (1995).

[32] P. Vonlanthen, J.L. Gavilano, B. Ambrosini, D. Heisenberg, F. Hulliger, and H.R. Ott, Z. Phys. B **102** 347 (1997).

[33] H.R. Ott, H. Rudigier, Z. Fisk, J.O. Willis, and G. Stewart, Solid State Commun. **53**, 235 (1985).

[34] H. v. Loehneysen, T. Pietrus, G. Portisch, H.G. Schlager, A. Schrder, M. Sieck, and T. Trappmann, Phys. Rev. Lett. **72** 3262 (1994).

[35] H. v. Loehneysen, J. Phys.: Condens. Matter **8** 9689 (1996).

[36] D.L. Cox, Phys. Rev. Lett. **59** 1240 (1987).

[37] E. Miranda, V. Dobrosavljevic, and G. Kotliar, Phys. Rev. Lett. **78** 290 (1997).

[38] See, e.g., M.B. Maple, J. Low Temp. Phys. **99** 223 (1995).

[39] H.R. Ott, H. Rudigier, Z. Fisk, and J.L. Smith, Phys. Rev. Lett. **50** 1595 (1983).

[40] G.R. Stewart, Z. Fisk, J.O. Willis, and J.L. Smith, Phys. Rev. Lett. **52** 679 (1984).

[41] P.W. Anderson, Phys. Rev. B **30** 1549 (1984).

[42] H.R. Ott, H. Rudigier, T.M. Rice, K. Ueda, Z. Fisk, and J.L. Smith, Phys. Rev. Lett. **52** 1915 (1984).

[43] D.E. MacLaughlin, C. Tien, W.G. Clark, M.D. Lan, Z. Fisk, J.L. Smith, and H.R. Ott, Phys. Rev. Lett. **53** 1833 (1984).

[44] F. Gross, B.S. Chandrasekhar, D. Einzel, K. Andres, P.J. Hirschfeld, H.R. Ott, Z. Fisk, and J.L. Smith, Z. Phys. B **64** 175 (1986).

[45] B.S. Shivaram, Y.H. Jeong, T.F. Rosenbaum, and D.G. Hinks, Phys. Rev. Lett. **56** 1078 (1986).

[46] H.R. Ott, E. Felder, C. Bruder, and T.M. Rice, Europhys. Lett. **3** 1123 (1987).

[47] R.H. Heffner, J.L. Smith, J.O. Willis, P. Birrer, C. Baines, F.N. Gygax, B. Hitti, E. Lippelt, H.R. Ott, A. Schenck, E.A. Knetsch, J.A. Mydosh, and D.E. MacLaughlin, Phys. Rev. Lett. **65** 2816 (1990).

[48] A. Schenstrom, M.F. Xu, Y. Hong, D. Bein, M. Levy, B.K. Sarma, S. Adenwalla, Z. Zhao, T. Tokuyasu, D.W. Hess, J. Ketterson, J. Sauls, and D.G. Hinks, Phys. Rev. Lett. **62** 332 (1989).

[49] N.D. Mathur, F.M. Grosche, S.R. Julian, I.R. Walker, D.M. Freye, R.K.W. Haselwimmer, and G.G. Lonzarich, Nature **394** 39 (1998).

[50] F.M. Grosche, S.R. Julian, N.D. Mathur, and G.G. Lonzarich, Physica B **224** 50 (1996).

[51] D. Jaccard, K. Behnia, and J. Sierro, Phys. Lett. A **163** 475 (1992).

[52] R. Movshovich, T. Graf, D. Mandrus, M.F. Hundley, J.D. Thompson, R.A. Fisher, N.E. Phillips, and J.L. Smith, Phys. Rev. B **53** 8241 (1996).

[53] E. Vargoz, P. Link and D. Jaccard, Physica B **230** 182 (1997).

[54] See, e.g., P.W. Anderson and Y. Ren, in *High Temperature Superconductivity*, The Los Alamos Symposium, eds. K.S. Bedell, D. Coffey, D.E. Meltzer, D. Pines, and J.R. Schrieffer, (Addison and Wesley, Redwood City 1990) p.3.

# CHAPTER 13

# SUPERCONDUCTIVITY AND MAGNETISM IN HEAVY-FERMIONS

F. STEGLICH[1], P. GEGENWART[1], C. GEIBEL[1], P. HINZE[1], M. LANG[1], C. LANGHAMMER[1], G. SPARN[1], T. TAYAMA[1], O. TROVARELLI[1], N. SATO[2], T. DAHM[3] AND G. VARELOGIANNIS[4]

[1]Max-Planck Institute for Chemical Physics of Solids
D-01187 Dresden, Germany
[2]Department of Physics, Graduate School of Science
Nagoya University, Nagoya 464-8602, Japan
[3]Institute for Theoretical Physics
University of Tübingen, D-72076 Tübingen, Germany
[4]Max-Planck Institute for Physics of Complex Systems
D-01187 Dresden, Germany

## ABSTRACT

Compared to the high-$T_c$ cuprates, the properties of the low-$T_c$ lanthanide and actinide-based 'heavy-fermion' (HF) superconductors share fewer 'universal' features. Except for $UBe_{13}$, the low-temperature normal state of all Uranium-based HF superconductors can be described as a heavy Landau Fermi liquid. The latter coexists with long-range antiferromagnetic order, which forms at a Néel temperature $T_N \gg T_c$ (the superconducting transition temperature). As an exemple, we discuss $UPd_2Al_3$, in which a large saturation moment is observed along with a large electronic specific heat $\gamma T$ at $T \geq T_c \approx 2$ K. The bulk of the normal-state properties hint at a 'two-component nature' of the U-based $5f$ configuration, namely, two '*localized*' $5f$ electrons together with one '*itinerant*' $5f$ electron or 'heavy fermion'. Analyzing recent neutron-scattering and tunneling data in this 'two-component model', we arrive at the proposal that superconductivity carried by the HFs is mediated by a magnetic exciton, i. e., a propagating crystal-field excitation of the 'localized' $5f^2$ configuration.

In several of the Cerium-based HF superconductors, so-called 'non-Fermi-liquid' (NFL) effects are observed in the low-temperature normal state. As an example, we discuss the prototypical system $CeCu_2Si_2$. Its NFL effects are ascribed to a 'magnetic instability', i. e. the disappearance of 'phase A' ($T_A \to 0$). 'Phase A' appears to be a spin-density wave with a very small ordered moment.

Finally, we discuss disparities between the resistivity and the specific heat found in the normal state of several stoichiometric HF metals, including the new NFL system $YbRh_2Si_2$, at low temperatures, $T < 0.3$ K. These observations make the applicability of a generalized Fermi-liquid concept to HF metals questionable.

## 13.1 INTRODUCTION

Magnetism and superconductivity (SC) are usually considered to be mutually exclusive phenomena: a tiny amount of local magnetic moments dissolved in a classical superconductor is sufficient to destroy the superconducting state [1, 2, 3]. Local magnetic moments are formed because of the strong repulsive Coulomb force between electrons on a partially filled core shell, particularly the 4f shell of a rare-earth element like Cerium [3]. On the other hand, SC in a classical superconductor like $LaAl_2$ [3] originates from a net attractive interaction between the delocalized ($s$-, $p$-, $d$-) conduction electrons which are well described as a low-density fluid, the 'Landau Fermi liquid' (LFL). According to the theory of Bardeen, Cooper and Schrieffer (BCS), the attractive interaction between conduction electrons is mediated by lattice vibrations.

The above antagonism between SC and magnetism appeared to be fundamentally violated in 1979, when SC was discovered in the tetragonal compound $CeCu_2Si_2$ [4]: Here, the $4f$ electrons give rise to the formation, not only of the local magnetic moments at elevated temperatures $T$, but also of the superconducting state at low $T$.

$CeCu_2Si_2$ belongs to the class of so-called *'heavy-fermion' metals*. These are three-dimensional intermetallic compounds containing a dense lattice of certain lanthanide or actinide ions. As mentioned, their high-$T$ properties are indicative of nearly independent local $4f$ or $5f$ magnetic moments. Owing to the mixing of the localized $f$-electron with the delocalized conduction-electron wave functions and the strong Coulomb correlations within the $f$ shells, a new metallic state is formed at low temperatures: it appears as if, in the static properties, the local moments are progressively reduced in magnitude on cooling. Simultaneously, the $f$ electrons become weakly itinerant. The quasiparticle excitations in the new state resemble the conduction electrons of a simple metal, but acquire a huge effective mass $m^*$, up to a thousand times greater than the free-electron mass. The mass is estimated from the large electronic specific heat $\gamma T$ at low $T$.

These quasiparticles, defined in a one-band model of itinerant fermions and called 'heavy fermions' (HF) or 'heavy electrons', are *complex* objects. They are dominated by the local $f$-degrees of freedom, with an admixture of itinerant conduction-electron contributions. In a few HF metals, both 'heavy' and 'light' charge carriers have, in fact, been identified via low-$T$ de Haas-van Alphen measurements on different parts of the renormalized Fermi surface [5].

For a long time, HF metals were considered to behave as LFLs at low temperatures: Nowadays, however, a *pure* heavy LFL state appears to be the exception rather than the rule: (i) Magnetic interactions between incompletely compensated local moments on different sites or between itinerant 'heavy electrons' may cause a phase transition at $T_N$ into an antiferromagnetically ordered state, usually with extremely small saturation moment ($\mu_s \approx 10^{-2}\mu_B$) [6]. In several systems, antiferromagnetic (AF) order *coexists* well below $T_N$ with a heavy LFL state. (ii) In the paramagnetic regime, close to a 'magnetic instability' at which $T_N \to 0$, extended and long-lived strong fluctuations of the order parameter (the staggered magnetization) lead to so-called 'non-Fermi-liquid' (NFL) phenomena in a second class of HF metals. According to spin-fluctuation (SF) theory [7, 8, 9], three-dimensional (3D) AF-SF should enhance the effective quasiparticle mass $m^*$ by a *finite* amount. This theory of the 'nearly antiferromagnetic Fermi liquid' (NAFFL), a generalization of the LFL theory, appears to be well applicable to transition-metal systems.

Except for $UBe_{13}$, *all* Uranium-based HF superconductors, i. e., $UPt_3$ [11], $URu_2Si_2$ [12], $UNi_2Al_3$ [13] and $UPd_2Al_3$ [14] belong to the first category of HF metals. On the other hand, $UBe_{13}$ [15] and $CeCu_2Si_2$ [16] exhibit normal-state (N-state) properties at low temperatures that are characteristic of the second category. While no ordering phenomena distinct from SC has been established to date in $UBe_{13}$ at ambient pressure, in $CeCu_2Si_2$, SC strongly interferes with an antiferromagnetically ordered state ('phase A'). Samples from one regime of its homogeneity range show weak AF order ('A type'), while those from another regime are superconductors ('S type'). In the intermediate regime, AF order and SC compete for stability ('AS type') [17]. At present, the nature of 'phase A' is not fully resolved: resistivity experiments hint at a spin-density wave (SDW) with a very small saturation moment and a nesting wave vector within the basal tetragonal plane [16]. However, to date, neutron diffractometry has not succeeded in resolving any magnetic Bragg peak. This has led to the proposal of an 'unconventional SDW' state (e. g., a spin-nematic) [18]. In Sec. 3 we will discuss the chemical and physical phase diagrams as well as two variants of $CeCu_2Si_2$ ('S' and 'AS' type). Our new results support the identification of 'phase A' as a conventional SDW.

Two compounds that have attracted much interest in the past two years are $CePd_2Si_2$ [19] and $CeIn_3$ [20]. At ambient pressure, both exhibit AF order which is suppressed by a hydrostatic pressure $p_c \approx 2.7$ GPa. SC occurs in the vicinity of $p_c$ at very low temperatures ($T \leq 0.4$ K and 0.2 K, respectively) and sufficiently low magnetic fields, if the quasiparticle mean free path $\ell$ is sufficiently long ($\ell \gg \xi_0$, the superconducting coherence length). For fields $B$ larger than the upper critical field $B_{c2} \approx$ 1-2 T, SC is suppressed, and an NFL power law $\rho - \rho_0 = \Delta\rho \sim T^{\varepsilon}$, with $\varepsilon < 2$, describes the low-$T$ resistivity over more than a decade in $T$ ($\rho_0$ is the residual resistivity). In the subsequently discovered HF superconductor $CeNi_2Ge_2$, all these phenomena are observed already at ambient pressure [15, 21, 22]. This compound has the advantage that one is not restricted to resistivity experiments, but can perform a full thermodynamic analysis. Since, at first glance, $CeNi_2Ge_2$ is a *paramagnetic* compound, the origin of its NFL behavior has been an open problem for some time. However, as was shown recently [23], 'phase A' seems to exist in $CeNi_2Ge_2$ as well, suggesting a similar microscopic origin of the NFL effects in this compound and its copper-silicide homologue.

Section 4 is devoted to 'disparities' between low-$T$, normal-state (N-state) results of the resistivity and the specific heat found in a number of stoichiometric Ce-, Yb- and U-based HF metals. The paper ends with a short outlook in Sec. 5.

In the following section, however, we focus on a member of the first category of HF metals, in which long-range AF order coexists with a heavy LFL state. Among them, $UPd_2Al_3$ is unique as it exhibits both a large saturation moment [24] and a pronounced electronic specific heat $\gamma T$ well below $T_N \approx 14$ K [14]. The LFL state becomes unstable against a superconducting transition below $T_c \approx 2$ K [14], the highest $T_c$ observed at $p = 0$ in any HF superconductor so far.

## 13.2 MAGNETIC-EXCITON MEDIATED SUPERCONDUCTIVITY

The simultaneous observation of a large saturation moment, $\mu_s \approx 0.85\,\mu_B$ [24], and a large Sommerfeld coefficient, $\gamma$ = 140 mJ/K$^2$mole [14] (Fig. 13.1), led to the assumption [25, 26, 27] that the average occupation number of the $5f$ shell of Uranium is only slightly smaller than three. Analyzing the magnetic susceptibility in the paramagnetic regime as a function of temperature and for different orientations, Sato *et al.* [28] confirmed the assumption [25, 26, 27] that two '*localized*' $5f$ electrons are responsible for the magnetic properties [29] and the additional '*itinerant*' one for the HF behavior. As the '*itinerant*' $5f$ electrons are more strongly hybridized with the Pd- and Al-derived $d$ and $p$ states (compared with the '*localized*' ones), they form the massive Cooper pairs below $T_c$. Measurements of the $^{27}$Al nuclear magnetic resonance (NMR) revealed a $T^3$ power law for the spin-lattice-relaxation rate $T_1^{-1}(T)$, with no Hebel-Slichter peak below $T_c$, and a ratio $2\Delta_0/k_BT_c \approx 5.5$ [30] where $2\Delta_0$ is the superconducting gap at $T = 0$. The large ratio indicates that $UPd_2Al_3$ is a *strong-coupling* superconductor.

The $T$ dependence of $T_1^{-1}$, together with a Pauli-limited and weakly anisotropic $B_{c2}(T)$ curve [31], indicate an even-parity $d$-wave state. According to angle-dependent $B_{c2}(T)$ measurements, line nodes exist *across* the $c^*$ axis [32]. Results of specific-heat experiments performed in the superconducting state [25] to

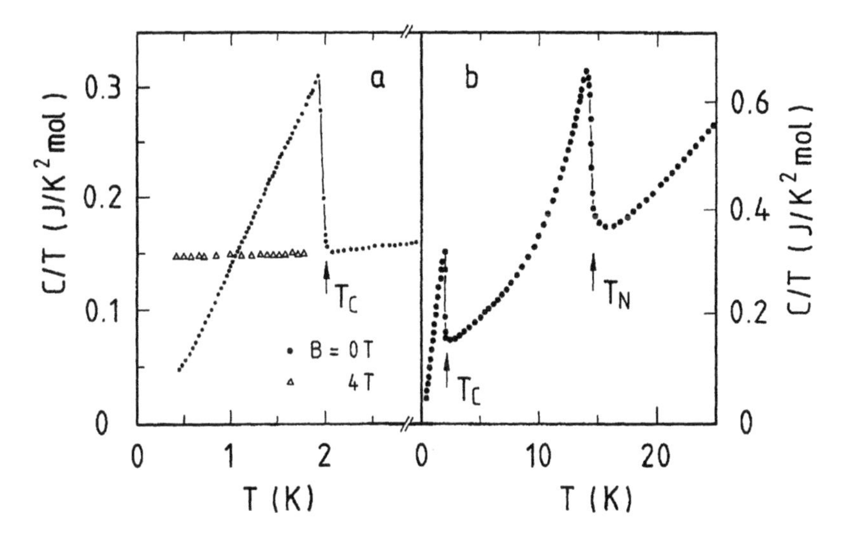


Figure 13.1: Specific heat of a $UPd_2Al_3$ polycrystal as $C/T$ *vs* $T$ for $T < 3$ K (a) and $T < 25$ K (b). The arrows indicate the phase transitions into the antiferromagnetically ordered ($T_N = 14.3$ K) and superconducting ($T_c = 2$ K) states.

substantially lower temperatures, when compared to those in Fig. 13.1, suggest the existence of an *intrinsic* term $\gamma_r T$, $\gamma_r = (13 \pm 5)$ mJ/K$^2$mole. Recalling the specific heat of antiferromagnetically ordered Kondo-lattice systems with (somewhat Kondo-reduced) localized $4f$ moments such as $CeAl_2$ [33] and $CeB_6$ [34], one may attribute the intrinsic 'residual' term $\gamma_r T$ in $UPd_2Al_3$ to the '*localized*' $5f$ states. Consequently, the remaining contribution $(\gamma - \gamma_r)T$ relates to the '*itinerant*' ones. The renormalized specific-heat jump $\Delta C/(\gamma - \gamma_r)T_c \approx 1.4$ at $T_c$ considerably exceeds the mean-field value 1.2 expected for a $d$-wave superconductor. Hence, it supports the strong-coupling nature of the SC state.

Recent inelastic neutron-scattering experiments by Sato *et al.* [28] confirm the *two-component* nature of the $5f$ electrons and a strong (intra-ionic) coupling between the 'localized' and 'itinerant' components. Figure 13.2 displays the intensity of the scattered neutrons as a function of energy transfer at different temperatures (in these figures, the momentum transfer $\vec{q}$ is fixed at the AF wave vector $\vec{Q}_0 = (0, 0, \frac{1}{2})$). The upper peak exhibits little intensity variation (not shown) but considerable dispersion near $\vec{Q}_0$. Thus, it is associated with a *collective mode* of the local-moment system due to intersite interaction. Sato *et al.* [28] concluded that a local crystal-field (CF) excitation of $\Delta_{CF} \approx 7$meV for the single $U^{4+}$ ($5f^2$) ion (with singlet ground state) develops dispersion in the molecular field of the antiferromagnet. Two magnetic-exciton modes are expected, but only one (the acoustic one) was observed in the low-energy neutron-scattering spectra. Its dispersion near the center of the AF Brillouin zone is shown in the inset of

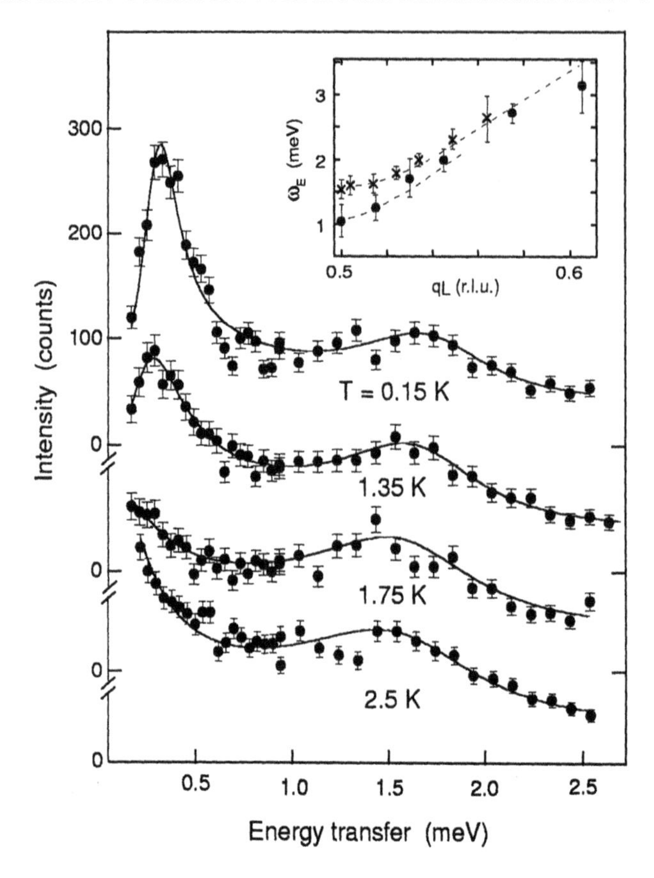


Figure 13.2: Temperature evolution of the inelastic neutron-scattering spectrum for a $UPd_2Al_3$ single crystal at $\vec{q} = (0, 0, q_L) = (0, 0, 0.5) = \vec{Q}_0$ through $T_c \approx 1.8\,K$. Solid lines represent fits using the microscopic 'two-component' model described in Ref. [28]. There is only one fit parameter at given values of the temperature and $\vec{q}$: the magnetic-exciton energy $\omega_E(\vec{q})$, shown in the inset for $T = 2.5\,K$ (crosses) and $0.15\,K$ (circles).

Fig. 13.2. The finite gap at $\vec{q} = \vec{Q}_0$ is caused by uniaxial exchange anisotropies. In addition, a remarkable 30% softening of this zone-center magnetic exciton due to the formation of the superconducting state is observed, apparently related to a change in the inter-site exchange interaction.

The low-energy peak in Fig. 13.2 ($\omega = 0.35\,\text{meV}$ at $T = 0.15\,K$) is progressively reduced and shifted downwards upon warming. It turns into a quasielastic line above $T_c \approx 1.8\,K$. This unique behavior indicates that the lower peak *must* be associated with the itinerant heavy quasiparticles. Since the 'localized' $5f$ electrons should dominate the magnetic neutron-scattering cross section (because of their large form factor), these data clearly demonstrate the *strong coupling* between '*itinerant*' and '*localized*' $5f$ states. The low-energy peak may be compared to the '41 meV resonance' observed in some of the high-$T_c$ cuprates [35]. However, in the latter case, the peak position is nearly *independent of temperature*. Moreover, its intensity vanishes as $T \rightarrow T_c^-$, in contrast to $UPd_2Al_3$ where the 'resonance mode' softens and still exists in the normal state, i. e. as a quasielastic line.

Recent tunneling experiments by Jourdan *et al.* [36] have revealed a distinct modulation in the tunneling density of states slightly above the gap structure of $UPd_2Al_3$. Inserting the spectrum of the acoustic magnetic exciton at $\vec{q} = \vec{Q}_0$ extracted from the neutron-scattering results into Eliashberg theory, Sato *et al.* [28] were able to reproduce the tunneling data surprisingly well. This analysis, thus, strongly suggests that the *bosons* identified by neutron scattering are responsible for SC. It should be stressed that the low-energy peak of Fig. 13.2 is related to the opening of the superconducting gap, but is *not* a direct measure of the superconducting gap energy $2\Delta_0$. From the existence of a strong, low-energy magnetic response at $\vec{q} = \vec{Q}_0$ observed well below $T_c$, one infers a non-vanishing coherence factor in the expression for the dynamical magnetic susceptibility. This implies the symmetry relation $\Delta(\vec{q} + \vec{Q}_0) = -\Delta(\vec{q})$ for the superconducting gap. The most obvious even-parity polar gap function obeying this relation has two line nodes at the intersection of $\vec{q} = \pm 0.5\,\vec{Q}_0$ with the Fermi surface.

To summarize, both inelastic neutron-scattering and tunneling experiments highlight a strong coupling of the charge carriers (i. e., '*itinerant*' $5f$ electrons or 'heavy fermions') to a magnetic exciton (i. e., a CF excitation within the '*localized*' $5f^2$ configuration, propagating in a molecular field). Among all unconventional superconductors, $UPd_2Al_3$ seems to be the first and only one, for which a dominating non-phononic pairing mechanism could be verified experimentally to such an extent.

## 13.3 ANTIFERROMAGNETISM AND SUPERCONDUCTIVITY

$CeCu_2Si_2$ shows an extreme sensitivity of its physical properties against variations of the stoichiometry. In a thorough investigation of the ternary chemical Ce-Cu-Si phase diagram [17, 25, 37], the Cu-Si composition, i. e., the average occupation of the non-$f$ sites surrounding Ce, was found to play a key role. Four different ground-state properties could be identified in different regions of the homogeneity range of $CeCu_2Si_2$, extending to approximately $\pm 0.5\,\%$ from the true stoichiometry point along the Cu-Si axis, see Fig. 13.3: SC is found for Cu-rich samples, 'phase A' for small and 'phase X' for larger Si excess, respectively. The latter has been associated with a disordered variant of static antiferromagnetism, and 'phase A' with a 'slowly fluctuating' ($\tau \approx 10^{-7}$ sec) SDW [38].

The magnetic/superconducting phase diagram of $CeCu_2Si_2$ is schematically shown in Fig. 13.4. A general coupling constant $g$, which reflects the strength of

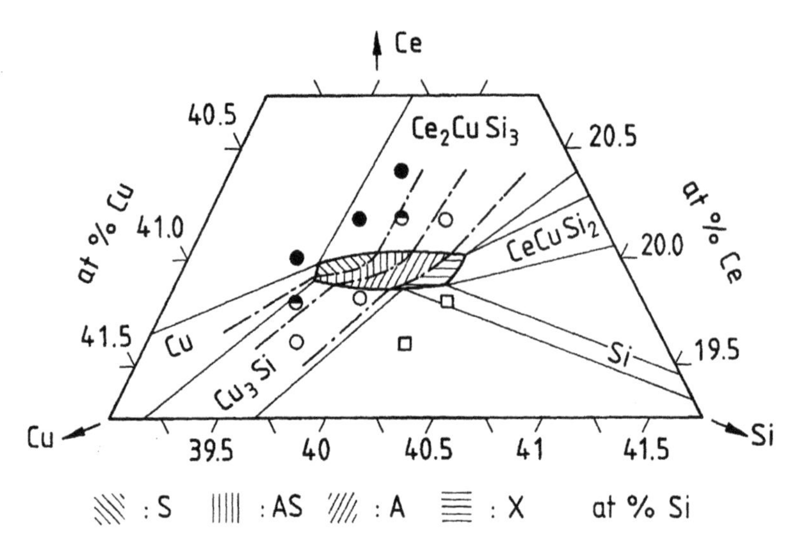


Figure 13.3: Ternary chemical Ce-Cu-Si phase diagram in the neighborhood of the homogeneity range of $CeCu_2Si_2$ (hatched). Note that the 'width' of this homogeneity range (associated with the Ce concentration) is limited by the resolution of the X-ray scattering experiments. Solid lines indicate foreign phases which form together with $CeCu_2Si_2$ in off-stoichiometric samples. Dash-dotted lines separate from each other different physical ground-state properties as determined with polycrystalline samples of varying composition: superconductivity 'S' (full circles), 'phase A' (open circles), competition between them 'AS' (half-filled circles) and 'phase X' (squares), see text. The true stoichiometric (1:2:2) composition is located in the 'AS sector'.

the $4f$-conduction electron hybridization, is used on the abscissa. Undoped samples from the homogeneity range are found in the hatched regime. The 'strong-coupling' (larger hybridization) sector of the phase diagram can be extended by application of hydrostatic pressure, the 'weak-coupling' sector by using partial substitution of Ge for Si. As shown by nuclear-quadrupole-resonance (NQR) experiments, the presence of the dopant atoms changes the nature of the low-$T$ antiferromagnetically ordered 'phase A' from slowly fluctuating to static [39]. The following cases can be distinguished in Fig. 13.4:

I. The 'A-phase' transition at $T_A$ is followed upon cooling by a superconducting one at $T_c \ll T_A$, below which SC and 'phase A' *coexist* ('A type') [40].

II. At $T_c \leq T_A$ and $B < B_{c2}$ ($\approx 1.5\,\mathrm{T}$ as $T \to 0$), 'phase A' is *replaced* by SC ('AS type') [41]. Note that the true stoichiometry point with the most perfect crystallinity residing in this sector is characterized by the lowest residual resistivities and is very close to the crossing of the $T_c(g)$ and $T_A(g)$ lines.

III. $T_c$ exceeds $T_A$, and the samples are superconducting ('S type'). However, in polycrystalline samples the 'A-phase' transition may be recovered at $B \geq B_{c2}$ (2-2.5 T, as $T \to 0$) [42]. *Single crystals* of 'S type' lack 'A-phase' signatures in

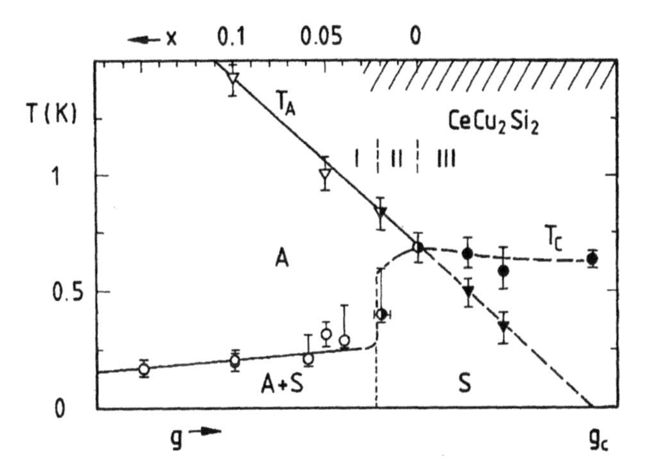


Figure 13.4: Generic phase diagram of $CeCu_2Si_2$ combining data obtained with undoped polycrystals from the homogeneity range (hatched) and from Ge-doped ones, $CeCu_2(Si_{1-x}Ge_x)_2$. Since $T_A$ increases linearly with the Ge concentration $x$, the coupling constant $g$ was *assumed* to be linear in $(1-x)$, i. e., $(1-T_A)$. Sectors I, II and III indicate samples of 'type A', 'AS' and 'S', respectively, see text.

their N state. They are, thus, located beyond the 'magnetic instability' at which $T_A \to 0$. With such single crystals, pronounced NFL effects have been observed [7]: At temperatures not too low ($T > 0.3$ K) and magnetic fields not too high ($B_{c2} < B < 7$ T), the resistivity obeys $\Delta\rho(T) = \beta T^{3/2}$, while the specific-heat coefficient $\gamma(T) = [C(T) - C_0(T)]/T = \Delta C(T)/T = \gamma_0 - \alpha T^{1/2}$ ($C_0$: specific heat of the non-$f$ reference compound $LaCu_2Si_2$). These power laws are in perfect agreement with the predictions of the SF theory [7, 8, 9] for an AF quantum-critical point (QCP) in 3D, with which the 'magnetic instability' at $g = g_c$ was, therefore, identified [16].

Before we comment in Sect. 4 on the behavior of $CeCu_2Si_2$ and a few other HF metals in the vicinity of a 'magnetic instability', we wish to address in the following the nature of 'phase A' as well as a possible scenario for the rich phase diagram shown in Fig. 13.4.

As already mentioned in the introduction, there is still some uncertainty as to whether 'phase A' denotes a conventional SDW (with a saturation moment $\mu_s$ too low to be resolved by neutron diffractometry). To shed more light on the nature of this state we have performed measurements of both the specific heat (Fig. 13.5) and the magnetization (Fig. 13.6) on a well-prepared $CeCu_2Si_2$ single crystal of the 'AS variety'. As shown in Fig. 13.5, the field dependences of the subsequent phase transitions at $T_A$ and $T_c$ are quite different. The upper one, from the paramagnetic state into 'phase A', is of second order, which is also evident from the isothermal ($T = 0.53$ K) magnetization curve in Fig. 13.6b. Further, the

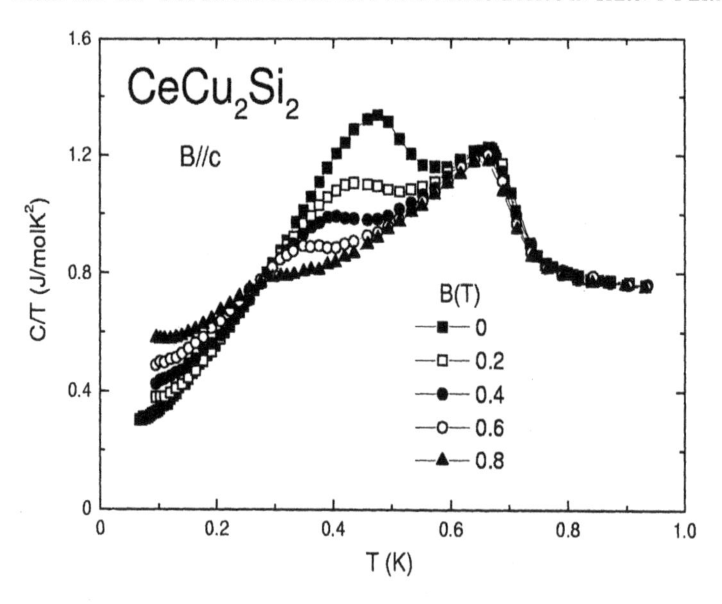


Figure 13.5: Specific heat of an 'AS-type' $CeCu_2Si_2$ single crystal as $C/T\ vs\ T$ at differing magnetic fields $B \leq 0.8$ T, applied along the $c$ axis. The upper phase transition marks the formation of 'phase A', the lower one the onset of superconductivity.

rapid drop at $T_A$ in the $M(T, B = \text{const.} > B_{c2})$ curves of Fig. 13.6 a clearly reflects the AF character of 'phase A'. Assuming, therefore, a SDW forming on parts of the renormalized Fermi surface (FS), the FS fraction involved can be estimated from the size of the discontinuity in the specific heat at $T_A$, $\Delta C_A$, normalized to the specific heat in the paramagnetic state, $\gamma T_A$. Assuming a mean-field ratio of $\Delta C_{SDW}/\gamma T_{SDW} = 1.43$ for a fully gapped FS, the experimental result $\Delta C_A/\gamma T_A \approx 0.55$ indicates that approximately 40 % of the FS takes part in the formation of 'phase A'.

The lower phase transition at $T_c$ from 'phase A' into the superconducting state is of first order, which cannot be recognized in Fig. 13.5, but is clearly resolved by our new high-resolution magnetization experiments (not shown) as well as by previous measurements of the sample length [43]. The magnetization curve taken at $T = 0.1$ K reveals a sharp metamagnetic-like transition at $B \approx 7$ T. This indicates a first-order transition from 'phase A' into the related high-field 'phase B' [41], which occurs for this sample below $T = 0.35$ K and corresponds to an increase of the saturation moment of only $2.5 \cdot 10^{-3} \mu_B/\text{Ce}$. De Haas-van Alphen oscillations at fields $B > 7$ T indicate the high quality of this single crystal.

After having substantiated the SDW nature of 'phase A', let us briefly discuss a model calculation that may explain the complex phase diagram of $CeCu_2Si_2$ (Fig. 13.4). Our theoretical result is presented in Fig. 13.7. Here, a simplified mean-field model for a SDW ground state competing with a $d$-wave superconduct-

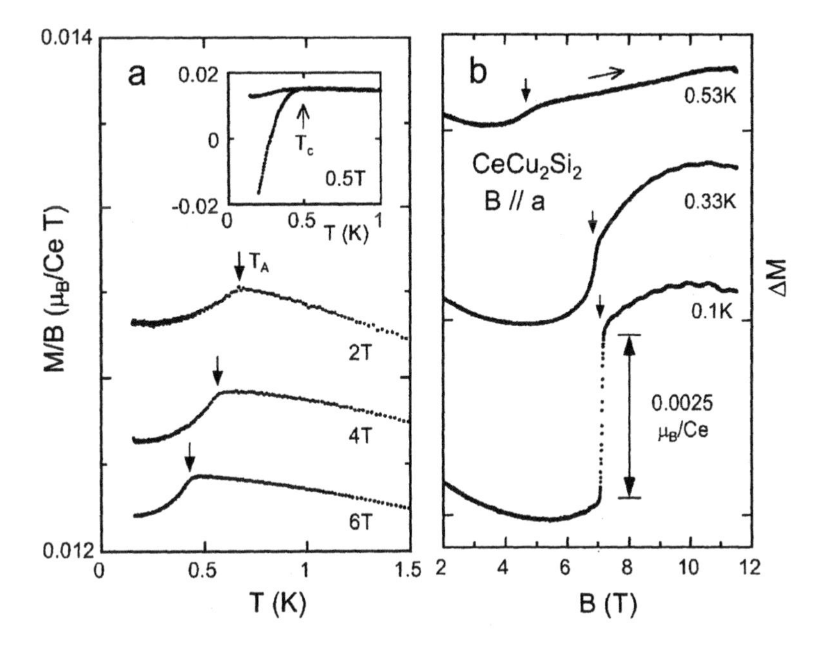


Figure 13.6: DC magnetization of an 'AS-type' single crystal for $B||a$ as (a) $M/B$ $vs$ $T$ for various fields and (b) $\Delta M(B, T = \text{const}) = M(B, T = \text{const}) - M(B, T = 0.95\,\text{K})$ at different temperatures. The curves in (b) are shifted by a constant value for clarity. Inset shows $M/B$ $vs$ $T$ at $B = 0.5$ T after zero-field (lower) and field cooling (upper curve).

ing state has been assumed. For the calculation we used a *two-dimensional* tight-binding band structure with an effective hopping integral $t_{eff}$, which can lead to interesting competition or coexistence of the magnetic and superconducting states, as has been shown earlier [44, 18]. In Ref. [44], a $d_{x^2-y^2}$ pairing state was studied. This state, in its pure form, was shown not to be stable in the presence of the SDW. Here, we have chosen a pairing state with $d_{xy}$ symmetry, having its nodes along the $\Gamma - X$ direction in the first Brillouin zone (see the inset in Fig. 13.7). This state is more competitive with the SDW and can become stable also *within* the SDW phase. We found a similar phase diagram also for an anisotropic $s$-wave state having its minima along the $\Gamma - X$ direction. NMR results showing a $T^3$ behavior of the $1/T_1$ relaxation rate seem to be consistent with such anisotropic pairing states [38].

In the calculation presented here, we included a second FS piece, well separated from the tight-binding FS (closed around the $\Gamma$-point), as shown in the inset. Such cases have been studied in connection with organic superconductors [45]. This second FS piece is closed around the M point and contains about 13% of the charge carriers in the actual calculation. The resulting phase diagram looks quite similar to the one observed in $CeCu_2Si_2$ (see Fig. 13.4). Note that the lower

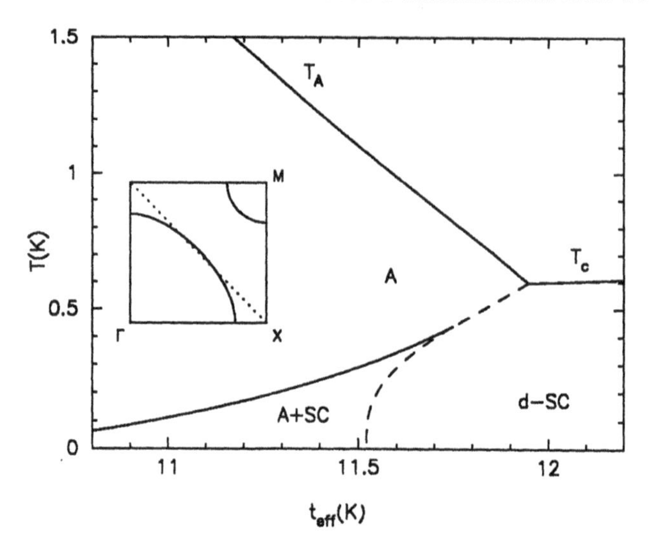


Figure 13.7: Phase diagram for the mean-field model described in the text. 'A' denotes a spin-density wave (SDW) phase, '*d*-SC' a superconducting phase with $d_{xy}$-symmetry and 'A+SC' a phase with coexistence of SDW and *d*-SC. The dashed line denotes a first-order phase transition, while the solid lines correspond to second-order phase transitions. $t_{eff}$ is the effective hopping integral of the charge carriers in the tight-binding band used for the calculation. The inset shows the Fermi-surface (FS) topology used. In addition to the tight-binding FS (closed around the Γ-point), a second FS piece, closed around the M-point and containing about 13% of the total charge carriers, has been included and leads to the long 'tail' of the (A+SC) phase at small values of $t_{eff}$. The dotted line in the inset shows the location where the SDW gap opens.

phase transition from 'phase A' into the superconducting state turns out to be of first order, in agreement with our observations mentioned above. The SDW gap which opens below $T_A$ along the dotted line (see inset) leads to a gap only on part of the FS, allowing the remaining charge carriers to form a superconducting state coexisting with the 'phase A' at low enough temperatures. Especially the charge carriers on the second FS piece remain mainly unaffected by the SDW gap, which makes the coexisting (A + SC) phase stable down to low hybridization strengths $t_{eff}$, leading to a long (A+SC) 'tail' in the phase diagram.

Although this two-dimensional model is certainly too simplified to fully account for all the properties of $CeCu_2Si_2$, it nevertheless suggests that charge carriers sitting on remote parts of the FS might give a possible explanation for the unusual stability of the (A+SC) phase. Such a scenario does not seem unreasonable in view of results of renormalized bandstructure calculations for this compound [46].

## 13.4 HEAVY-FERMION METALS NEAR A MAGNETIC INSTABILITY

As already pointed out in the introduction, SF theory [7, 8, 9] predicts a 3D system in the vicinity of an AF-QCP to behave as a NAFFL, with asymptotic $T$ dependences $\Delta\rho(T) = \beta T^{3/2}$ and $\gamma(T) = \gamma_0 - \alpha T^{1/2}$. In Figs. 13.8a and b we present N-state resistivity and specific-heat results for several $CeNi_2Ge_2$ polycrystals and for an 'S-type' $CeCu_2Si_2$ single crystal [16]. $\gamma(T)$ obeys the predicted $T^{1/2}$ law within more than half a decade in $T$ *above* $T \approx 0.3$ K for *all* samples. Two of them also show the predicted $T^{3/2}$ dependence of $\Delta\rho(T)$: the 'standard-quality' sample of $CeNi_2Ge_2$ ($\rho_0 = 2.7\,\mu\Omega$cm) measured at $B$ = 1 T (necessary to suppress the 'A-phase' transition) and the $CeCu_2Si_2$ crystal measured at moderate fields, $2\,T < B < 7\,T$. For the $CeNi_2Ge_2$ samples with much lower residual resistivities ($\leq 0.4\,\mu\Omega$cm), the exponent $\varepsilon$ in the power law $\Delta\rho \sim T^{\varepsilon}$ appears somewhat reduced, in accord with what was recently calculated for a clean NAFFL [47].

In the following we briefly address observations which seem to complicate the interpretation of our results in terms of SF theory:

(i) For $B = 14\,T$, the low-$T$ resistivity of $CeCu_2Si_2$ obeys $\Delta\rho(T) = a(B)T^2$ with a gigantic coefficient $a \approx 10\,\mu\Omega\text{cmK}^{-2}$, becoming somewhat reduced on increasing the field. The dramatic change [16] in the resistivity exponent from 1.5 ($B < 7\,T$) to 2 ($B > 7\,T$) has *no* correspondence in the $T$-dependence of the specific-heat coefficient $\gamma(T)$, see Fig. 13.8b. For $CeCu_2Si_2$ samples of both 'A' and 'AS type', such a quadratic temperature dependence of the resistivity is preceding not only the 'B-phase' transition for $B$ >7 T, but is also found [16, 48] precursive to the 'A-phase' transition below 7 T. Based upon $\Delta\rho(T, B$ = const.) measurements alone one might, therefore, be inclined to ascribe the 'A' and 'B-phase' transitions to instabilities of a heavy LFL. However, a strongly temperature-dependent Sommerfeld coefficient in either case (Ref. [48] and Fig. 13.8b) argues against such a scenario.

(ii) Considering the low-field results for the polycrystalline $CeNi_2Ge_2$ samples ($B \leq 1\,T$) as well as for the 'S-type' $CeCu_2Si_2$ single crystal ($B < 7\,T$) in Figs. 13.8a and b, we find that *below* $T \approx 0.3$ K the N-state resistivity follows the predicted [7, 8, 9, 47] quantum-critical power law. On the other hand, for one of the $CeNi_2Ge_2$ polycrystals of Fig. 13.8b (measured at $T_c < T < 0.3$K and $B = 0$) as well as for the $CeCu_2Si_2$ single crystal in its normal state ($B = 4\,T$ and $14\,T$), $\gamma(T)$ is found to *deviate distinctly* from the quantum-critical $T^{1/2}$ dependence obeyed at temperatures $T > 0.3\,K$. Note that an 'A-phase' transition is not detected in either of these two samples. Also, quadrupole and/or Zeeman splitting of nuclear-spin states may be safely discarded in both cases [49]. This means that the anomalous low-$T$ upturn of $\gamma(T)$ is very likely of *electronic* origin. In addition, this feature is extremely dependent on the stoichiometry, cf. the two $CeNi_2Ge_2$ samples in Fig. 13.8b. Such 'disparities' between $\Delta\rho(T)$ and $\gamma(T)$ which have also been reported for N-state $UBe_{13}$ [15] and, more recently,

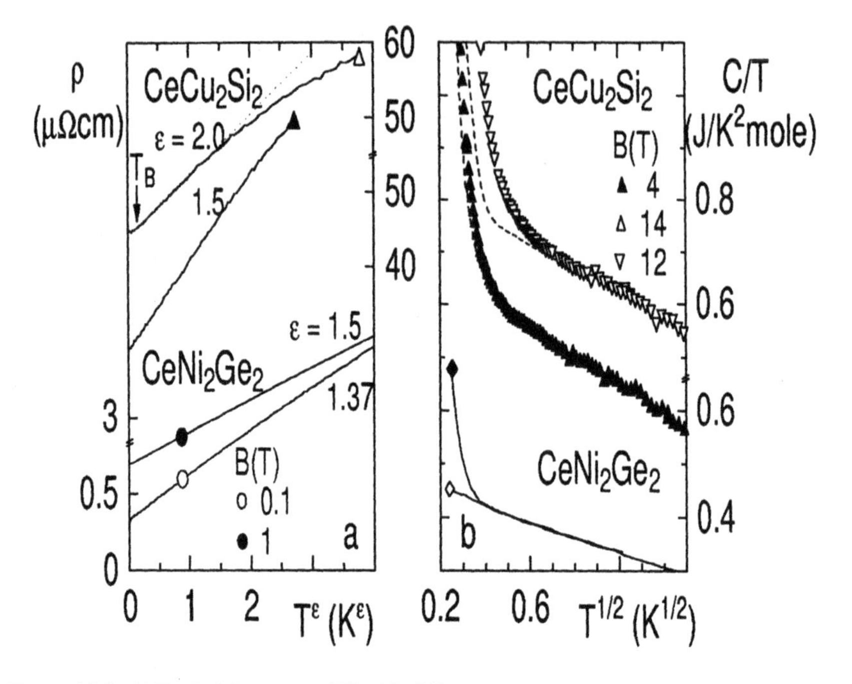


Figure 13.8: (a) Resistivity as $\rho\ vs\ T^{\varepsilon}$ with different exponents $\varepsilon$ for an 'S-type' $CeCu_2Si_2$ single crystal at $B = 4\,T$ (full) and $14\,T$ (open triangle) as well as for two $CeNi_2Ge_2$ polycrystals: a 'standard-quality' sample ($\rho_0 \approx 2.7\,\mu\Omega cm$) measured at $B = 1\,T$ (full) and a very pure one ($\rho_0 \approx 0.3\,\mu\Omega cm$) measured at $B = 0.1\,T$ (open circle). Arrow indicates 'B-phase' transition at $T_B$ in $CeCu_2Si_2$. (b) Specific heat as $C/T\ vs\ T^{1/2}$ for the same single crystal as in (a) studied at $B = 4\,T$ and $12\,T$, respectively, as well as of two polycrystalline $CeNi_2Ge_2$ samples. The latter exhibit $\rho_0 \leq 0.4\,\mu\Omega cm$ and $\Delta\rho \sim T^{\varepsilon}$ ($\varepsilon \approx 1.4$), very similar to the 'cleaner' sample in (a). Dashed lines display $C(T)/T$ data after subtraction of nuclear hyperfine contributions due to the applied $B$-fields.

for $CeRu_4Sb_{12}$ [50] may signal a 'break up' of the 'heavy quasiparticles' when approaching a QCP [16], cf. also Refs. [51] and [52].

In Fig. 13.9 we summarize our observations for $YbRh_2Si_2$, a new NFL compound [53] crystallizing in the same $ThCr_2Si_2$ structure as the two Ce homologues addressed above. According to susceptibility measurements, this material exhibits a strong easy-plane anisotropy and undergoes a weak AF phase transition at $T_N = 70\,mK$ which cannot be resolved in $\rho(T)$, measured along the tetragonal basal plane (Fig. 13.9a). Rather, the resistivity shows a nearly $T$-linear dependence from $T \approx 10\,K$ down to 10 mK. Not shown in Fig. 13.9b, a broad maximum in the specific-heat coefficient $\gamma(T)$ occurs near 60 K which indicates CF splitting of the $Yb^{3+}$ ($J$ = 7/2) Hund's rule ground state. Results for the entropy gain hint at a low-lying Kramers doublet, well separated from the excited CF states. $\gamma(T)$ diverges logarithmically in the range of $0.3\,K < T < 10\,K$, see Fig. 13.9b.

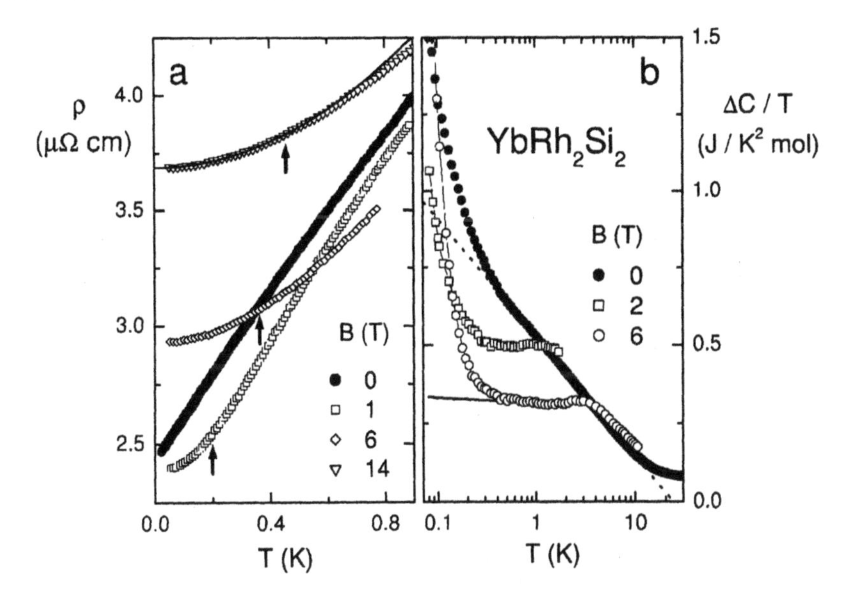


Figure 13.9: (a) Low-temperature resistivity of $YbRh_2Si_2$, measured along the $a$ axis for $B = 0$ and differing magnetic fields applied along the $c$ axis. Arrows indicate the crossover temperature below which a $T^2$ law is recovered (cf. solid line for $B = 14\,\mathrm{T}$). (b) Yb increment to the specific heat as $\Delta C/T$ *vs* $T$ (on a logarithmic scale) at different fields applied along the $a$ axis. The thin solid lines are guides to the eye. The dotted line representing $\Delta C/T = \gamma_0' \ell \mathrm{N}(T_0/T)$ is a guide to the eye, too. The thick solid line represents the $B = 6\,\mathrm{T}$ data after subtraction of the hyperfine contribution.

Application of a magnetic field induces a low-$T$ LFL state below a crossover temperature which increases with the field. Similar observations have been made earlier with other NFL systems, e. g., $CeCu_{6-x}Au_x$ ($x \approx 0.1$) [54] and $CeNi_2Ge_2$ [21]. At $B = 0$, the low-$T$ part of $\gamma(T)$ deviates strongly from the quantum-critical ($-\ln T$) behavior found for $T > 0.3$ K, whereas $\Delta\rho(T)$ remains linear in the whole temperature range. This $\gamma(T)$ upturn might be related to the AF transition which takes place just below the temperature limit of our specific-heat experiment and/or to the unexplained upturns in the *nonmagnetic* HF compounds mentioned before.

The observation of $\Delta\rho(T) \sim T$ and $\gamma(T) \sim (-\ln T)$ in a temperature window much larger than a decade has not been made with any stoichiometric HF metal before. This phenomenology is in apparent contradiction to the asymptotic ($T \to 0$) dependences at an AF-QCP [7, 8, 9, 47] in 3D. However, similar behavior has been found in the alloy system $CeCu_{6-x}Au_x$ ($x \approx 0.1$) [54], where it was ascribed [55] to quasi-2D AF-SF coupled to a 3D system of charge carriers. We speculate that geometrical frustration of the spin-exchange coupling along the $c$ direction, inherent to the tetragonal $ThCr_2Si_2$ structure [53], may lead to a decoupling of adjacent Yb planes and, thus, to a quasi-2D SF spectrum in this Yb compound as well.

## 13.5 OUTLOOK

HF phenomena usually occur in the neighborhood of some (mostly antiferro-) magnetic ordering. Thus it is relatively easy to 'tune' a given HF metal through a 'magnetic instability' either by application of hydrostatic pressure, if one starts at $p = 0$ from a magnetically ordered state like in $CePd_2Si_2$ [19] and $CeIn_3$ [20], or by application of negative ('lattice') pressure, i. e., by alloying with sizeable dopants like in $CeCu_{6-x}Au_x$ [54] or $Ce_{1-x}La_xRh_2Si_2$ [56]. In order to be able to perform a full thermodynamic analysis and to avoid disorder as much as possible it is preferable, however, to investigate such *stoichiometric* HF metals that are accidentally close to a 'magnetic instability' already at *ambient* pressure.

The existence of a magnetic instability gives rise to several interesting observations, namely

(i) complex phase diagrams with several different ground-state properties like in $CeCu_2Si_2$ and $CeNi_2Ge_2$,

(ii) N-state NFL effects,

(iii) HF-SC, often found to be extremely sensitive to disorder [19, 20, 21] suggesting strongly anisotropic superconducting order parameters that are efficiently destroyed by already weak potential scattering.

A few final remarks concerning points (ii) and (iii) are in order. Unlike the high-$T_c$ cuprates, HF superconductors behave rather non-universally: SC can form either out of a LFL state (e.g., $CeRh_2Si_2$ [57]) which may coexist with AF order (e. g., $UPd_2Al_3$ [14]) or, alternatively, out of a NFL state. Among the latter systems some, like $CeCu_2Si_2$ [58] and $UBe_{13}$ [10], may exhibit quasiparticle mean free paths *shorter* than the superconducting coherence length. A major mystery in HF physics is the observation that *Th impurities* in $UBe_{13}$ give rise to a *multiphase diagram* [59] usually ascribed to an *unconventional*, i.e. *strongly anisotropic, superconducting order parameter*! This striking non-universal behavior has led to the proposal [60] that more than just one pairing mechanism may have to be considered in these materials. In the case of $UPd_2Al_3$, the dominating coupling mechanism is, no doubt, of electronic nature. But, different from the assumption made by the SF theory, which attributes the formation of Cooper pairs to *itinerant* AF-SFs with sufficiently high frequencies [61], *local* magnetic excitations appear to be essential in $UPd_2Al_3$.

*Local degrees of freedom* also seem to markedly manifest themselves in the low-$T$ N-state properties of HF metals near a 'magnetic instability'. While the electrical resistivity, probing the 'light (itinerant) component' of the complex quasiparticles, behaves as predicted by SF theory [7, 8, 9], if the amount of impurity scattering [47] and the effective dimensionality of the SF spectrum [62] are properly taken into account, the specific-heat coefficient, probing the 'heavy (local-$f$) component', does not. Obviously, these observations present a challenge that invites further experimental and theoretical investigations.

## ACKNOWLEDGMENTS

We thank P. Thalmeier for numerous conversations and a careful reading of, as well as C. Lang for the expert preparation of this manuscript. We are grateful to W. Assmus for supplying the $CeCu_2Si_2$ single crystal. Work in Dresden was supported by Fonds der Chemischen Industrie, Grant No. 700171, and the ESF Scientific Programme 'Fermi-liquid instabilities in correlated metals' (FERLIN).

## BIBLIOGRAPHY

[1] A. Abrikosov and L.P. Gor'kov, Sov. Phys. JETP **12** 1234 (1961).

[2] E. Müller-Hartmann and J. Zittartz, Phys. Rev. Lett. **26** 428 (1971).

[3] G. Riblet and K. Winzer, Solid State Commun. **9** 1663 (1971). M.B. Maple, Appl. Phys. **9** 179 (1976).

[4] F. Steglich *et al.*, Phys. Rev. Lett. **43** 1892 (1979).

[5] For early experiments, see, e. g., P. H. P. Reinders *et al.*, Phys. Rev. Lett. **57** 1631 (1986). L. Taillefer and G. G. Lonzarich, Phys. Rev. Lett. **60** 1570 (1988).

[6] See, e. g., C. Broholm *et al.*, Phys. Rev. Lett. **58** 1467 (1987). G. Aeppli *et al.*, J. Magn. Magn. Mat. **76 & 77** 385 (1988).

[7] A. J. Millis, Phys. Rev. B **48** 7183 (1993).

[8] G.G. Lonzarich, College on Quantum Phases (ICTP Trieste, 1994, unpublished).

[9] T. Moriya and T. Takimoto, J. Phys. Soc. Jpn. **64** 960 (1995).

[10] H. R. Ott *et al.*, Phys. Rev. Lett. **50** 1585 (1983).

[11] G.R. Stewart *et al.*, Phys. Rev. Lett. **52** 679 (1984).

[12] W. Schlabitz *et al.*, Abstract ICVF, Cologne, August 1984 (unpublished). Z. Phys. B **62** 171 (1986).

[13] C. Geibel *et al.*, Z. Phys. B **83** 305 (1991).

[14] C. Geibel *et al.*, Z. Phys. B **84** 1 (1991).

[15] F. Steglich *et al.*, Z. Phys. B **103** 235 (1997).

[16] P. Gegenwart *et al.*, Phys. Rev. Lett. **81** 1501 (1998).

[17] F. Steglich *et al.*, Physica B **223-224** 1 (1996).

[18] P. Thalmeier, Z. Phys. B **95** 39 (1994).

[19] F. M. Grosche *et al.*, Physica B **223-224** 50 (1996).

[20] N. D. Mathur *et al.*, Nature **394** 39 (1998). For $CeRhIn_5$, a quasi-2D variant of $CeIn_3$, see: H. Hegger, Phys. Rev. Lett. **84** 4986 (2000).

[21] P. Gegenwart *et al.*, Phys. Rev. Lett. **82** 1293 (1999).

[22] F. M. Grosche *et al.*, cond-mat 98 12 133. D. Braithwaite *et al.*, J. Phys.: Condens. Matter **12** 1339 (2000).

[23] F. Steglich *et al.*, Physica B **280** 349 (2000). P. Hinze (unpublished results).

[24] A. Krimmel *et al.*, Z. Phys. B **86** 161 (1992).

[25] F. Steglich *et al.*, in: *Physical Phenomena at High Magnetic Fields-II*, Z. Fisk, L. Gor'kov, D. Meltzer and R. Schrieffer, eds. (World Scientific, Singapore, 1996) p. 125.

[26] T. Takahashi *et al.*, J. Phys. Soc. Jpn. **65** 156 (1996).

[27] A. Yaounc *et al.*, Phys. Rev. B **58** 8793 (1998).

[28] N. Sato *et al.* (submitted for publication).

[29] The magnetic structure consists of ferromagnetically ordered basal planes with the moments aligned in the planes and an AF stacking of adjacent planes with a modulation wave vector $(0, 0, \frac{1}{2})$.

[30] M. Kyougaku *et al.*, J. Phys. Soc. Jpn. **62** 4016 (1993).

[31] N. Sato *et al.*, J. Phys. Soc. Jpn. **61** 32 (1992).

[32] J. Hessert *et al.*, Physica B **230-232** 373 (1997).

[33] C. D. Bredl *et al.*, Z. Phys. B **29** 327 (1978).

[34] C. D. Bredl, J. Magn. Magn. Mat. **63 & 64** 355 (1987).

[35] H. F. Fong *et al.*, Phys. Rev. Lett. **75** 316 (1995).

[36] M. Jourdan *et al.*, Nature **398** 47 (1999).

[37] R. Müller-Reisener, Diploma Thesis, TU Darmstadt 1995 (unpublished). C. Geibel (to be published).

[38] K. Ishida *et al.*, Phys. Rev. Lett. **82** 5353 (1999).

[39] K. Ishida (unpublished results). See also, O. Trovarelli *et al.*, Phys. Rev. B **56** 678 (1997).

[40] C. Geibel, Phys. Bl. **53** 689 (1997). C. Geibel *et al.*, *to be published.*

[41] G. Bruls *et al.*, Phys. Rev. Lett. **72** 1754 (1994).

[42] P. Gegenwart *et al.*, Physica B **259-261** 403 (1999).

[43] M. Lang *et al.*, in *Electron Correlations and Materials Properties*, Gonis *et al.*, eds. (Kluwer Academics/Plenum Publishers, 1999) p. 153.

[44] M. Kato and K. Machida, Phys. Rev. B **37** 1510 (1988).

[45] K. Machida and T. Matsubara, J. Phys. Soc. Jpn. **50** 3231 (1981).

[46] G. Zwicknagl and U. Pulst, Physica B **186-188** 895 (1993).

[47] A. Rosch, Phys. Rev. Lett. **82** 4280 (1999).

[48] G. Sparn *et al.*, Rev. High Pressure Sci. Technol. **7** 431 (1998).

[49] C. Langhammer, Dissertation, TU Dresden 2000 (unpublished).

[50] N. Takeda and M. Ishikawa, Physica B **259-261** 92 (1999).

[51] P. W. Anderson, Adv. Phys. **46** 3 (1997).

[52] P. Coleman and A.M. Tsvelik, Phys. Rev. B **57** 12757 (1998).

[53] O. Trovarelli *et al.* Phys. Rev. Lett. **85** 626 (2000).

[54] H. von Löhneysen, J. Phys. Condens. Matter **8** 9689 (1996).

[55] O. Stockert *et al.*, Phys. Rev. Lett. **80** 5627 (1998).

[56] S. Kambe *et al.*, Physica B **223-224** 135 (1996).

[57] T. Graf *et al.*, Phys. Rev. Lett. **78** 3769 (1997).

[58] U. Rauchschwalbe *et al.*, Phys. Rev. Lett. **49** 1448 (1982).

[59] H. R. Ott *et al.*, Phys. Rev. B **31** 1651 (1985).

[60] N. Grewe and F. Steglich, in: *Handbook on the Physics and Chemistry of Rare Earths*, Vol. 14, K.A. Gschneidner Jr. and L. Eyring, eds. (North Holland, Amsterdam, 1991) p. 343.

[61] A. J. Millis *et al.*, Phys. Rev. B **37** 4975 (1988).

[62] A. Rosch *et al.*, Phys. Rev. Lett. **79** 159 (1997).

# CHAPTER 14

# THE MOTT TRANSITION

GABRIEL KOTLIAR
Serin Physics Laboratory
Rutgers University, Piscataway, NJ 08855-0849

ABSTRACT

We review some aspects of the Mott transition from the point of view of dynamical mean-field theory, a generalization of the dynamical mean-field construction from classical to quantum systems. Comparison with experiments suggests that the approach captures some of the essential physics of three-dimensional transition-metal compounds. We conclude with directions for further extensions.

## 14.1 INTRODUCTION

The Mott transition [1], i.e. the interaction-driven metal-to-insulator transition, has many experimental realizations starting with the famous $V_2O_3$ system. Recently these systems have been studied with much higher resolution, and new compounds displaying similar physics have been synthesized [2].

From a theoretical point of view, this is a fundamental problem in the theory of strongly correlated electron systems that requires a coherent framework to describe the dual character of the electron: wave and particle, itinerant and localized. The two limits of well-localized and fully-itinerant are fairly well understood. In the weakly-correlated limit, a wave-like description of the electron is appropriate. Fermi-liquid theory explains why, at low energies, systems such as the alkali metals behave as weakly interacting fermions. Their transport properties are well described by Boltzmann theory applied to long-lived quasiparticles, an approach that works well as long as $k_F l \gg 1$ ($k_F$ is the Fermi wavevector and $l$ the mean-free-path).

In insulators, far away from the Mott transition, the electron is well-described in real space. A solid is viewed as a regular array of atoms that bind an integer number of electrons. Transport occurs via activation, with the creation of vacancies and doubly-occupied sites. Atomic-physics calculations, together with perturbation theory around the atomic limit, allows us to derive accurate spin Hamiltonians. The spectrum is composed of atomic excitations that are broadened to form states that have no single-particle character. These are known as Hubbard bands.

These two limits, well-separated atoms and strongly-overlapping bands, are easily described in real space and **k**-space, respectively. The strong correlation problem is the description of the electronic structure of solids away from these well-understood limits. The challenge is to develop new concepts and new computational methods, capable of describing situations where both itinerancy and localization are simultaneously important. The Mott transition problem forces us to confront these issues head-on.

Strongly correlated electron systems have anomalous properties resulting from the proximity to a localization-delocalization boundary. We mention, among other things, resistivities that far exceed the Ioffe-Regel limit ${\rho_{Mott}}^{-1} \approx e^2/hk_F$, non-Drude like optical conductivities, and spectral functions that are not well-described by band theory [2]. To address this anomalous behavior, one needs a technique capable of treating quasiparticle bands and Hubbard bands on the same footing, as well as providing a continuous interpolation between the atomic and band limits.

A recently developed approach to strongly correlated electron systems, the Dynamical Mean-Field Theory (DMFT) in the local-impurity self-consistent approximation, satisfies these requirements, and has advanced our understanding of the Mott transition problem [3]. It is a great pleasure to present a contribution in the area of correlated electron systems at a conference in honor of P. W. Anderson. Phil introduced me to this problem when I was a graduate student. His early work permeates this entire subject, and has been very influential in the developments I will present.

## 14.2 MODEL HAMILTONIAN

A minimal Hamiltonian to describe transition-metal oxides was introduced by Anderson in a classic paper. The electron moves via hopping integrals $t_{ij}$ between localized states. The localizing influences are described by a matrix of on-site Coulomb interactions [4]. If the model is simplified further by ignoring the orbital degeneracy, one obtains the Hamiltonian of the one-band Hubbard model, viz.

$$H = -\sum_{\langle ij\rangle,\sigma}(t_{ij}+\mu\delta_{ij})(c^{\dagger}_{i\sigma}c_{j\sigma}+c^{\dagger}_{j\sigma}c_{i\sigma}) + U\sum_{i} n_{i\uparrow}n_{i\downarrow}, \tag{14.1}$$

which plays the role of the "Ising model" of strongly-correlated electrons. It is the simplest model describing the competition between localization and itinerancy.

Here, $U$ is the Coulomb energy of two electrons occupying the same site, $c^\dagger_{i\sigma}$, $c_{j\sigma}$ are creation and annihilation operators for an electron on a site $i$, and $n_{i\sigma} = c^\dagger_{i\sigma}c_{j\sigma}$ is the density of electrons at site $i$ with spin $\sigma =\uparrow,\downarrow$ [5].

The essential parameters in the model are the doping $\delta$ (or the chemical potential $\mu$), the temperature $T$, and the ratio $U/t$. It turns out that the nature of the Mott transition depends on the degree of magnetic frustration in the insulating phase. The degree of magnetic frustration is thus another important parameter that is controlled by the lattice structure and the hopping integrals. Frustrated models are constructed, for example, by assigning comparable values to $t$ and $t'$ (the nearest-neighbor and next-nearest-neighbor hopping amplitudes, respectively).

The solution of the Hamiltonian in Eq. 14.1 (for dimensions $d = 2$ and 3) is a central problem in condensed matter physics that has resisted exact solution [6]. The opposite limit of infinite lattice dimensionality ($d \to \infty$) was formulated by Metzner and Vollhardt [7]. They pointed out that this limit is well-defined if the hopping matrix elements are scaled as $t_{ij} = (\frac{1}{\sqrt{d}})^{|i-j|}$ in the limit $d \to \infty$, where $|i-j|$ is the minimum number of links one has to traverse in going from sites $i$ to $j$. In this limit, the kinetic energy and the potential energy per site are of the same order. This retains the main effects of the competition between the two terms.

Müller-Hartmann pointed out an essential simplification of the limit of large lattice coördination. While the Greens function depends on the wave vector **k** via the Fourier transform of the hopping matrix elements $\epsilon_k$, the self-energy becomes **k**-independent [8], viz.

$$G(k, i\omega_n) = \frac{1}{i\omega_n + \mu - \epsilon_k - \Sigma(i\omega_n)}. \tag{14.2}$$

The one-particle spectral function of the Hubbard in the limit of large lattice coördination then requires the computation of a function of a single variable, namely $\Sigma(i\omega_n)$.

The limit of large lattice coördination was used to simplify several approximation schemes, such as perturbative expansions in the skeleton series [9], and variational studies of the Gutzwiller type [10]. Brandt and Mielsch [11] used it to solve the Falicov-Kimball model, a simplified version of the Hubbard model.

A different direction came about with the introduction of dynamical mean-field ideas, which provided a clear link between the Hubbard model (a model defined on a lattice) and the Anderson impurity model (AIM), a quantum-impurity problem introduced by P. W. Anderson nearly twenty years ago to describe magnetic moments in a metal [12]. In Section 14.3 I present a pedagogical discussion of this connection done in collaboration with A. Georges [13]. For related work, see Refs. [14]. Consequences of this insight into strongly correlated electron systems are described in the following sections. Anderson's early work on the formation of local moments greatly facilitated the direct solution of the Hubbard model in the limit of large lattice coördination, which in turn gave rise to a deeper understanding of the correlation-induced localization-delocalization transition.

## 14.3 MEAN FIELD THEORY

The goal of a mean-field theory is to approximate a lattice problem with many degrees of freedom by a *single-site effective problem.* The underlying physical idea is that the dynamics at a given site can be thought of as the interaction of the degrees of freedom at this site with an external bath created by all other degrees of freedom on other sites. We illustrate this idea with the Ising model with ferromagnetic couplings $J_{ij} > 0$ between the nearest-neighbor sites of a lattice of coördination $z$, viz.

$$H = -\sum_{(ij)} J_{ij} S_i S_j - h \sum_i S_i. \tag{14.3}$$

The Weiss mean-field theory views each given site (say, $o$) as governed by the effective Hamiltonian

$$H_{eff} = -h_{eff} S_o. \tag{14.4}$$

Formally, the effective Hamiltonian $H_{eff}$ is obtained by performing a partial trace over all the spin variables except for the spin at site $o$ in the partition factor

$$\frac{\int_{i \neq o} dS_i e^{-\beta H}}{\int_i dS_i e^{-\beta H}} \equiv e^{-H_{eff}[S_o]}. \tag{14.5}$$

All interactions with the other degrees of freedom are contained in the effective field $h_{eff}$ given by

$$h_{eff} = h + \sum_i J_{oi} m_i = h + zJm, \tag{14.6}$$

where $m_i = \langle S_i \rangle$ is the magnetization at site $i$, and translation invariance has been assumed ($J_{ij} = J$ for nearest-neighbor sites, $m_i = m$). Hence, $h_{eff}$ has been related to a local quantity that,in turn, may be computed from the single-site effective model $H_{eff}$. For the simple case at hand, this reads:

$$m = \tanh(\beta h_{eff}), \tag{14.7}$$

which may be combined with Eq. 14.6 to yield the well-known mean-field equation for the magnetization, viz.

$$m = \tanh(\beta h + z\beta J m). \tag{14.8}$$

These mean-field equations are, in general, an approximation to the true solution of the Ising model. They become *exact* in the limit where the *coördination of* *the lattice becomes large.* It is quite intuitive indeed that the neighbors of a given site can be treated globally as an external bath when their number becomes large, and that the spatial fluctuations of the local field become negligible. As is clear from Eq. 14.6, the coupling $J$ must be scaled as $J = J^*/z$ to yield a sensible limit $z \to \infty$ (this scaling ensures that both the entropy and the internal energy per site remain finite, so as to maintain a finite $T_c$).

These ideas can be directly extended to quantum many-body systems, and will be illustrated with the model Hamiltonian in Eq. 14.1 [15, 3]. We assume that no symmetry breaking occurs, i.e. that one deals with the translation-invariant paramagnetic phase. The generalization to phases with broken symmetry is straightforward [3].

A mean-field description associates with the *lattice* Hamiltonian (Eq. 14.1) a single-site effective dynamics, which is conveniently described in terms of an imaginary-time action for the fermionic degrees of freedom $(c_{o\sigma}, c^{\dagger}_{o\sigma})$ at that site:

$$
\begin{aligned}
S_{eff} \quad &= \quad -\int_0^{\beta} d\tau \int_0^{\beta} d\tau' \sum_{\sigma} c^{\dagger}_{o\sigma}(\tau)\mathcal{G}_0^{-1}(\tau-\tau')c_{o\sigma}(\tau') \\
&\quad + U\int_0^{\beta} d\tau\, n_{o\uparrow}(\tau)n_{o\downarrow}(\tau). \qquad (14.9)
\end{aligned}
$$

Formally, we imagine integrating out all the degrees of freedom except for those living at that specific site in a path-integral formulation. This step is the quantum analog of Eq. 14.5. The role of the effective local Hamiltonian $H_{eff}$ is played by an effective local action $S_{eff}$.

The effect on the selected site of the other sites that have been integrated out is contained in $\mathcal{G}_0(\tau - \tau')$, which plays the role of the Weiss effective field. Its physical content is that of an effective amplitude for a fermion created on the isolated site at time $\tau$ (coming from the "external bath") and destroyed at time $\tau'$ (going back to the bath). The main difference with the classical case is that this generalized "Weiss function" is a *function of time* instead of a single number. This, of course, is required to take into account *local quantum fluctuations*. Indeed, the mean-field theory presented here freezes spatial fluctuations but takes full account of local temporal fluctuations (hence the name 'dynamical'). $\mathcal{G}_0$ plays the role of a bare Green's function for the effective action $S_{eff}$, but it should not be confused with the non-interacting local Green's function of the original lattice model.

Notice that knowledge of $S_{eff}$ allows us to calculate *all the local* correlation functions of the original Hubbard model.

A closed set of mean-field equations is obtained by supplementing Eq. 14.9 with the expression relating $\mathcal{G}_0$ to local quantities computable from $S_{eff}$ itself, in complete analogy with Eq. 14.6 above. It reads:

$$
\mathcal{G}_0^{-1}(i\omega_n) = i\omega_n + \mu + G^{-1}(i\omega_n) - R[G(i\omega_n)]. \qquad (14.10)
$$

In this expression, $G(i\omega_n)$ denotes the on-site interacting Green's function calculated from the effective action $S_{eff}$:

$$G(\tau - \tau') \equiv -\langle Tc(\tau)c^\dagger(\tau')\rangle_{S_{eff}}, \tag{14.11}$$

$$G(i\omega_n) = \int_0^\beta d\tau G(\tau) e^{i\omega_n\tau}, \qquad \omega_n \equiv \frac{(2n+1)\pi}{\beta}, \tag{14.12}$$

and $R(G)$ is the reciprocal function of the Hilbert transform of the density-of-states corresponding to the lattice at hand. Explicitly, given the non-interacting density-of-states (DOS) defined by

$$D(\epsilon) = \sum_k \delta(\epsilon - \epsilon_k), \text{ with } \qquad \epsilon_k \equiv \sum_{ij} t_{ij} e^{ik.(R_i - R_j)},$$

the Hilbert transform $\tilde{D}(\zeta)$ and its reciprocal function $R(G)$ are defined by

$$\tilde{D}(\zeta) \equiv \int_{-\infty}^{+\infty} d\epsilon \frac{D(\epsilon)}{\zeta - \epsilon}, \qquad R[\tilde{D}(\zeta)] = \zeta. \tag{14.13}$$

For a rigorous derivation of these equations, the reader is referred to Refs. [15, 3]. Since $G$ may in principle be computed as a functional of $\mathcal{G}_0$ using the impurity action $S_{eff}$, Eqs. 14.9,14.10, and 14.11 form a closed system of functional equations for the on-site Green's function $G$ (analogous to the magnetization) and the Weiss function $\mathcal{G}_0$ which contains the effects of the surrounding medium on the selected site. It is instructive to check these equations in two simple limits [13]:

- In the *non-interacting limit* $U = 0$, solving Eq. 14.9 yields $G(i\omega_n) = \mathcal{G}_0(i\omega_n)$ and hence from Eq. 14.10, $G(i\omega_n) = \tilde{D}(i\omega_n + \mu)$ reduces to the free on-site Green's function.

- In the *atomic limit* $t_{ij} = 0$, one just has a collection of disconnected sites and $D(\epsilon)$ becomes a $\delta$-function, with $\tilde{D}(\zeta) = 1/\zeta$. Then, Eq. 14.10 implies $\mathcal{G}_0(i\omega_n)^{-1} = i\omega_n + \mu$, and the effective action $S_{eff}$ becomes essentially local in time and describes a four-state Hamiltonian which yields: $G(i\omega_n)_{at} = (1 - n/2)/(i\omega_n + \mu) + n/2(i\omega_n + \mu - U)$, with $n/2 = (e^{\beta\mu} + e^{\beta(2\mu - U)})/(1 + 2e^{\beta\mu} + e^{\beta(2\mu-U)})$.

Solving the coupled equations above not only yields *local quantities* but also allows us to reconstruct all the **k**-dependent correlation functions of the original lattice Hubbard model. For this purpose one needs to calculate the self-energy from

$$\Sigma(i\omega_n) = \mathcal{G}_0^{-1}(i\omega_n) - G^{-1}(i\omega_n), \tag{14.14}$$

and substitute it in Eq. 14.2.

The dynamical mean-field equations can be studied on many different lattices. The lattice structure enters the single-particle properties only via the DOS. There are many useful examples [3], but in the following we will focus on the Bethe lattice (Cayley tree) with coördination $z \to \infty$ and nearest-neighbor hopping $t_{ij} = t/\sqrt{z}$. In this case, the DOS is semicircular, viz.

$$D(\epsilon) = \frac{1}{2\pi t^2}\sqrt{4t^2 - \epsilon^2}, \qquad |\epsilon| < 2t. \tag{14.15}$$

The Hilbert transform and its reciprocal function take very simple forms, viz.

$$\tilde{D}(\zeta) = (\zeta - s\sqrt{\zeta^2 - 4t^2})/2t^2, \qquad R(G) = t^2 G + 1/G, \tag{14.16}$$

so that the self-consistency relation between the Weiss function and the local Green's function has the explicit form

$$\mathcal{G}_0^{-1}(i\omega_n) = i\omega_n + \mu - t^2 G(i\omega_n). \tag{14.17}$$

The same DOS is also realized for a random Hubbard model on a fully-connected lattice (all $N$ sites pairwise connected) where the hoppings are independent random variables with variance $\overline{t_{ij}^2} = t^2/N$.

The structure of the dynamical mean-field theory is that of a functional equation for the local Green's function $G(i\omega_n)$ and the 'Weiss function' $\mathcal{G}_0(i\omega_n)$. In contrast to the classical case, the on-site effective action $S_{eff}$ remains a *many-body problem.* This is because the present approach freezes *spatial fluctuations* but fully retains *local quantum fluctuations.* As a function of imaginary time, each site undergoes transitions between the four possible quantum states

$$|0\rangle,\ |\uparrow\rangle,\ |\downarrow\rangle,\ |\uparrow,\downarrow\rangle$$

by exchanging electrons with the rest of the lattice described as an external bath. The dynamics of these processes is encoded in the Weiss function $\mathcal{G}_0(\tau - \tau')$.

For these reasons, no Hamiltonian form involving *only* the on-site degrees of freedom $(c_{o\sigma}, c_{o\sigma}^\dagger)$ can be found for the effective on-site model: Once the bath has been eliminated, $S_{eff}$ necessarily includes retardation effects. In order to gain physical intuition and also to make some practical calculations with $S_{eff}$, it is useful to have a Hamiltonian formulation [13]. This is possible upon reintroducing auxiliary degrees of freedom describing the 'bath'. For example, one can view $(c_{o\sigma}, c_{o\sigma}^\dagger)$ as an 'impurity orbital', and the bath as a 'conduction band' described by operators $(a_{l\sigma}, a_{l\sigma}^\dagger)$, and consider the Hamiltonian

$$H_{AIM} = \sum_{l\sigma} \tilde{\epsilon}_l a_{l\sigma}^\dagger a_{l\sigma} + \sum_{l\sigma} V_l (a_{l\sigma}^\dagger c_{o\sigma} + c_{o\sigma}^\dagger a_{l\sigma}) - \mu \sum_{\sigma} c_{o\sigma}^\dagger c_{o\sigma} + U n_{o\uparrow} n_{o\downarrow}. \tag{14.18}$$

This Hamiltonian is quadratic in $a_{l\sigma}^{\dagger}$, $a_{l\sigma}$: Integrating these out gives rise to an action of the form Eq. 14.9, with

$$\mathcal{G}_0^{-1}(i\omega_n) = i\omega_n + \mu - \int_{-\infty}^{+\infty} d\omega \frac{\Delta(\omega)}{i\omega_n - \omega}, \qquad \Delta(\omega) \equiv \sum_{l\sigma} V_l^2 \delta(\omega - \widetilde{\epsilon}_l). \tag{14.19}$$

Hence, Eq. 14.18 can be viewed as a Hamiltonian representation of $S_{eff}$, provided the $\Delta(\omega)$ (i.e the parameters $V_l$, $\widetilde{\epsilon}_l$) are chosen to reproduce the actual solution $\mathcal{G}_0$ of the mean-field equations. The spectral representation Eq. 14.19 is general enough to permit this in all cases. Note that the $\widetilde{\epsilon}_l$'s are *effective* parameters that should not be confused with the single-particle energies $\epsilon_k$ of the original lattice model. The Hamiltonian Eq. 14.18 is the Anderson model of a magnetic impurity coupled to a conduction bath [12]. Note, however, that the shape of the hybridization function $\Delta(\omega)$ is *not known à priori* in the present context, and must be found by solving the self-consistent problem. The isolated site *o* plays the role of the impurity orbital, and the conduction bath is built out of all other sites. Comparing Eq. 14.19 with 14.17 allows us to rewrite the mean-field equations in terms of the hybridization function, now written as a function of Matsubara frequencies

$$\Delta(i\omega_n) = \sum_l \frac{V_l^2}{(i\omega_n - \widetilde{\epsilon}_l)},$$

viz.

$$t^2 G_{imp}(i\omega_n)[\Delta, \alpha] = \Delta(i\omega_n). \tag{14.20}$$

The index $\alpha$ denotes all the parameters of the problem such as $T$, $U$ etc. The quantity $G_{imp}$ is the local Greens function of the Anderson impurity model.

The reduction of a lattice problem to a single-site problem with *effective* parameters is a common feature of both the classical and the quantum mean-field constructions. The main difference is that the Weiss field is a number in the classical case, and a function in the quantum case. Physically, this reflects the existence of *many energy scales* in strongly correlated fermion models. (We note in passing that this also occurs in the mean-field theory of some classical problems with many energy scales such as spin glasses). This points out to the limitations of other 'mean-field' approaches, such as the Hartree-Fock approximation or slave-boson methods, where one attempts to parametrize the whole mean-field function by a single *number* (or a few of them). This, in effect, amounts to freezing local quantum fluctuations by replacing the problem with a purely classical one, and can only be reasonable when a single low-energy scale is important. This is the case, for instance, for a Fermi-liquid phase. However, even in such cases, parametrizing the Weiss field by a single number can only be satisfactory at low energy, and misses the high-energy incoherent features associated with other energy scales in the problem. When no characteristic low-energy scale is present, a single-number parametrization fails completely.

Finally, besides its intuitive appeal, the mapping onto impurity models has proven to be useful for two reasons. First, it allows us to understand qualitatively the physics of the Hubbard model in the limit of large lattice coördination by exploiting the known analytic structure of the Anderson impurity model. The key insight is that the impurity model is generally an analytic function of the interaction $U$. Hence, all the non-analyticities in the problem result from the self-consistency condition Eq. 14.20. Secondly, the Anderson impurity model can be accurately solved with a variety of techniques [16], which, in combination with self-consistency conditions such as 14.20, result in a quantitative solution of many models of lattice correlated electrons [3].

## 14.4 SPECTRAL FUNCTIONS OF STRONGLY CORRELATED STATES

A simple local picture of the various regimes of the Hubbard model follows from the early work of Anderson and Yuval, and Anderson, Yuval and Hamman [17]. Their results are instrumental for generating the qualitative picture of the spectra of the Hubbard model in large dimensions [13]. They propose to express the partition function of the Anderson-impurity model as a sum-over-histories of the local-impurity spin, as is currently done in modern numerical evaluations of the partition function using the Hirsch-Fye algorithm [18]. Then, the partition function is expressed in terms of the positions of the spin-flips, or kinks, in the spin-impurity history.

In the DMFT context, the kinks represent transitions between the degenerate ground states of the paramagnetic phase of the Mott insulator, i.e. *local* spin fluctuations. The partition function is

$$Z = \sum_n \sum_{\tau_1 \dots \tau_n} y_1 \dots y_n e^{-A[\tau_1 \dots \tau_n]}. \tag{14.21}$$

The $y_i$'s in Eq. (14.21) are the Anderson-Yuval fugacities for flipping the spin, and $A$ is the action that governs interactions between the kinks, i.e. the energy of the fermionic degrees of freedom in the presence of a sequence of spin-flips at times $\tau_1 \dots \tau_n$. It results from integrating out the bath electrons that hybridize with the impurity.

When the system is metallic, the bath function has a finite density of states at low energies $[\Delta(i\omega_n) \propto -i \ \mathrm{sign}(\omega_n)]$. The interaction between the spin-flips is logarithmic, and the number of spin-flips proliferates at long times (low energies). The local spin-spin autocorrelation function then decays as $\tau^{-2}$ at long times.

On the other hand, in the Mott insulating state, the self-consistent hybridization function has zero density-of-states at low frequencies $[\Delta(i\omega_n) \propto -i\ \omega_n]$. This gives rise to a confining interaction among the kinks, since now $\mathcal{G}_0$ does not decay. The local impurity spin now has long-range order in time. In this case, the interaction between the Fermions and the spin fluctuations remains finite at low frequencies. A gap opens in the one-particle excitation spectrum.

The mapping onto an Anderson impurity model offers an intuitive picture of both metallic states and Mott insulating states. A correlated metal is described locally as an Anderson impurity model in a metallic bath: The Kondo effect gives rise to strongly renormalized quasiparticles when the interactions are strong, and to a broad band when the interactions are weak. A Mott insulator is locally described as an Anderson impurity model in an insulating bath. The charge degrees of freedom are gapped, but the spin degrees of freedom are not quenched; they dominate the low-energy physics. When there is one electron per site, the Mott transition takes place as one goes from the first regime to the second by increasing the strength of the interaction $U$ [19].

A sketch of the evolution of the spectral function $-\mathrm{Im}\, G(i\Omega + i\delta)$ of the half-filled Hubbard model is displayed in Fig. 14.1. For interactions $U$ close to, but smaller than the critical value denoted by $U_{c2}$, the one-electron spectral function of the Hubbard model in the strongly correlated metallic region contains both atomic features (i.e. Hubbard bands) and quasiparticle features in its spectra [13]. This may be understood intuitively from the Anderson-Yuval-Hamman path-integral representation. In the regime of strong correlations, paths that are nearly constant in imaginary time, as well as those that fluctuate strongly, have substantial weight in the path integral [17]. The former give rise to the Hubbard bands, while the latter are responsible for the low-frequency Kondo resonance. Two features in this spectra are suprpising. First, we have the narrow central-peak before the Mott transition, which results from quasiparticle states formed in the backround of the coherent Kondo tunneling of the local spin fluctuations. Secondly, atomic physics leaves a signature on the one-particle spectral properties in the form of well-formed Hubbard band at higher frequency, in the strongly correlated metallic state.

As the transition at zero temperature is approached, there is a substantial transfer of spectral weight from the low-lying quasiparticles to the Hubbard bands. The Mott transition at zero temperature takes place at the critical value $U_{c2}$ at which the integrated spectral weight at low frequency vanishes, as shown in Fig. 14.1. This results in a Mott transition-point at which the quasiparticle mass diverges, but a discontinuous gap opens in the quasiparticle spectra [20]. These results are in agreement with the early work of Fujimori *et al.* [21], who arrived at essentially the same picture on the basis of experiment. It is worth remarking that the spectral function in the strongly correlated metallic region is better regarded as composed of three components: Hubbard bands centered at $U/2$, a low-energy quasiparticle peak with a height of order unity, and a total intensity proportional to $(U_{c2}-U)$ distributed over an energy range $U_{c2}-U$, and an incoherent background connecting the high-energy and low-energy regions. The last feature ensures that there is no real gap between the quasiparticle features and the Hubbard bands, as long as one stays in the metallic regime. The intensity in the background between the quasiparticles and the Hubbard bands can be estimated to vanish as $(U_{c2} - U)^{1/2}$.

The Anderson-Yuval Coulomb gas was extended to the study of the mixed-valence systems by Haldane [22]. The local configurations contain low-lying charge and spin fluctuations. Surprisingly the mixed-valence regime occurs in a wide region of parameters when the Anderson impurity model describes the local physics of a lattice system [23]. A general, extended Hubbard model [24] was investigated extensively using the Anderson-Yuval Coulomb gas approach. There is a regime of parameters where the charge fluctuations are *incoherent*.

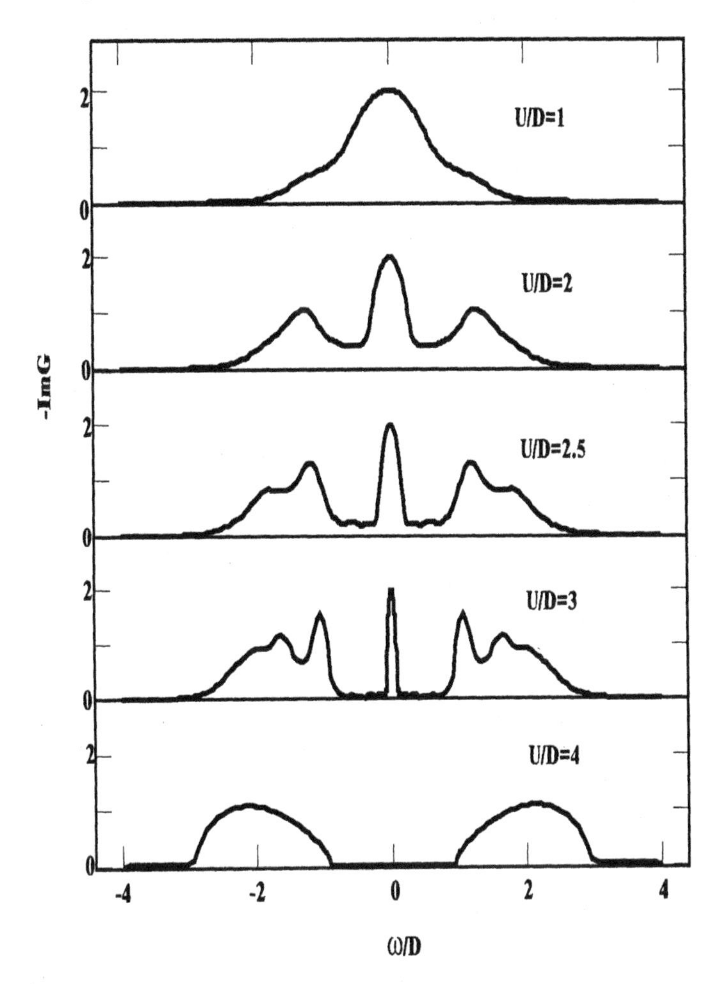


Figure 14.1: Evolution of the spectral function at zero temperature as a function of $U$, from Ref. [20].

In this regime, the physics is similar to that of the Falicov-Kimball [25] model. The physics is described by Fermions that are strongly scattered off quasiclassical, low-energy local modes (these modes cannot tunnel coherently between the degenerate states, but instead flip among them incoherently) [23, 24]. Systems of this kind, Falicov-Kimball liquids, are accurately described by Hubbard III-like approximations, which are now viewed in a new light as accurate solutions of the dynamical mean-field equations in a regime where the Coulomb gas fugacities remain small. The Hubbard III approximation describes very low-lying local-spin fluctuations in a regime where they are not quenched by the fermionic degrees of freedom. Other charge or orbital fluctuations can play the role of the spin fluctuation in more complex (multiorbital) systems.

The coupling of these unquenched collective modes to quasiparticles gives rise to the incoherent regime.

## 14.5 ANOMALOUS RESISTIVITY AND SPECTRAL WEIGHT

The results described in the previous section require a large degree of magnetic frustration to suppress the long-range order and to stabilize a finite-temperature, paramagnetic insulator with finite entropy. In systems without magnetic frustration, the onset of magnetism prevents us from accessing the strongly correlated regime. Indeed, in the absence of frustration, the quasiparticle residue in the antiferromagnetic metal-to-antiferromangetic insulator transition remains finite, while it is vanishingly small in the paramagnetic metal-to-paramagnetic insulator transition [26].

However, in three-dimensional transition metal oxides [27], orbital degeneracy introduces magnetic frustration which weakens the strength of magnetic correlations enough to make the strongly correlated regime observable. This can be understood in very simple physical terms: Exchange among identical orbitals is antiferromagnetic while exchange among orthogonal orbitals has ferromagnetic character. In the absence of orbital ordering, cancellation between these two processes takes place, resulting in reduction of the magnetic interactions between neighboring spins.

The phase diagram of a *partially frustrated* Hubbard model is shown in Fig. 14.2. This dynamical mean-field phase diagram has the same topology as the observed phase diagram of $V_2O_3$ and $NiSe_{2-x}S_x$ [2]. In such cases, the change from the paramagentic insulator (localized ) to the paramagentic metal (extended) regime proceeds as a function of $U/t$, and occurs as a first-order phase transition [28, 29].

There is a remarkable degree of universality in the high temperature part of the phase diagram describing the localization-delocalization phase transtion (below $T_c$), and the localization-delocalization crossover (above $T_c$ ) that follows from the Landau analysis discussed in the next section. This phase diagram is *generic*. While concrete calculations have been performed mostly for the one-band model, the first-order phase transition takes place for arbitrary orbital degeneracy, as long as magnetic order is sufficiently suppressed. The basic physical ingredients for obtaining such a first-order line are on-site repulsion comparable with the bandwidth and magnetic frustration to suppress the onset of magnetic long-range order.

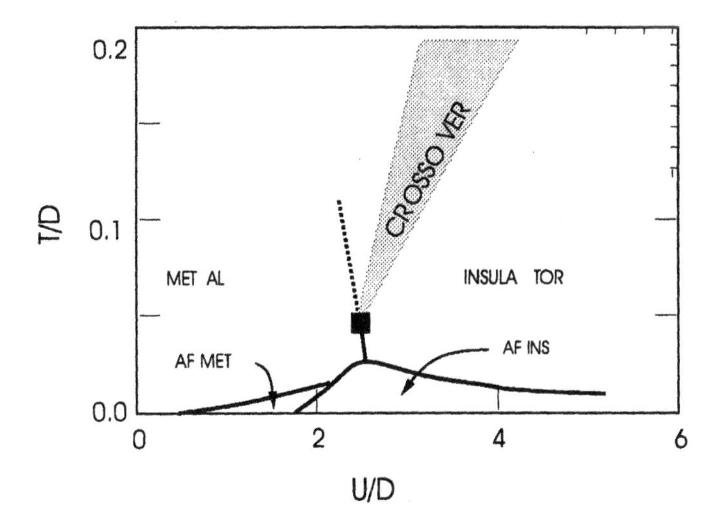


Figure 14.2: Schematic phase diagram of partially frustrated Hubbard model, from Ref. [29]. The square dot denotes $(U_c, T_c)$, the critical endpoint of the first-order line. The dotted line continues the first-order line above the second-order endpoint. It denotes the location of the sudden rise in the resistivity shown in Fig. 14.3.

The phase diagram in Fig. 14.2 displays two crossover lines. The dotted line is a coherence-to-incoherence crossover (i.e. the continuation of the $U_{c2}$ line where metallicity is lost). The shaded area is a continuation of the $U_{c1}$ line, where the temperature becomes comparable with the gap. Figure 14.3 describes the anomalous resistivities near the crossover region. They are qualitatively similar to the behavior of the resisitivities of $V_2O_3$ and the $NiSe_{2-x}S_x$ mixtures in the corresponding region of the phase diagram [30] [31], as well as some organic materials [32]. This should be contrasted with the low temperature ordered phases and the phase transitions into these ordered states which depend on the detailed orbital structure of the material [33][26].

Notice the anomalously large metallic resistivity that is typical of many oxides [34]. While the curves in this figure far exceed the Ioffe-Regel limit (using estimates of $k_F$ from $T = 0$ calculations), there is no violation of any physical principle. At low temperatures, a **k**-space based Fermi-liquid theory description works, but in this regime the resistivity is low (below the Ioffe-Regel limit). Above a certain temperature, the resistivity exceeds the Ioffe-Regel limit, and the quasiparticle description becomes inadequate. There is, however, no breakdown or singularities in our formalism. The spectral functions remain smooth; only the physical picture changes. At high temperatures, we have an incoherent regime in which the Ioffe-Regel criterion does not apply, because there are no long-lived excitations with well-defined crystal momentum in the spectra. The electron is strongly scattered, and is better described in real space. In this regime, there is no simple description in terms of **k**-space elementary excitations, but one can construct a simple description and perform quantitative calculations if one uses a theory based on the spectral function as a fundamental object. The temperature dependence of the transport in the high-temperature, incoherent regime depends on whether the system is at integer-filling or doped, as may be seen from the analysis in Ref. [35].

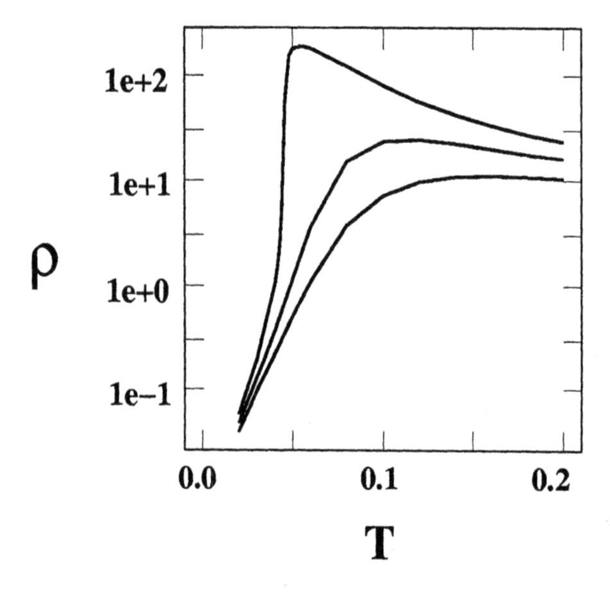


Figure 14.3: Resistivity $\rho_{dc}(T)$ versus $T$ above the critical endpoint of the line of first-order phase transitions in Fig. 14.2. Values of $U/D$ are 2.1, 2.3, 2.5 from bottom to top. From Ref. [29].

Another manifestation of the interplay of itinerancy and localization is the anomalous transfer of spectral weight observed in the one-electron spectrum and in the optical spectra of correlated systems, as the doping concentration or pressure is varied. When one proceeds from a metallic state to an insulating state, one expects spectral weight to be removed from the low-frequency region. What is anomalous and unexpected is that this spectral weight reappears at much higher frequencies (of the order of $U$ or $D$) which is very different from what is observed in weak-coupling superconductors and spin-density-wave systems. In weak-coupling

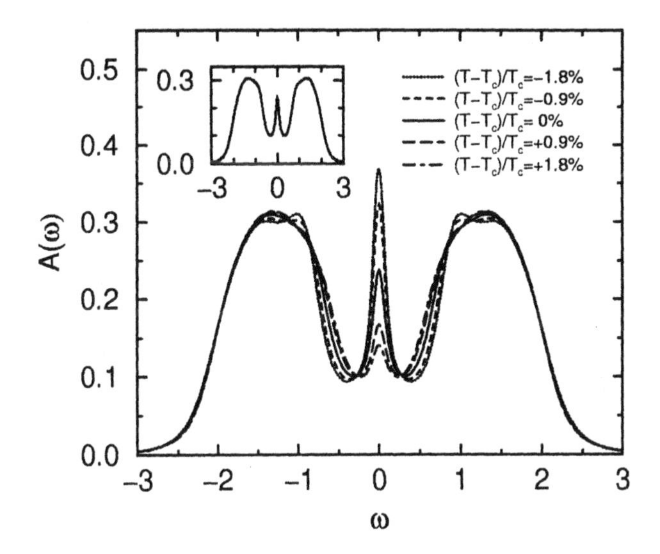


Figure 14.4: Evolution of the spectral function as function of temperature (bottom to top), in the vicinity of the critical temperature $T_c$ (Ref. [36]). The inset shows the spectral function at the critical point.

situations, when a small gap opens in the single-particle spectra, the lost spectral density reappears in the immediate vicinity of the gap.

This surprising aspect of strong correlation physics was noted and emphasized by many authors [37]. Transfer of spectral weight can also take place as a function of temperature. For example, the 'kinetic energy' that appears in the low-energy optical sum-rule can have sizeable temperature dependence, an effect that was discovered experimentally [38], and explained theoretically by DMFT calculations [39].

Once again, thinking about this problem in terms of well-defined quasiparticles is not useful in the finite-temperature, strongly-correlated region. It is more fruitful to formulate the problem in terms of spectral functions describing, on the same footing, coherent and incoherent excitations.

The relative weights of the coherent and incoherent components of the spectra evolve rapidly with temperature, and lead to sizeable variations in the integrated optical intensity. The rapid variation of the low-energy photoemission and optical spectral weight predicted by the theory has already been seen in $V_2O_3$ and in the system $NiSe_{2-x}S_x$ [29, 31, 41, 42]. These effects are most pronounced near the Mott transition endpoint, as shown in Fig. 14.5. Finally, the fundamental role of the spectral function becomes even more prominent in the Landau-theory approach to the Mott transition discussed in the next section, in which

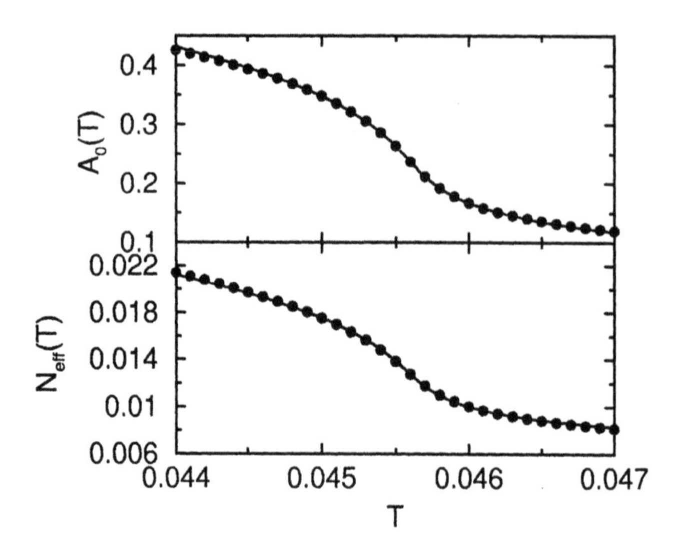


Figure 14.5: Integrated optical spectral weight $N_{eff}$, and quasiparticle weight, $A_0$, as a function of temperature for $U$ near $U_c$.

the Greens function is allowed to fluctuate away from its physical saddle-point value to explore non-perturbative states that may not be accessed in a perturbation calculation in the interaction strength.

## 14.6 THE MOTT TRANSITION AS A BIFURCATION

A new perspective on the Mott transition is obtained by viewing the dynamical mean-field equations as a saddle-point of a Landau Ginzburg-like functional [40], viz.

$$F_{LG}[\Delta] = -T\sum_{\omega} \frac{\Delta(i\omega)^2}{t^2} + F_{imp}[\Delta], \tag{14.22}$$

where

$$-\beta F_{imp} = \int df^\dagger df e^{-L_{loc}[f^\dagger,f]-\sum_{\omega,\sigma} f^\dagger{}_\sigma(i\omega)\Delta(i\omega)f_\sigma(i\omega)}, \tag{14.23}$$

and

$$L_{loc}[f^\dagger, f] = \int_0^\beta f^\dagger[\frac{d}{d\tau} + \varepsilon_f]f + U f^\dagger{}_\uparrow f_\uparrow f^\dagger{}_\downarrow f_\downarrow.$$

This functional may be understood by analogy with the Hubbard-Stratonovich construction of the free energy of the Ising model which (in the absence of a magnetic field) is given by

$$\beta F_{LG}[h] = \beta \frac{h^2}{2Jz} - \log[\coth[\beta h]]. \tag{14.24}$$

The first term in Eqs. 14.22 and 14.24 represents the energy cost of forming the Weiss field, while the second term is the energy gain of the local degree of freedom in the presence of the Weiss field (spin or electron in the classical and quantum cases, respectively). Differentiating Eq. 14.24 gives the mean-field theory described by Eqs. 14.6 and 14.7.

A critical insight in this problem is the fact that the Mott transition point appears as a bifurcation point in the Landau functional Eq. 14.22. Even in infinite dimensions, we already know of several different classes of bifurcations, depending on whether we are at finite or zero temperature, and whether the transition takes place in the presence of magnetic long-range order. Zero-temperature bifurcation points are discussed in Ref. [40]. The bifurcation that takes place at the finite-temperature endpoint $(U_c, T_c)$ of the line of first-order phase transitions (Fig. 14.2) was understood only very recently [36]. It is a simple Ising-like bifurcation, characterized by the fact that the fluctuation matrix

$$M_{nm} = \frac{t^2}{2T} \left. \frac{\delta^2 F_{\text{LG}}[\Delta]}{\delta\Delta(i\omega_n)\delta\Delta(i\omega_m)} \right|_{\text{critical point}} \tag{14.25}$$

acquires a single zero-energy mode $\phi_0\ (i\omega_n)$. The form of the critical theory is derived by expanding around the critical point $\alpha_c = (U_c, T_c)$ up to third order in the deviation of the hybridization function from its value at the critical point, $\delta\Delta = \Delta(\alpha_c + \delta\alpha) - \Delta(\alpha_c)$, and to first order in $\delta\alpha = (U - U_c, T - T_c)$. This expansion is well-behaved because the impurity model depends smoothly on $\alpha$ and $\Delta(i\omega_n)$. The singular part of the component of the spectral function along the zero-energy mode is denoted by $\eta$, and plays the role of the order parameter of the finite temperature Mott transition endpoint. It obeys the equation [36]

$$c_1\eta + c_2\eta^3 = h. \tag{14.26}$$

Here, $c_1$ and $c_2$ are smooth functions of $U$ and $T$ that model pressure and temperature in real physical systems. The coefficient $c_1$ goes to zero linearly at the endpoint of the first-order Mott transition line, while $c_2$ is non-vanishing.

Since $\delta\Delta \sim \eta\phi_0(i\omega_n)$, the quantity $\eta$ may be regarded as the *singular* part of the spectral function. It is observable in photoemission and inverse photoemission experiments. Almost all experimental probes, optical conductivity, transport coefficients etc., couple directly to the order parameter. Therefore, all these physical quantities should have an Ising-like singularity, which is generically driven by a magnetic field-like Ising variable with $T - T_c$ or $p - p_c$ playing the role of magnetic field in the Ising model (here, $p$ is pressure). There is also a special combination of pressure and temperature that plays the role of an Ising-model temperature-like variable. Therefore, we expect that, near criticality, $\eta \approx (\delta\alpha)^{\frac{1}{\delta}}$. In mean-field, $\delta = 3$, whereas $\delta = 5$, 15 in three and two dimensions, respectively. However, the critical region of the finite-temperature Mott transition might be fairly small if, as in BCS theory, the coherence length for the fluctuation quantity $\eta$ is inversely proportional to a power of $T_c/\epsilon_F$. The predictions of an Ising-like transition should be tested against careful experiments near the critical region of the compounds $V_2O_3$ and $NiSe_{2-x}S_x$, or in other systems exhibiting a finite-temperature Mott endpoint to refine our understanding of the anomalous transfer of spectral weight discussed in the previous section.

The analysis of Ref. [36] reconciles two very different viewpoints of the Mott transition. Ideas put forward by Castellani *et al.* [43], in which the Mott transition appears as a condensation of doubly-occupied sites, are tied to a dual picture in which the Mott transition is associated with a rapid variation of the low-energy part of the one-electron spectral function, as described in slave-boson, mean-field-theories [44].

An important quantity that generically couples to the order parameter $\eta$ is the density $\delta n \approx a\eta \approx a\delta\mu^{\frac{1}{\delta}}$. This dependence leads to a compressibility that diverges as

$$\frac{\delta n}{\delta \mu} = \delta\mu^{1-\frac{1}{\delta}} \approx \frac{1}{\delta n^{\delta-1}}. \tag{14.27}$$

Notice, however, that in particle-hole symmetric situations, the coefficient $a$ is zero. The divergence in the compressibility can be observed at the second-order endpoint of the chemical potential-driven Mott transition. It can also be observed at the endpoints of the interaction-driven Mott transition for integer-fillings which do not correspond to half-filled shells. The compressibility is not singular in interaction-driven Mott transition in the one-band, half-filled Hubbard model because of its perfect particle-hole symmetry.

The Landau functional in Eq. 14.22, viewed as a functional of the Weiss field or as a functional of the local Greens function, has several interesting properties. Its stationary point gives the physical Greens function, and consideration of its bifurcations give interesting insights into the corresponding Mott transitions.

Generalizations of this construction to finite dimensions are highly desirable, and some progress was recently made by R. Chitra and the author [45]. A simple consequence of viewing the Mott transition as a smooth bifurcation is the connection between the Mott transition, viewed as a bifurcation in the one-electron Greens function (a one particle property), and the divergence of the compressibility (a two-particle property). This is interesting in light of the important work of Furukawa and Imada [46] who discovered that the approach to the Mott transition in two dimensions, as a function of doping, is characterized by a *divergence* of the charge susceptibility. Reference [45] proceeds by constructing a functional $\Gamma$ of the one-electron Greens function that is stationary at the correct Greens function.

The Mott transition is a bifurcation of the stationary solutions of $\delta\Gamma[G,\alpha]/\delta G = 0$, as a control parameter $\alpha$ is varied ($\alpha$ may represent $\mu$, $T$ or $U$). For a smooth bifurcation to take place at a critical value of the control parameter $\alpha_c$, the matrix

$$\chi = \frac{\delta^2\Gamma}{\delta G(x,y)\delta G(u,v)}$$

has to be non-invertible when evaluated at $\alpha_c$. Otherwise, the equation

$$\delta G = -\left(\frac{\delta^2\Gamma}{\delta G\,\delta G}\right)^{-1}_{\alpha_c}\left(\frac{\delta^2\Gamma}{\delta G\delta\alpha}\right)_{\alpha_c}\delta\alpha \tag{14.28}$$

would have a unique solution for $\delta\alpha+\alpha_c$, contradicting the starting assumption of a bifurcation point at $\alpha_c$. The smoothness assumption is crucial for this argument. In infinite dimensions at zero temperature, the necessary smoothness conditions are not satisfied. As a result, the compressibility vanishes at $U_{c2}$ and remains finite when the Mott transition is approached as a function of chemical potential at $\mu_{c2}$ [47]. The results in Ref. [46] and the previous arguments support the conjecture of Ref. [40] that the zero-temperature Mott transition in finite dimensions is smoother than its infinite-dimensional counterpart at $T = 0$ and $U = U_{c2}$, perhaps closer in spirit to what was found at bifurcation point at the top of the coexistence region of the DMFT phase diagram (Fig. 14.2) i.e. at $(U_c, T_c)$.

The finite-temperature Mott transition is really a transition between a 'good' and a 'bad' metal. The Landau analysis, which is based on an order-parameter description, is applicable to other localization-delocalization electronic transitions, such as the $\alpha$-to-$\gamma$ transition observed in the lanthanide series, and related transitions observed in some actinides [48]. To describe Ce, we adopt the Anderson lattice model, which introduces the hybridization of the $f$ electrons with a broad $s-d$ band. Within the Anderson lattice Hamiltonian, one may prove that the first-order phase-transition line shown in Fig. 14.2 persists if the hybridization term is sufficiently small. This is a direct consequence of the smoothness of the Landau functional at finite $T$.

The Landau analysis demonstrates that the introduction of particle-hole asymmetry is a very relevant perturbation. Hence, it must be introduced as an essential ingredient in a model of $f$ electrons such as Ce. Quantitative estimates of the phase diagram and the physical properties of the $\alpha$-to-$\gamma$ transition [49] requires a combination of realistic DMFT (as described in Ref. [50]) and explicit coupling of the DMFT energy to the lattice displacements (as in Ref. [51]). Nonetheless, we can already make exact statements about the critical behavior based on the Landau theory approach. For example, we predict an Ising-like divergence in the specific heat and in the bulk modulus (compressibility) of Ce at the finite-temperature endpoint of the $\alpha$-to-$\gamma$ transition line.

## 14.7 EXTENSIONS OF DYNAMICAL MEAN FIELD METHODS

The dynamical mean-field approach may be extended in several directions with a view towards realistic applications. To account for spin-spin and charge-charge interactions that are non-local, an extended dynamical mean-field approach (EDMFT) can be used [52, 53]. An imaginary time formulation starts from the action

$$\begin{aligned} \mathcal{S} &= \sum_{ij} \int \int d\tau dv \left[ c^{\dagger}_{i\sigma}(\tau) F^{-1}(\tau i, v j) c_{\sigma j}(v) \right. \\ &\quad \left. + \phi_{i\sigma}(\tau) B_{\sigma}{}^{-1}(\tau i, v j) \phi_{j\sigma}(v) \right] \\ &\quad + \sum_{i} \int d\tau \left[ U n_{i\uparrow} n_{i\downarrow} + \phi_{i\sigma}(\tau) n_{\sigma}(\tau) \right]. \end{aligned} \tag{14.29}$$

This action describes a Fermion system interacting with a Bose field. The bare Bose and Fermi propagators are represented by $B_{\sigma}{}^{-1}(\tau i, v j)$ and $F^{-1}(\tau i, v j)$, respectively. The case of long-range Coulomb interaction, corresponding to the addition of a term $\sum_{i \neq j, \sigma\sigma'} V_{ij} : n_{i\sigma} n_{j\sigma'} :$, was considered in Ref. [54] [this case corresponds to $B_{\sigma}(q, i\omega_n) = V(q)$ in Eq. 14.29]. It produces qualitatively new effects that drive the Mott transition first-order even at zero temperature, where the conventional DMFT predicts a second-order transition [55]. DMFT treats both electrons and the collective excitations (spin and charge fluctuations) on equal footing. The fermion (boson) propagators $G$ and $\Pi_{\sigma}$ are expressed in terms of the self-energies $\Sigma(i\omega_n)$, and $\tilde{\Pi}_{\sigma}(i\omega_n)$, which are assumed to be momentum independent, viz.

$$\begin{aligned} G^{-1}(i\omega_n, q) &= i\omega_n - \epsilon_q - \Sigma(i\omega_n), \\ &\quad -\Pi_{\sigma}(q, i\omega_n) = \tilde{\Pi}_{\sigma} - B_{\sigma}^{-1}(q, i\omega_n). \end{aligned}$$

The self-energies $\tilde{\Pi}$ and $\Sigma$, as well as other local quantities, are computed from the local action

$$\begin{aligned} S_{loc} &= \int d\tau d\tau' \sum_{\sigma} c^{\dagger}_{\sigma}(\tau) \mathcal{G}_{0\sigma}^{-1}(\tau - \tau') c_{\sigma}(\tau') + \frac{U}{\beta} n_{\uparrow}(\tau) n_{\downarrow}(\tau) \delta(\tau - \tau') \\ &\quad - \sum_{\sigma} \phi_{\sigma}(\tau) \Pi_{0\sigma}^{-1}(\tau - \tau') \phi_{\sigma}(\tau'). \end{aligned} \tag{14.30}$$

The parameters of the local action are determined by solving the EDMFT self-consistency conditions which in the spinless case read:

$$\begin{aligned} \Pi_0^{-1}(i\omega_n) &= \left[ \sum_{q} \frac{1}{-\tilde{\Pi}^{-1}\{\Pi_0, \mathcal{G}_0\}(i\omega_n) + B(q, i\omega_n)} \right]^{-1} \\ &\quad + \tilde{\Pi}^{-1}\{\Pi_0, \mathcal{G}_0\}(i\omega_n), \end{aligned} \tag{14.31}$$

and

$$\mathcal{G}_0^{-1}(i\omega_n) = \left[\sum_q \frac{1}{i\omega_n - \epsilon_q - \Sigma\{\Pi_0, \mathcal{G}_0\}}\right]^{-1} + \Sigma\{\Pi_0, \mathcal{G}_0\}(i\omega_n). \quad (14.32)$$

A full solution of extended dynamical mean-field equations [53] for a spinless electron-phonon system, using the Quantum Monte-Carlo method, was obtained recently [56].

A second generalization is needed to deal with the electronic structure problem which starts with the atomic positions and the unit cell, but no *à priori* model Hamiltonian. It has been noted, by analogy with the Legendre transform construction of Density Functional Theory [57], that one may construct formally a functional of the local one-electron Greens function, which is stationary at the physical local spectral function. This construction was carried out to all orders in the interaction and was given a concrete diagrammatic interpretation [58], but is not unique since there are several possible definitions of what is meant by the local Greens function in the solid. In this approach DMFT is viewed as an exact theory.

While the density-functional theory in the local approximation is closely related to the idea that the solid can be viewed as a perturbation around the homogeneous electron gas, Dynamical Mean-Field Theory, based on the concept of a Weiss field, offers a complementary perspective, the possibility of viewing the solid as a perturbation around a periodic collection of atoms. Formal developments of these ideas, combined with state-of-the-art electronic calculations, are now producing first-principles calculations on materials for which density-functional calculations were unsuccessful (e.g. $\delta$ plutonium [60]). Viewed from this perspective, the Mott transition problem has proved to be a valuable training ground for developing techniques for calculating electronic structure. These new methods handle the dual nature of the electrons (itinerant and localized) in a unified framework. A full-fledged realistic dynamical mean-field theory, including orbital degeneracy and band structure, is a new research tool in the science of complex materials with many possible applications.

An alternative approach focuses on functionals of the full one-electron particle, or its conjugate field as generalizations of the functional, and view dynamical mean-field as an approximate solution to the quantum many-body problem. In this view DMFT is an approximate theory. For example, the functional of the one-particle Greens function [45]

$$\Gamma[G] = -\mathrm{Tr}\log[G_0^{-1} - \frac{\delta\Phi}{\delta G}] - \mathrm{Tr}G\frac{\delta\Phi}{\delta G} + \Phi,$$

gives the dynamical mean-field equations when restricted to local Greens functions. Here, $\Phi[G]$ is defined as the sum of all two-particle irreducible graphs computed with the full propagator $G$ .

A third generalization involves extensions of DMFT to clusters [3]. Conceptually, the simplest generalization consists of dividing the lattice into supercells, and viewing each supercell as a complex 'site' to which one can apply ordinary DMFT. Denoting by $R_n$ the supercell position and by $\rho$ and $\varsigma$ the sites within the unit cell, one may rewrite the kinetic energy term of the Hamiltonian in Eq. 14.1 in terms of a matrix $\widehat{t}_{\rho\,\varsigma}(\mathbf{k})$, where $\mathbf{k}$ is now a vector in the Brillouin zone (reduced by the size of the cluster due to the supercell construction). The local term in Eq. 14.1 is left unchanged. In supercell notation, the Hamiltonian reads

$$H = -\sum_{\rho\,\varsigma\sigma} \widehat{t}_{\rho\,\varsigma}(k) c^{\dagger}_{\rho\sigma}(k) c_{\varsigma\sigma}(k) + U \sum_{n\rho} n_{R_n\rho\uparrow} n_{R_n\rho\downarrow}. \tag{14.33}$$

The DMFT equations now become matrix equations relating the cavity Greens function $\widehat{G}_0$ and the self-energy matrix $\widehat{\Sigma}$ (both matrices are indexed by the cluster sites):

$$\widehat{G}_0^{-1} = [\sum_k \, [i\omega - \widehat{t}(k) - \widehat{\Sigma}]]^{-1} + \widehat{\Sigma}.$$

In the presence of long-range order, the supercell scheme reduces to the ordinary DMFT, if in addition we ignore the intersite self-energies. In the absence of long-range order, it contains additional information on short-range order. The division of the lattice into supercells is artificial when spatial symmetries are not broken. Jarrell [59] and collaborators have suggested an alternative cluster scheme (the dynamic cluster approximation or DCA) to take into account the lattice periodicity. That approach relies on clusters with periodic boundary conditions. Lichtenstein and Katsenelson [61] have applied cluster methods together with an interpolation scheme in $\mathbf{k}$-space to produce continuous self energies. The interplay of superconductivity and antiferromagnetism is currently being studied with these periodic cluster methods [61] [62].

## 14.8 CONCLUSIONS

Dynamical mean-field methods provide a zeroth-order starting point for describing the strongly correlated regime of many three-dimensional transition-metal oxides. Further generalizations, as the ones mentioned in this section and related schemes, are likely to be useful for attacking the major open problem in this field, how to incorporate the effects of magnetic correlations (with no magnetic long-range-order) in the theory of the Mott transition.

From a broader perspective, we have simple physical pictures as well as powerful computational tools to understand complex materials containing $f$ and $d$ electrons. The future of the theory of strongly correlated electrons, a field that Phil Anderson pioneered with his unique, in-depth creativity and insight, seems very bright and will continue to thrive in this new century.

## BIBLIOGRAPHY

[1] N. F. Mott, Proc. Roy. Soc. A **62** 416 (1949).

[2] For a recent review and for further references, see M. Imada, A. Fujimori, and Y. Tokura, Rev. Mod. Phys. **70** 1039 (1998).

[3] For a recent review and further references, see A. Georges, G. Kotliar, W. Krauth, and M. Rozenberg, Rev. Mod. Phys. **68** 13 (1996).

[4] P. W. Anderson, Phys. Rev. B **115** 2 (1959).

[5] J. Hubbard, Proc. Roy. Soc. (London) **A281** 401 (1964).

[6] For a review, see *The Mathematics and Phyics of the Hubbard model.* D. Bareswyl *et al.*, eds. (Plenum Press, New York) (1995).

[7] W. Metzner and D. Vollhardt, Phys. Rev. Lett. **62** 324 (1989).

[8] E. Müller-Hartmann, Z Phys. B **74** 507 (1989).

[9] E. Müller-Hartmann, Z. Phys. B **76** 211 (1989).

[10] F. Gebhardt Phys. Rev. B **41** 9452 (1990), and **44** 992 (1991).

[11] U. Brandt and C. Mielsch, Z. Phys. **75** 365 (1989).

[12] P. W. Anderson, Phys. Rev. **124** 41 (1969).

[13] A. Georges and G. Kotliar, Phys. Rev. B **15** 6479 (1992).

[14] Y. Kuramoto and T. Watanabe, Physica **148B** (1987); F. J. Ohkawa, J. Phys. Soc. Jpn **60** 3218 (1991); V. Janiš, Z. Phys. B **83** 227 (1991). [3] M. Jarrell, Phys. Rev. Lett. **69** 168 (1992).

[15] G. Kotliar in *Strongly Correlated Electronic Materials*, K. Bedell, Z. Wang, D. Meltzer, A. Balatzky, E. Abrahams, eds, (Adison-Wesley) (1993).

[16] A. Hewson, *The Kondo Problem*, Cambridge Studies in Magnetism Vol 2. (Cambridge University Press, U.K.) (1993).

[17] P. W. Anderson and G. Yuval, Phys. Rev. Lett. **23** 89 (1969); P. W. Anderson, G. Yuval, and D. Hamman, Phys. Rev. B **1** 4464 (1970).

[18] J. E. Hirsch, and R. M. Fye, Phys. Rev. Lett. **56** 2521 (1986).

[19] M. Rozenberg, X. Y. Zhang and G. Kotliar, Phys. Rev. Lett. **69** 1236 (1992); A. Georges, and W. Krauth, Phys. Rev. Lett. **69** 1240 (1992).

[20] X. Y. Zhang, M. Rozenberg, and G. Kotliar, Phys. Rev. Lett. **70** 1666 (1993).

[21] A. Fujimori *et al.*, Phys. Rev. Lett. **69** 1796 (1992).

[22] F.D.M. Haldane, J. Phys. C **11** 5015 (1978).

[23] Q. Si, G. Kotliar and A. Georges, Phys. Rev. B **46** 1261 (1992).

[24] Q. Si and G. Kotliar, Phys. Rev. Lett **70** 3143 (1993).

[25] L.M. Falicov and J. C. Kimball, Phys. Rev. Lett 22, 997 (1969).

[26] R. Chitra and G. Kotliar, Phys. Rev. Lett **83** 2386 (1999).

[27] H. Kajueter and G. Kotliar, Int. J. Mod. Phys. **11** 729 (1997).

[28] A. Georges and W. Krauth, Phys. Rev. B **48** 7167 (1993).

[29] M. Rozenberg, G. Kotliar, and X. Y. Zhang, Phys Rev. B **49** 10181 (1994); M. Rozenberg *et al.*, Phys. Rev. Lett. **75** 105 (1995).

[30] D. B. McWhan, J. Remeika, W. Brinkman, and T. M. Rice, Phys. Rev. B **7** 1920 (1973); H. Kuwamoto, J. Honig and J. Appel, Phys. Rev. B **22** 2626 (1980).

[31] A. Miyasaka and H. Takagi, *unpublished.*

[32] H. Ito *et al.*, J. Phys. Soc. Jpn **65** 2987 (1996); K. Kanoda, Physica C **282-287** 299 (1987).

[33] G. Kotliar, Physica B **259-261** 711 (1999).

[34] V. J. Emery and S. Kivelson, Phys. Rev. Lett. **74** 3253 (1995).

[35] G. Palsson and G. Kotliar, Phys. Rev. Lett. **80** 4775 (1998).

[36] G. Kotliar, E. Lange, and M. Rozenberg, Phys. Rev. Lett. **84** 5180 (2000).

[37] For an early discussion, see H. Eskes, M. B. J. Meinders, and G. A. Sawatzky, Phys. Rev. Lett. **67** 1035 (1991).

[38] Z. Schlesinger *et al.*, Phys. Rev. Lett. **71** 1748 (1993).

[39] M. Rozenberg, G. Kotliar, and H. Kajueter, Phys. Rev. B **54** 8452 (1996).

[40] G. Kotliar, Euro. J. Phys. B, **11** 27 (1999).

[41] A. Matsuura *et al.*, Phys. Rev. B **58** 3690 (1998).

[42] S. Watanabe and S. Doniach, Phys. Rev. B **58** 3690 (1998).

[43] C. Castellani, C. D. Castro, D. Feinberg, and J. Ranninger, Phys. Rev. Lett. **43** 1957 (1979).

[44] G. Kotliar and A. Ruckenstein, Phys. Rev. Lett. **57** 1362 (1986).

[45] R. Chitra and G. Kotliar, cond-mat/9911223.

[46] N. Furukawa and M. Imada, J. Phys. Soc. Jpn. **61** 331 (1992); **62** 2557 (1993).

[47] H. Kajueter, G. Kotliar, and G. Moeller, Phys. Rev. B **53** 16214 (1996).

[48] A. McMahan *et al.*, J. Comput-Aided Mater. Des. **5** 131 (1998).

[49] J. W. Allen and R. M. Martin, Phys. Rev. Lett **49** 1106 (1982); L. Z. Liu *et al.*, Phys. Rev. B **45** 8934 (1992).

[50] V. Anisimov *et al.*, J. Phys. Cond. Matt. **9** 7359 (1997).

[51] P. Majumdar and H. R. Krishnamurthy, Phys. Rev. Lett. **73** 1525 (1994).

[52] S. Sachdev and Y. Ye, Phys. Rev. Lett. **70** 339 (1993).

[53] Q. Si and J. L. Smith, Phys. Rev. Lett. **77** 3391 (1997); H. Kajueter, *Ph.D. thesis, Rutgers University* (1996).

[54] R. Chitra and G. Kotliar, Phys. Rev. Lett **84** 3678 (2000).

[55] G. Moeller, Q. Si, G. Kotliar, M. Rozenberg and D. S Fisher, Phys. Rev. Lett. **74** 2082 (1995).

[56] Y. Motome and G. Kotliar, Phys. Rev. B, *submitted.*

[57] M. Valiev and G. Fernando, Phys. Lett. A **227** 265 (1997).

[58] R. Chitra and G. Kotliar, cond-mat/9911056.

[59] C. Huscroft *et al.*, cond-mat/9910226; T. Maier, M. Jarrell, T. Pruschke, J. Keller, Euro. Phys. Jnl. **B 13** 613 (2000); M. H. Hettler *et al.*, Phys. Rev. B **58** 7475 (1998).

[60] S. Savrasov and G. Kotliar, Phys. Rev. Lett. **84** 3670 (2000).

[61] L. Lichtenstein and M. Katsenelson, cond-mat/9911320; Phys. Rev. B, *in press.*

[62] Th. Maier, M. Jarrell, Th. Pruschke, J. Keller, cond-mat/0002352.

# CHAPTER 15

# FIRST STEPS IN GLASS THEORY

MARC MÉZARD [1]

Laboratoire de Physique Théorique de l'Ecole Normale Supérieure
24 rue Lhomond, F-75231 Paris Cedex 05, France

## 15.1 INTRODUCTION

Nearly all liquids become a glass, when quenched fast enough to avoid the crystallization transition [1]. This means that the density profile is not flat as in a liquid. It contains some peaks as in a crystal, but the peaks are not located at the nodes of a periodic or quasi-periodic lattice. The understanding of such amorphous 'solid' states has been recognized for a long time to be a major question in Condensed Matter Physics. The sentence by Phil Anderson "... there are still fascinating questions of principle about glasses and other amorphous phases..." [1], written nearly thirty years ago, was once again visionary in that it foresaw the wonderful developments in glassy systems, especially spin glasses. The progress in these areas has been difficult, particularly in the study of structural glasses.

## 15.2 MATHEMATICS

The first question that comes to mind is whether the glass is a new state of matter. It is not distinguished by any obvious symmetry (a non-obvious symmetry will be discussed later) from the liquid state, and one might think (as many people do) that the density profile should actually become flat on time scales longer than the experimental ones: the glass is just a liquid with a long relaxation time.

From the statistical physics point of view, one wants to start from a microscopic Hamiltonian. The simplest situation is that of $N$ point-like particles in a

[1] Present Address: Universite Paris Sud Laboratoire de Physique Theorique et Modeles Statistiques, Batiment 100, 91405 Orsay Cedex, France

volume $V$, with a pair-interaction potential

$$H = \sum_{i<j} V_{ij}(r_i - r_j). \tag{15.1}$$

A simple case is a homogeneous system in which $V_{ij}$ is, for instance, either a hard-sphere potential, a 'soft-sphere' potential ($V_{ij}(r) = A/r^{12}$), or a Lennard-Jones potential ($V_{ij}(r) = A/r^{12} - B/r^6$). Another class of systems (well-studied numerically because crystallization is more easily avoided) are the binary mixtures with two types of particles [3]: particle $i$ has $\epsilon_i \in \{\pm 1\}$ and $V_{ij}(r) = V_{\epsilon_i \epsilon_j}(r)$, where $V_{++}$, $V_{--}$, and $V_{+-} = V_{-+}$ are three potentials of the same type as before, but with different $A$ and $B$ parameters corresponding to particles $+$ and $-$ with different radii.

Does there exist, in any such case, an independent state of matter that is the glass state? Does it exist as a long-lived metastable state (like the diamond phase of carbon)? Nobody knows the rigorous mathematical answer to these questions. Actually, simpler, related questions remain unanswered (e.g. proving the existence of a spin-glass phase in a finite dimensional short-range system [4]). Others have only been solved after great effort (e.g. proving Kepler's conjecture that the densest three-dimensional packing of hard spheres is the fcc/hcp lattice [5]).

## 15.3 EXPERIMENTS

Experimentally, the liquid falls out of equilibrium on experimental time scales and becomes a 'glass' at a temperature $T_g$ called the glass temperature [1]. This glass temperature is conventionnally defined as the one at which the relaxation time $\tau$ of the liquid, as obtained e.g. from viscosity or from susceptibility measurements, becomes of the order of $10^3$ seconds. Angell's plot of $\log(\tau/1s)$ versus $T_g/T$ allows us to distinguish several types of behavior (fig. 15.1). So-called strong glasses like $SiO_2$ have a typical Arrhenius behavior with one well-defined free energy barrier. On the other hand, some glasses, called fragile, show a dramatic increase of the relaxation time with decreasing temperature $T$ that is much faster than Arrhenius: the typical free energy barrier thus increases when $T$ decreases. This implies a collective behavior involving more and more particles. An increase of the dynamic correlation, characteristic of the mobile particles (rather than the more natural correlation of frozen particles), has been found in recent simulations [7]. A popular fit of the relaxation time versus $T$ is the Vogel-Fulcher rule,

$$\tau \sim \tau_0 \exp\left(\frac{A}{T - T_{VF}}\right) \tag{15.2}$$

which would predict a phase transition at a temperature $T_{VF}$ that is not accessible experimentally (while staying at equilibrium). The more fragile the glass, the closer is $T_{VF}$ to $T_g$, while strong glass formers have a $T_{VF}$ close to zero.

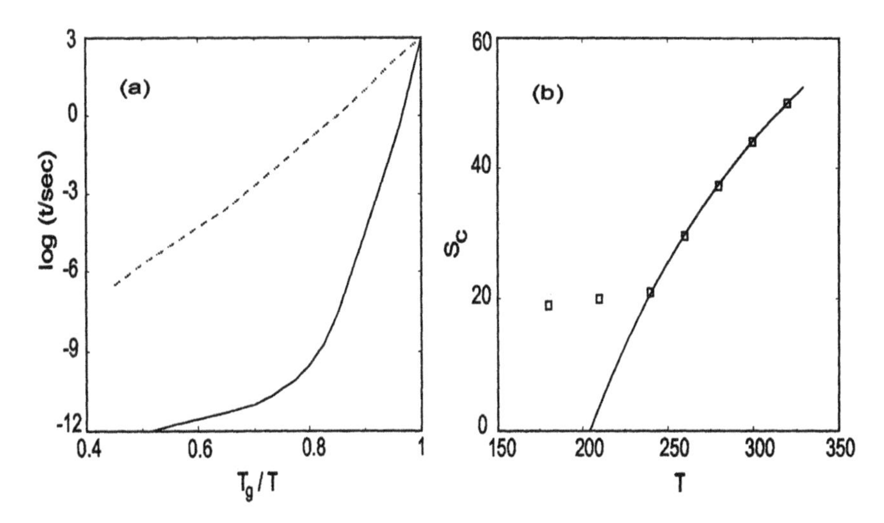


Figure 15.1: The left-hand figure (a) shows the behavior of the logarithm (to base 10) of the relaxation time in seconds, versus $T_g/T$, for two extreme cases of glass formers. The dashed line represents $GeO_2$, a strong glass former displaying Arrhenius-like behavior. The full line is the measurement from OTP, which is a fragile glass former with a relaxation time that diverges much faster than the Arrhenius law. The right-hand figure (b) shows the configurational entropy $S_c$ of OTP (in $JK^{-1}$/mol) versus $T$. The configurational entropy, defined as the difference between the entropy of the supercooled liquid and that of the crystal, is measured through an integral of the specific heat difference. The squares are the experimental values. The glass temperature is $T_g$ = 246 K. The full line is a fit to the equilibrated data, of the type $S_c = S_\infty(1 - T_K/T)$. The Kauzmann temperature is $T_K$ = 204 K, while the fusion temperature is 331 K. The data are taken from [6].

Another interesting experimental signature is that of the specific heat. When one cools the liquid slowly, at a cooling rate $\Gamma = -dT/dt$, it freezes into a glass at a temperature that decreases slightly when $\Gamma$ decreases. When this freezing occurs, the specific heat jumps downward, from its value in the equilibrated supercooled liquid state to a glass value that is close to that of the crystal. From the specific heat one can compute the entropy. The configurational entropy, defined experimentally as the difference $S_c = S_{liq} - S_{crystal}$, behaves smoothly in the supercooled liquid phase, until the system becomes a glass (see fig.15.1). It was noted by Kauzmann long ago that, if extrapolated, $S_c(T)$ vanishes at a finite temperature $T_K$. If cooled more slowly, the system follows the smooth $S_c(T)$ curve down to slightly lower temperatures, but then freezes again. One can wonder what could happen at infinitely slow cooling. As a negative $S_c$ does not make sense (except for pure hard spheres, where there is no energy), something must happen at $T > T_K$. The curve $S_c(T)$ could flatten down smoothly, or there might be a phase transition, which in the simplest scenario would lead to

| Substance | $T_K(K)$ | $T_{VF}(K)$ | $T_g(K)$ |
|---|---|---|---|
| o-terphenyl | 204.2 | 202.4 | 246 |
| salol | 175.2 | - | 220 |
| 2-MTHF | 69.3 | 69.6 | 91 |
| n-propanol | 72.2 | 70.2 | 97 |
| 3-bromopentane | 82.5 | 82.9 | 108 |

Table 15.1: Comparison of $T_K$ and $T_{VF}$ in various glass-formers (from [6])

$S_c(T) = 0$ at $T < T_K$. This idea of an underlying 'ideal' phase transition, which could be obtained only at infinitely slow cooling, receives some support from the following observation: the two temperatures $T_{VF}$ and $T_K$, at which the *extrapolated* experimental behavior has a singularity, turn out to be amazingly close to each other (see Table 15.1) [6]. The first phenomenological attempts to explain this fact were by Kauzmann [8]. They were further developed by, among others, Adam, Gibbs and Di-Marzio [9], who identified the glass transition as a 'bona fide' thermodynamic transition blurred by some dynamical effects.

If there exists a true thermodynamic glass transition at $T = T_K = T_{VF}$, it is a transition of a strange type. On the one hand, it is of second-order because the entropy and internal energy are continuous. On the other hand, the order parameter is discontinuous at the transition, as in first-order transitions: the modulation of the microscopic density profile in the glass does not appear continuously from the flat profile of the liquid. As soon as the system freezes, there is a finite jump in this modulation (A more precise definition of the order parameter is given below).

## 15.4 A MEAN-FIELD SPIN-GLASS ANALOGY

A totally different class of systems, in which a similar 1st- to 2nd-order type of transition has been studied in great detail, is a certain category of mean-field spin glasses. A few years after the replica symmetry breaking (RSB) solution of the mean-field theory of spin glasses [10], it was realized that there exists another category of mean-field spin glasses in which the static phase transition exists and is due to an entropy crisis [11]. These are now called discontinuous spin glasses because their phase transition has a discontinuous order parameter, despite being second-order in the Ehrenfest sense [12]. Another name often found in the literature is 'one-step RSB' spin glasses, because of the special pattern of symmetry breaking involved in their solution. These are spin glasses with infinite-range interactions involving a coupling between triplets (or higher order groups) of spins. The simplest among them is the random-energy model, which is the $p \to \infty$ limit version of the $p$-spin models described by the Hamiltonian

$$H = - \sum_{i_1 < \ldots < i_p} J_{i_1 \ldots i_p} s_{i_1} \ldots s_{i_p} , \tag{15.3}$$

where the $J$'s are (appropriately scaled) quenched random couplings, and the spins can be either of Ising or spherical type [12, 13, 14].

The analogy between the phase transition of discontinuous spin glasses and the thermodynamic glass transition was first noticed by Kirkpatrick, Thirumalai and Wolynes in a series of inspired papers of the mid-eighties [13]. While some of the basic ideas of the present development were around at that time, a few crucial ingredients were still missing. On the one hand, one needed to get more confidence that this analogy was not just fortuitous. The big obstacle was the existence (in spin glasses) versus the absence (in structural glasses) of quenched disorder. The discovery of discontinuous spin glasses without any quenched disorder [15, 16, 17] provided an important new piece of information: contrary to what had long been believed, quenched disorder is not necessary for the existence of a spin-glass phase (but frustration is).

It is important to analyze critically this analogy from the point of view of the dynamical behavior. In discontinuous, mean-field spin glasses, there exists a 'dynamical transition' temperature $T_c$ that is higher than the equilibrium transition $T_K$. When $T$ decreases and approaches $T_c$, the correlation function relaxes with a characteristic two-step form: a fast $\beta$-relaxation leading to a plateau takes place on a characteristic time that grows slowly, while the $\alpha$-relaxation from the plateau takes place on a time scale that diverges when $T \to T_c$ (see fig. 15.2). This dynamic transition is exactly described by the schematic mode-coupling equations.

However, the existence of a dynamic relaxation at a temperature above the true thermodynamic one is possible only in mean-field, and the conjecture [13] is that, in a realistic system like a glass, the region between $T_K$ and $T_c$ will have instead a finite, but very rapidly increasing, relaxation time, as explained in fig. 15.2. A similar behavior has been found in finite-size, mean-field models [21].

Another very interesting dynamical regime is the one in which the system is out of equilibrium ($T < T_g$). The system is then no longer stationary: it ages. This is well known, for instance, from studies in polymeric glasses. If one measures the response of one's favorite plastic ruler to stress, it will behave differently depending on its age. Schematically, new relaxation processes come into play on a time scale comparable to the age of the system: the older the system, the longer the time needed for this 'aging' relaxation to take place. Recent years have seen important developments in the out-of-equilibrium dynamics of the glassy phases [22], initiated by the exact solution of the dynamics in a discontinuous spin glass by Cugliandolo and Kurchan [23]. It has become clear that, in realistic systems with short-range interactions, the pattern of replica symmetry breaking can be deduced from measurements of the violation of the fluctuation-dissipation theorem [24]. These measurements are difficult. However, numerical simulations performed on different types of glass-forming systems have provided an independent and spectacular confirmation of their 'one-step-RSB' structure [25, 26, 27, 28] on the (short) time scales that are accessible. Experimental results

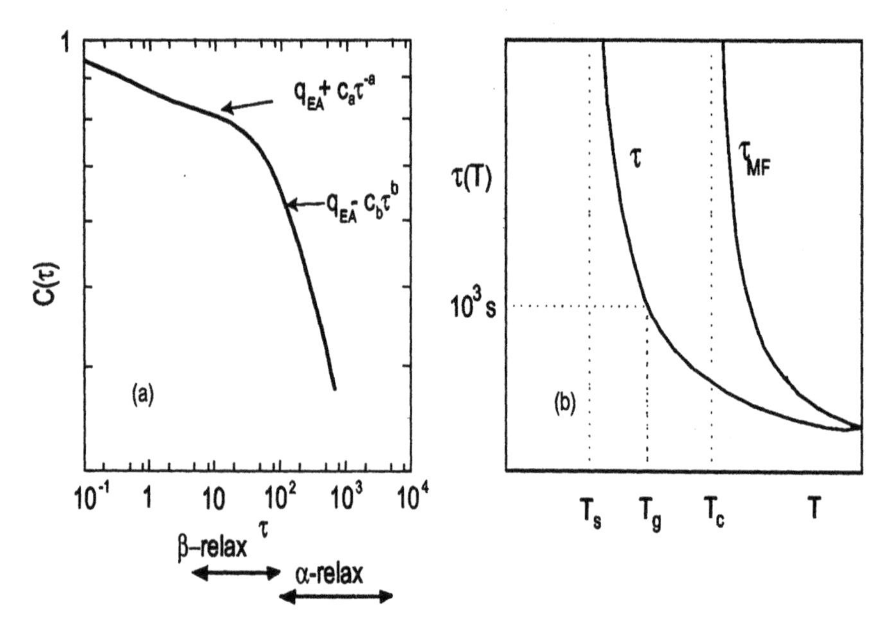


Figure 15.2: The left-hand figure (a) shows the schematic behavior of the correlation function found in mean-field discontinuous spin glasses and observed in structural glasses. The typical two-step relaxation consists of a fast $\beta$-relaxation leading to a plateau, followed by an $\alpha$-relaxation from the plateau, whose typical time scale increases rapidly as $T$ decreases, and diverges at $T = T_c$, which is equal to the mode-coupling transition temperature. The right-hand figure (b) shows the behavior of the relaxation time versus $T$. The right-hand curve is the prediction of mode-coupling theory without any activated processes: it is a mean-field prediction that is exact for instance in the discontinuous mean-field spin glasses [18]. The left-hand curve is the observed relaxation time in a glass. The mode-coupling theory provides a quantitative prediction for the increase of the relaxation time with decreasing $T$, at high temperatures (well above the mode coupling transition $T_c$) [19, 20]. The departure from the mean-field prediction at lower temperatures is usually attributed to 'hopping' or 'activated' processes, in which the system is trapped for a long time in a valley, but can eventually jump out of it. The ideal glass transition, which takes place at $T_K$, cannot be observed directly since the system falls out of equilibrium on laboratory time scales at the 'glass temperature' $T_g$.

have not yet settled the issue, but the first measurements of effective temperatures in the fluctuation-dissipation relation were made recently [29].

To summarize, the analogy between the phenomenology of fragile glass formers and discontinuous mean-field spin glasses accounts for:

- The discontinuity of the order parameter
- The continuity of the energy and the entropy

- The jump in specific heat (and the sign of the jump)
- Kauzmann's "entropy crisis"
- The two-step relaxation of the dynamics, and the success of Mode-Coupling Theory at relatively high temperatures.
- The aging phenomenon and the pattern of modification of the fluctuation dissipation relation in the low-$T$ phase

## 15.5 A LESSON FROM MEAN-FIELD: MANY VALLEYS

The successes of the above analogy suggest that we should have a closer look at the mean-field models to understand, at least at the mean-field level, the basic ingredients that are at work in the glass transition. In mean-field spin glasses, at temperatures $T_K < T < T_c$, the phase space breaks up into ergodic components that are well-separated, the so-called free-energy valleys or TAP states [30, 31]. Each valley $\alpha$ has a free energy $F_\alpha$ and a free-energy density $f_\alpha = F_\alpha/N$. The number of free-energy minima with free-energy density $f$ is found to be exponentially large:

$$\mathcal{N}(f,T,N) \approx \exp(N\Sigma(f,T)), \tag{15.4}$$

where the function $\Sigma$ is called the complexity. The total free energy of the system, $\Phi$, may be well-approximated by:

$$e^{-\beta N\Phi} \simeq \sum_\alpha e^{-\beta N f_\alpha(T)} = \int_{f_{min}}^{f_{max}} df \, \exp\left(N[\Sigma(f,T) - \beta f]\right) , \tag{15.5}$$

where $\beta = 1/T$. The minima that dominate the sum are those with a free-energy density $f^*$ that minimizes the quantity $\Phi(f) = f - T\Sigma(f,T)$. At large enough $T$, the saddle point is at $f > f_{min}(T)$. When one decreases $T$, the saddle-point free-energy decreases (see Fig. 15.3, with $m = 1$). The Kauzman temperature $T_K$ is the temperature below which the saddle point sticks to the minimum: $f^* = f_{min}(T)$. It is a genuine phase transition [11, 12, 13]. However, because $T_c > T_K$, the phase space is actually separated into non-ergodic components (valleys) at $T < T_c$ (actually, some non-ergodic components also exist above $T_c$, but they are not accessed when the system starts from random initial conditions [32]). The total equilibrium free energy is analytic at $T_c$: in spite of the ergodicity breaking, the system has the same free energy as that of the liquid, as if transitions were allowed between valleys.

What remains of this mean-field picture in finite-dimensional glasses? When one decreases the temperature, there is a well-defined separation of time scales between the $\alpha$ and the $\beta$-relaxations. This suggests that we should regard the

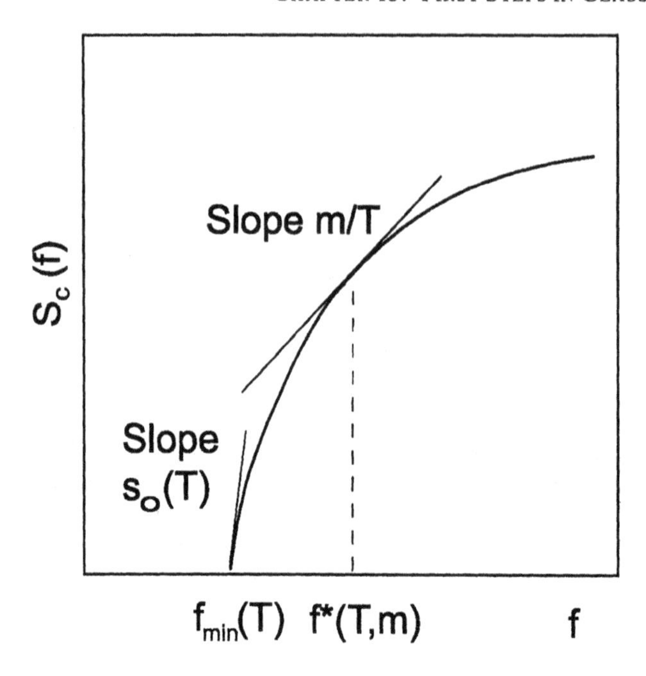


Figure 15.3: Qualitative shape of the complexity versus free energy in mean-field discontinuous spin glasses. The whole curve depends on the temperature. The saddle point that dominates the partition function, for $m$ constrained replicas, is the point $f^*$ such that the slope of the curve equals $m/T$ (in the usual unreplicated system, $m = 1$). If $T$is small enough, the saddle point sticks to the minimum $f = f_{min}$, and the system is in its glass phase. For $m = 1$, this equilibrium phase transition occurs at $T = T_K$.

dynamical evolution of system in phase space as a superposition of two processes: an intravalley relaxation that is relatively fast, and an intervalley hopping process that is slow, and becoming dramatically slower with decreasing $T$.

A popular way of making this statement more precise (hence facilitating numerical studies) is to introduce 'inherent structures' (IS) [33]. Given a configuration of the system characterized by its phase space position $x = \{\vec{x}_1, ..., \vec{x}_N\}$, the corresponding inherent structure $s(x)$ is another point in phase space that is the local minimum of the Hamiltonian reached from the starting configuration via a steepest-descent dynamics. The inherent structures are readily identified numerically. A given trajectory $x(t)$ of the glass through phase space maps onto the corresponding trajectory $s(t)$ in the space of inherent structures. Looking at the dynamical evolution in the space of IS [34] makes the valley structure slightly more apparent, since one gets rid of the small thermal excitations around each valley minimum. Calling $\mathcal{D}_s$ the set of those configurations that are mapped into the coherent structure $s$, a natural definition of the IS entropy density $\Sigma_{is}$ is $N\Sigma_{is}(T) = -\sum_s P(s)\ln(P(s))$, where the weight of the inherent structure $s$

is

$$P(s) = Z(s)/\sum_b Z(b) \quad ; \quad Z(s) \equiv \int_{x\in\mathcal{D}_s} dx \exp(-\beta H(x)) \,. \tag{15.6}$$

In a system with short-range interactions, it is reasonable to expect that one may have two distinct IS that differ by a local rearrangement of a finite number of atoms. It is then easy to show that the slope of configurational entropy versus free energy is infinite around $f_{min}$ [35], which does not agree with the general scenario, except when the Kauzmann temperature vanishes. This problem arises because the IS are objects that are too simple to be identified with the free energy valleys. The difference is very easily seen in spin systems [36]: IS are nothing but configurations that are stable against one spin flip. Zero-temperature free-energy valleys, defined as TAP states, are stable against the flip of an arbitrarily large number $k$ of spins (but the limit $N \to \infty$ must be taken before the limit $k \to \infty$). In continuous systems, the generalization is clear: IS are local minima of the energy, so that any infinitesimal move of the positions of all $N$ particles raises the energy. Let us generalize the notion of a minimum as follows: define a $k$-th order local minimum as a configuration of particles such that any infinitesimal move of all $N$ particles, together with a move of *arbitrary size* of $k$ particles, raises the energy. The limit $k \to \infty$ gives the proper definition of a zero-temperature, free-energy valley. The proper definition at finite $T$ is slightly more involved [37]. Let us summarize it here briefly. Given two configurations $x$ and $y$ we define their overlap as before as $q(x,y) = -1/N\sum_{i,k=1,N} w(x_i - y_k)$, where $w(x)$ = -1 for $x$ small, $w(x)$ = 0 for $x$ larger than the typical interatomic distance. We add an extra term to the Hamiltonian: we define

$$\exp(-N\beta F(y,\epsilon)) = \int dx \exp(-H(x) + \beta\epsilon N q(x,y)), \tag{15.7}$$

$$F(\epsilon) = \langle F(y,\epsilon)\rangle, \tag{15.8}$$

where $\langle f(y)\rangle$ denotes the average value of $f$ over equilibrium configurations $y$ thermalized at temperature $\beta^{-1}$. Taking the thermodynamic limit before the limit $\epsilon \to 0$ allows us to identify the valley around any generic equilibrium configuration $y$ [37, 39].

In a nutshell, two configurations that differ by the (arbitrarily large) displacement of a finite number of atoms are in the same thermodynamic valley. This definition of the valleys also suffers from some difficulties: Nucleation arguments then forbid the existence of a non-trivial complexity versus free energy curve in a finite-dimensional system. The solution consists in noticing that there exist many more metastable valleys that have a finite but very long lifetime. These can be identified by taking the Legendre transform $W(q)$ of the free energy $F(\epsilon)$:

$$W(q) = F(\epsilon) + \epsilon q \quad ; \quad q = \frac{-\partial F}{\partial \epsilon} \,. \tag{15.9}$$

Analytic computation in mean-field models [37], as well as in glass-forming liquids, using the replicated HNC approximation [38], show that $W(q)$ is minimal at $q = 0$, but has a secondary minimum at a certain $q = q_{EA}$ in the temperature range $T_K < T < T_c$. The behavior around this second, metastable minimum corresponds to phenomena that can be observed on time scales shorter than the lifetime of the metastable state. The thermodynamic configurational entropy is the value of the potential $W(q)$ at the secondary minimum with $q \neq 0$ [37], and it can be defined only if the minimum exists (i.e. for $T < T_c$). Of course, the secondary minimum for $T > T_k$ is always in the metastable region. However, if one starts from a large value of $\epsilon$ and decreases $\epsilon$ to zero (not too slowly), the system should not escape from the metastable region. One then obtains a proper definition of the thermodynamic configurational entropy in this region $T > T_K$. In a similar way one may compute $q(\epsilon)$ in the region ($\epsilon > \epsilon_c$) where the high-$q$ phase is thermodynamically stable, and extrapolate it to $\epsilon \to 0$. The ambiguity in the definition of the thermodynamic configurational entropy at temperatures above $T_k$ becomes larger and larger as $T$ increases. It cannot be defined for $T > T_c$.

## 15.6 BEYOND THE ANALOGY: FIRST-PRINCIPLES COMPUTATION

In recent years, it has become possible to go beyond the simple analogy between structural glasses and mean-field discontinuous spin glasses. One can actually use the concepts and the techniques that are suggested by this analogy to start a systematic, first-principles study of the glass phase [41, 42, 39]. So far, we have focused on the equilibrium study of the low-$T$ phase. The reason is that the direct study of out-of-equilibrium dynamics is more difficult, and one might be able to make progress by a careful analysis of the landscape [40]. One assumes that there exists a phase transition, and that it is of the same type as observed in discontinuous mean-field spin glasses. Within this framework, one tries to compute the properties of the glass phase. This involves several quantitites like the Kauzmann temperature, the radius of the cage that confines the particles in the glass phase, the configurational entropy, and so on. The validity of the scenario is checked from the comparison of various predictions with numerical simulations of well-equilibrated systems.

The first task is to define an order parameter. This is not trivial in an equilibrium theory in which we have no notion of time-persistent correlations. The best way is to introduce two copies of the system, with a weak interaction. The two sets of particles have positions $x_i$ and $y_i$, respectively, the total Hamiltonian is

$$E = \sum_{1 \leq i \leq j \leq N} (v(x_i - x_j) + v(y_i - y_j)) + \epsilon \sum_{i,j} w(x_i - y_j) \qquad (15.10)$$

where we have introduced a small attractive potential $w(r)$ between the two systems. The precise shape of $w$ is irrelevant, insofar as we shall be interested in the limit $\epsilon \to 0$, but its range should be of order or smaller than the typical interparticle distance. The order parameter is then the correlation function between the two systems:

$$g_{xy}(r) = \lim_{\epsilon \to 0} \lim_{N \to \infty} \frac{1}{\rho N} \sum_{ij} < \delta(x_i - y_j - r) > . \tag{15.11}$$

In the liquid phase this correlation function is identically equal to one, while it has a nontrivial structure in the glass phase, reminiscent of the pair-correlation in a dense liquid, but with an extra peak around $r \simeq 0$. Let us notice that we expect a discontinuous jump of this order parameter at the transition, despite its second-order nature (in the thermodynamic sense). The existence of a non-trivial order parameter is associated with the spontaneous breaking of a symmetry: For $\epsilon = 0$, with periodic boundary conditions, the system is symmetric under a global translation of the $x$ particles with respect to the $y$ particles. This symmetry is spontaneously broken in the low-$T$ phase, where the particles of each subsystem tend to sit in front of each other.

Generalizing this approach to a system of $m$ coupled replicas, sometimes named 'clones' in this context (the order parameter used only $m = 2$), provides a wonderful method for studying analytically the thermodynamics of the glass phase [43, 44]. In the glass phase, the attraction will force all $m$ systems to fall into the same glass state, so that the partition function is:

$$Z_m = \sum_{\alpha} e^{-\beta N m f_\alpha(T)} = \int_{f_{min}}^{f_{max}} df \, \exp\left(N[\Sigma(f,T) - m\beta f]\right). \tag{15.12}$$

In the limit $m \to 1$ the corresponding partition function $Z_m$ is dominated by the correct saddle point $f^*$ for $T > T_K$. The interesting regime is when $T < T_K$, and the number $m$ is allowed to become smaller than one. The saddle point $f^*(m,T)$ in the expression (15.12) is the solution of $\partial\Sigma(f,T)/\partial f = m/T$. Because of the convexity of $\Sigma$ as function of $f$, the saddle point is at $f > f_{min}(T)$ when $m$ is small enough, and it sticks at $f^* = f_{min}(T)$ when $m$ exceeds a certain value $m = m^*(T)$, which is smaller than one if $T < T_K$ (see fig. 15.3). The free energy in the glass phase, $F(m = 1, T)$, is equal to $F(m^*(T), T)$. As the free energy is continuous along the transition line $m = m^*(T)$, one can compute $F(m^*(T), T)$ from the region $m \le m^*(T)$, where the replicated system is in the liquid phase. This is the clue to the explicit computation of the free energy in the glass phase. It may sound a bit strange because one is tempted to think of $m$ as an integer number. However, the computation is much clearer if one regards $m$ as a real parameter in (15.12). As one considers low temperatures $T < T_K$ the $m$ coupled replicas fall into the same glass state. Hence, they assemble some molecules of $m$ atoms, each molecule being built from one atom of each 'color'. Now, the strength of the interaction between the molecules is basically rescaled by a factor $m$ (this statement becomes exact in the limit of zero temperature where the molecules are point-like). If $m$ is small enough, this interaction is small, and the system of molecules is a liquid. When $m$ increases, the molecular fluid freezes into a glass state at the value $m = m^*(T)$. Thus, the method requires us to estimate the replicated free energy, $F(m,T) = -\log(Z_m)/(\beta m N)$, in a molecular liquid phase, where the molecules consist of $m$ atoms, with $m$ smaller than one. For $T < T_K$, $F(m,T)$ is maximum at a value $m = m^*$ smaller than one, whereas for $T > T_K$, the maximum is reached at a value $m^*$ larger than one. The knowledge of $F_m$ as a function of $m$ allows us to reconstruct the configurational entropy function $Sc(f)$ at a given $T$ via a Legendre transform, using the parametric representation

$$f = \frac{\partial\,[mF(m,T)]}{\partial m} \quad ; \quad \Sigma(f) = \frac{m^2}{T}\frac{\partial F(m,T)}{\partial m}\,. \tag{15.13}$$

(This is easily deduced from a saddle-point evaluation of Eq. 15.12 [43].) The Kauzmann temperature ('ideal glass-temperature') is that at which $m^*(T_K) = 1$. For $T < T_K$ the equilibrium configurational entropy vanishes. Above $T_K$ one obtains the equilibrium configurational entropy $\Sigma(T)$ by solving Eq. 15.13 at $m = 1$.

This gives the main idea that enables the free energy in the glass phase for $T < T_K$ to be computed from first principles. It is equal to the free energy of a molecular liquid at the same $T$, in which each molecule is built of $m$ atoms, and an appropriate analytic continuation to $m = m^*(T) < 1$ has been taken. The whole problem is reduced to a computation in a liquid. This is not a trivial task, and requires specific approximations to be adopted. I refer the reader to the original papers [42, 45, 46, 47]. The basic idea of the approximation is that the size of the molecules is directly related to the thermal wandering of an atom in its cage. Therefore, at low temperatures one can use some small-cage approximation. It is natural to write the partition function in terms of the center-of-mass and relative coordinates $\{r_i, u_i^a\}$, with $x_i^a = r_i + u_i^a$ and $\sum_a u_i^a = 0$, and to expand the interaction in powers of the relative displacements $u$. Keeping only the term quadratic in $u$ (harmonic vibrations of the molecules), and integrating over these vibration modes, one gets the 'harmonic resummation' approximation in which the partition function is given by:

$$Z_m = Z_m^0 \int dr \exp\left(-\beta m H(r) - \frac{m-1}{2} Tr \log M\right), \tag{15.14}$$

where $Z_m^0 = m^{Nd/2}\sqrt{2\pi T}^{Nd(m-1)}/N!$, and the matrix $M$, of dimension $dN \times dN$, is given by

$$M_{(i\mu)(j\nu)} = \frac{\partial^2 H(r)}{\partial r_i^\mu \partial r_j^\nu} = \delta_{ij}\sum_k v_{\mu\nu}(r_i - r_k) - v_{\mu\nu}(r_i - r_j), \tag{15.15}$$

and $v_{\mu\nu}(r) = \partial^2 v/\partial r_\mu \partial r_\nu$ (the indices $\mu$ and $\nu$ denote space directions). Now we are back to a real problem of liquid theory, since we have only $d$ degrees of freedom per molecule (the center of mass coordinates), and the number of clones, $m$, appears as a parameter in (15.14).

Once one has derived an expression for the replicated free energy, one can deduce from it the whole thermodynamics, as described above (notice that the 'technical' approximation of neglecting the exchange of atoms between different molecules, as well as using a harmonic model, means that one really studies the IS in this computation, rather than the real free energy valleys). In all three cases, one finds an estimate of the Kauzmann temperature that is in reasonable agreement with simulations, with a jump in specific heat, from a liquid value at $T > T_K$ to the Dulong-Petit value $C = 3/2$ below $T_K$ (we have included only positional degrees of freedom). This is similar to the experimental result, where the glass specific-heat jumps down to the crystal value as $T$ decreases (our approximations so far are similar to the Einstein approximation of independent vibrations of atoms, in which case the contribution of positional degrees of freedom to the crystal specific heat is $C = 3/2$). The parameter $m^*(T)$ and the cage sizes are nearly linear in $T$ in the whole glass phase. This means, in particular, that the effective temperature $T/m$ is always close to $T_K$, so in our theoretical computation we need only to evaluate the expectation values of observables in the liquid phase at temperatures where the HNC approximation for the liquid still works quite well.

A more detailed numerical check of these analytical predictions involves the measurement of the complexity,

$$\Sigma_t = S(T) - S_{valley}(T). \tag{15.16}$$

The liquid entropy is estimated by a thermodynamic integration of the specific heat from the very dilute (ideal gas) limit. It turns out that in the deeply supercooled region the $T$ dependence of the liquid entropy is well-fitted by the law predicted in [48]: $S_{liq}(T) = aT^{-2/5} + b$, which presumably allows for a good extrapolation to temperatures that cannot be simulated. As for the 'valley' entropy, it may be estimated as that of an harmonic solid. One needs, however, the vibration frequencies of the solid. These have been approximated by several methods, most of which are based on some evaluation of the Instantaneous Normal Modes (INM) [50] in the liquid phase, and the assumption that the spectrum of frequencies does not depend much on $T$ below $T_K$.

Starting from a typical configuration of the liquid, one can look at the INM around it. In general, there exist some negative eigenvalues (the liquid is not a local minimum of the energy) that one must take care of. Several methods have been tried: either one keeps only the positive eigenvalues, or one considers the absolute values of the eigenvalues [45, 46, 47]. Alternatively, one may also consider the INM around the nearest inherent structure that has, by definition, a

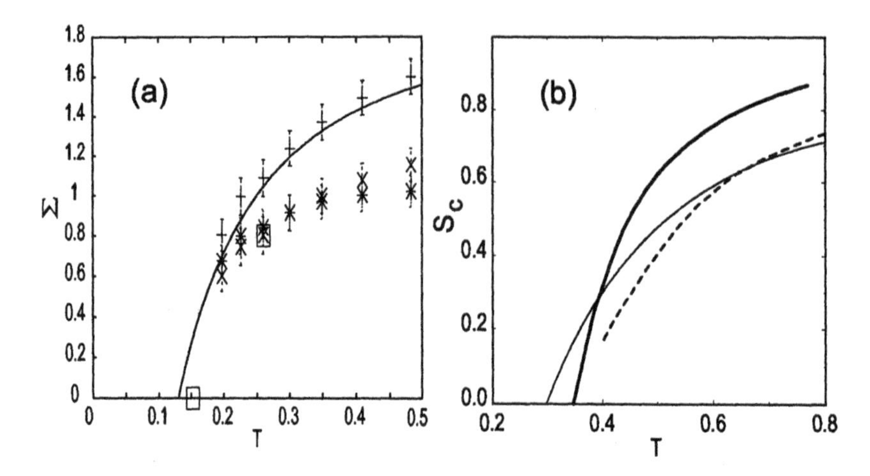


Figure 15.4: The configurational entropy versus $T$ in binary mixtures of soft spheres (Panel (a)) and of Lennard-Jones particles (b). The soft-sphere result (a), from [45], compares the analytical prediction obtained within the harmonic resummation scheme (full line), to simulation estimates of $S_{liq} - S_{valley}$, where the valley entropy is that of a harmonic solid with INM eigenvalues projected onto positive eigenvalues (+), taken in absolute values (×), or taken around the nearest inherent structure (∗). The squares correspond to the numerical estimate of the thermodynamic configurational entropy obtained by studying the system coupled to a reference configuration (see text, and Ref. [45] for details). In the Lennard-Jones system (b), the prediction obtained from the cloned molecular liquid approach [46, 47] is displayed as a thick, solid line. The broken line is the result from the simulations in Ref. [46, 47], while the thin, solid line is from simulations in Ref. [51]. Both simulations use the $S_{liq} - S_{valley}$ estimate in which the harmonic solid vibration modes are approximated by the ones of the nearest inherent structure.

positive spectrum [45, 46, 47, 51]. This procedure really measures the configurational entropy rather than the thermodynamic complexity. The computation of the thermodynamic complexity, using its definition as a system coupled to a reference thermalized configuration, has also been computed in Ref. [45], and turns out to be not very different from the configurational entropy, on the time and temperature scales that have been studied so far (they must differ on infinitely long time scales, as we discussed in the previous section).

The results for the configurational entropy as a function of $T$ are shown in Fig. 15.4, for binary mixtures of soft spheres and of Lennard-Jones particles. The agreement with the analytical result obtained from the replicated fluid system is rather satisfactory, considering the approximations in both analytical and numerical approaches.

## 15.7 CONCLUSION

Our knowledge of first-principle computations of glasses is still rather primitive. Basically, we have obtained for the glass the equivalent of the Einstein approximation for the crystal. Even within this simple scheme, doing the actual computation for the glass turns out to be rather formidable. For further progress, we need: on the analytical side, better approximations of the molecular liquid state that allow us to go beyond the small cage expansions, and reliable estimate of time scales in the regime $T_K < T < T_c$; on the numerical side, some precise results in the glass phase at equilibrium [52]. On the experimental side, we need more measurements of the fluctuation-dissipation ratio in the out-of-equilibrium dynamics. No doubt, "... there are still fascinating questions of principle about glasses and other amorphous phases..." [1].

## ACKNOWLEDGMENTS

It is a great pleasure to thank Giorgio Parisi for the collaboration that led to the work described here, as well as G. Biroli and R. Monasson for useful discussions.

## BIBLIOGRAPHY

[1] Recent reviews can be found in: C. A. Angell, Science **267** 1924 (1995); P. De Benedetti, *Metastable liquids*, Princeton University Press (1997); G. Tarjus and D. Kivelson, to appear in "Jamming" volume, A. Lui and S. Nagel eds.; J. Jäckle, Rep. Prog. Phys. **49** 171 (1986).

[2] P. W. Anderson, Science **177** 393 (1972).

[3] See, for e.g., the review by W. Kob, J. Phys.: Cond. Matter **11** R85 (1999).

[4] A review of recent developments is *Spin glasses and random fields*, A. P. Young ed., (World Scientific, Singapore) 1998.

[5] For recent work on Kepler's conjecture, see www.math.lsa.umich.edu/ hales/countdown/.

[6] R. Richert and C. A. Angell, J. Chem. Phys. **108** 9016 (1999).

[7] C. Benneman, C. Donati, J. Baschnagel and S. C. Glotzer, Nature **399** 246 (1999).

[8] A. W. Kauzman, Chem. Rev. **43** 219 (1948).

[9] G. Adams and J. H. Gibbs, J. Chem. Phys **43** 139 (1965); J. H. Gibbs and E. A. Di Marzio, J. Chem. Phys. **28** 373 (1958).

[10] For a review, see M. Mézard, G. Parisi and M. A. Virasoro, *Spin glass theory and beyond*, (World Scientific, Singapore 1987).

[11] B. Derrida, Phys. Rev. B **24** 2613 (1981).

[12] D. J. Gross and M. Mézard, Nucl. Phys. B **240** 431 (1984).

[13] T. R. Kirkpatrick and P. G. Wolynes, Phys. Rev. A **34** 1045 (1986); T. R. Kirkpatrick and D. Thirumalai, Phys. Rev. Lett. **58** 2091 (1987); T. R. Kirkpatrick and D. Thirumalai, Phys. Rev. B **36** 5388 (1987); T. R. Kirkpatrick, D. Thirumalai and P. G. Wolynes, Phys. Rev. A **40** 1045 (1989).

[14] A. Crisanti, H. Horner and H. J. Sommers, Z. Physik B **92** 257 (1993).

[15] J.-P. Bouchaud and M. Mézard; J. Physique I (France) **4** 1109 (1994). E. Marinari, G. Parisi and F. Ritort; J. Phys. A **27** 7615 (1994); J. Phys. A **27** 7647 (1994).

[16] P. Chandra, L. B. Ioffe and D. Sherrington, Phys. Rev. Lett. **75** 713 (1995); P. Chandra *et al.*, cond-mat/9809417. P. Chandra, M. V. Feigelman and L. B. Ioffe, Phys. Rev. Lett. **76** 4805 (1996).

[17] E. Marinari, G. Parisi and F. Ritort, cond-mat/9410089. S. Franz and J. Hertz, Phys. Rev. Lett. **74** 2114 (1995).

[18] J. P. Bouchaud, L. F. Cugliandolo, J. Kurchan and M. Mézard, Physica A **226** 243 (1996).

[19] For a review on mode couping theory, see W. Götze, in *Liquid, freezing and the Glass transition*, Les Houches (1989), J. P. Hansen, D. Levesque, J. Zinn-Justin editors, North Holland.

[20] Some discussion of the experimental situation can be found in H. Z. Cummins and G. Li, Phys. Rev. E **50** 1720 (1994); H. Z. Cummins, W. M. Du, M. Fuchs, W. Götze, S. Hildebrand, A. Latz, G. Li and N. J. Tao, Phys. Rev. B **47** 4223 (1993); P. K. Dixon, N. Menon and S. R. Nagel, Phys. Rev. E **50** 1717 (1994). For an enlightening introduction to the experimental controversy, see the series of Comments in Phys. Rev. E: X. C. Zeng, D. Kivelson and G. Tarjus, Phys. Rev. E **50** 1711 (1994).

[21] A. Crisanti and F. Ritort, cond-mat/9911226 and cond-mat/9911351.

[22] For a review, see J.-P. Bouchaud, L. Cugliandolo, J. Kurchan., M. Mézard, in *Spin glasses and random fields*, A. P. Young editor, (World Scientific, Singapore) 1998.

[23] L. F. Cugliandolo and J. Kurchan, Phys. Rev. Lett. **71** 1 (1993).

[24] S. Franz, M. Mézard, G. Parisi and L. Peliti, Phys. Rev. Lett. **81** 1758 (1998); S. Franz, M. Mézard, G. Parisi and L. Peliti, J. Stat Phys **97** 459 (1999).

[25] G. Parisi, Phys. Rev. Lett. **78** 4581 (1997).

[26] W. Kob and J.-L. Barrat, Phys. Rev. Lett. **79** 3660 (1997).

[27] J.-L. Barrat and W. Kob, Europhys. Lett. **46**, 637 (1999).

[28] R. Di Leonardo, L. Angelani, G. Parisi and G. Ruocco, cond-mat/0001311.

[29] T. S. Grigera and N. E. Israeloff, cond-mat/9904351; S. Ciliberto *et al.*, private communication.

[30] D. J. Thouless, P. W. Anderson and R. G. Palmer, Phil. Mag. **35** 593 (1977).

[31] A. Crisanti and H.-J. Sommers, J. Phys. I (France) **5** 805 (1995); A. Crisanti, H. Horner and H-J Sommers, *Z. Phys.* B **92** 257 (1993)

[32] A. Barrat, R. Burioni, and M. Mézard, J. Phys. A **29** L81 (1996).

[33] M. Goldstein, J. Chem. Phys. **51** 3728 (1969); F. H. Stillinger and T. A. Weber,Science **225** (1984) 983. For a recent review , see F. H. Stillinger, Science **267** 1935 (1995).

[34] T. B. Schroder, S. Sastry, J. C. Dyre and S. C. Glotzer, cond-mat/9901271.

[35] F. H. Stillinger, J. Chem. Phys. **88** 7818 (1988).

[36] G. Biroli and R. Monasson, cond-mat/9912061.

[37] S. Franz ang G. Parisi, J. Physique I **5** 1401 (1995); Phys. Rev. Lett. **79** 2486 (1997).

[38] M. Cardenas, S. Franz and G. Parisi, cond-mat/9712099.

[39] This is reviewed in more details in M. Mézard and G. Parisi, Proceedings of the Trieste conference on *Unifying Concepts in Glass Physics*, cond-mat/0002128, to appear in J. Phys.

[40] L. Angelani, G. Parisi, G. Ruocco and G. Viliani, cond-mat/9904125.

[41] M. Mézard and G. Parisi, Phys. Rev. Lett. **82** 747 (1998).

[42] M. Mézard and G. Parisi, J. Chem. Phys. **111** 1076 (1999).

[43] R. Monasson, Phys. Rev. Lett. **75** 2847 (1995).

[44] M. Mézard, Physica A **265** 352 (1999).

[45] B. Coluzzi, M. Mézard, G. Parisi and P. Verrocchio, J. Chem. Phys. **111** 9039 (1999).

[46] B. Coluzzi, G. Parisi and P. Verrocchio, *Lennard-Jones binary mixture: a thermodynamical approach to glass transition*, cond-mat/9904124.

[47] B. Coluzzi, G. Parisi and P. Verrocchio, *The thermodynamical liquid-glass transition in a Lennard-Jones binary mixture*, cond-mat/9906124.

[48] Y. Rosenfeld and P. Tarazona, Mol. Phys. **95** 141 (1998).

[49] Some recent references are S. Sastry, P. G. Debenedetti and F. H. Stillinger, Nature **393** 554 (1998); W. Kob, F. Sciortino, and P. Tartaglia, Europhys. Lett. **49** 590 (2000), *ibid.* cond-mat/9910476; S. Büchner and A. Heuer, cond-mat/9906280.

[50] T. Keyes, J. Chem. Phys. A **101** 2921 (1997).

[51] F. Sciortino, W. Kob, and P. Tartaglia, Phys. Rev. Lett. **83** 3214 (1999); F. Sciortino, W. Kob, and P. Tartaglia, cond-mat/9911062.

[52] A first step in this direction has been taken recently in L. Santen and W. Krauth, cond-mat/9912182.

# CHAPTER 16

# GEOMETRICAL FRUSTRATION AND MARGINAL CONSTRAINT

A. P. RAMIREZ [1]

Bell Laboratories, Lucent Technologies
600 Mountain Ave., Murray Hill, New Jersey 07974-0636, U.S.A.

## 16.1 INTRODUCTION

In the early 70's, Phil Anderson articulated the key to understanding spin glasses when he wrote on a blackboard in Aspen "frustration is the name of the game". Spin glasses were already known to possess quenched disorder, so this observation quickly became part of a central tenet - that both disorder and frustration are the necessary microscopic ingredients of spin glasses [1, 2]. We now know that these ingredients give rise to a low energy spectrum consisting of a rough landscape of nearly degenerate states. When disorder is removed from the system, but frustration still exists on a periodic lattice, one has a qualitatively different problem. In the context of magnetism, this problem was first discussed in 1950 by Wannier [3] and Houtappel [4] in the realization of Ising spins on a triangular lattice interacting antiferromagnetically (TIA) with Hamiltonian $\mathcal{H} = J\sum_{i,j} \mathbf{S}_i \cdot \mathbf{S}_j$, where $J$ is a nearest-neighbor exchange energy. In Wannier's paper, an argument attributed to Phil Anderson describes a set of ground states with zero modes the number of which grows exponentially with system size. These modes therefore yield a macroscopic entropy at temperature $T = 0$, or in other words, spectral weight shifts down to energies much lower than the two-body interaction energy $J$. It is rare to find good experimental realizations of model problems but the spectral weight shift predicted for the TIA is robust and observed in a variety of different guises in a class of materials known as *Geometrically Frustrated Magnets* (GFMs) [5]. Phil Anderson didn't publish a great deal on this topic. Nevertheless, his influence is felt in this area of Condensed Matter Physics, as in many others. In this review, we discuss experimental realizations of ground

[1] Present address: Condensed Matter and Thermal Physics, Los Alamos National Laboratories, Los Alamos, NM 87545, U.S.A.

state entropy, present a broader class of GFMs, and discuss what geometrical frustration would look like outside of a magnetic context, in particular for a lattice dynamics problem.

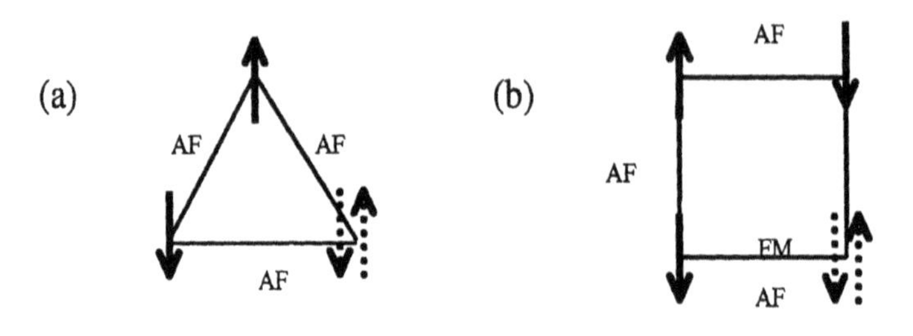


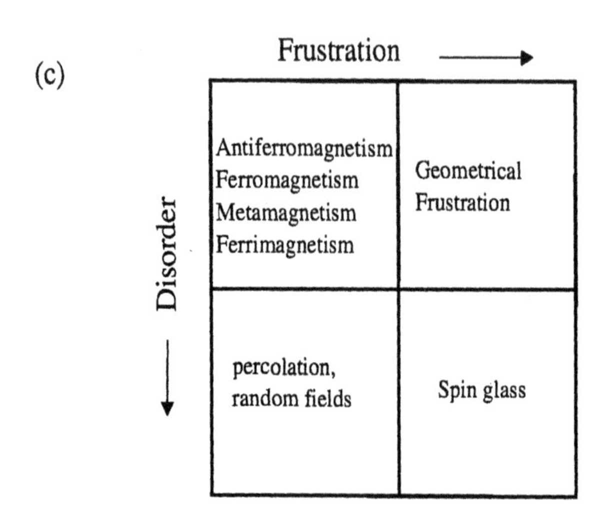


Figure 16.1: (a) Geometrical frustration on a triangular plaquette. There are six equivalent energy configurations per plaquette. (b) Frustration due to a defect. The ferromagnetic bond is representative of a ligand defect in an otherwise periodic antiferromagnetic lattice. (c) The different magnetic ground states as a function of disorder and frustration.

## 16.2 GEOMETRICAL FRUSTRATION

To describe the basic features of geometrical frustration we appeal to a triangular plaquette, as shown in figure 1a. Clearly for antiferromagnetic (AF) interactions, it is not possible to satisfy all the bonds simultaneously. Thus, the lowest-energy state is one which has a higher energy per spin than if either i) the interaction were ferromagnetic (FM) or ii) the spins were arrayed on a square lattice. Thus, there is an incompatibility between the symmetry of the interaction in the AF case, and the spatial symmetry of the lattice where a unique doubling of the unit cell is not permitted (i.e., it is nonbipartite). When the triangular plaquette is replicated to form an extended periodic lattice, the ground state of the entire spin system will have a higher energy than, say, the FM case, as was shown by Wannier [3]. This situation is to be contrasted with the situation in Fig. 1b where a square plaquette of AF spins has an 'impurity' FM bond introduced. Here the ground state is also higher in energy than if the FM bond were not present. If one considers an extended square-plaquette system, then wherever FM bonds are introduced, frustration will exist locally - frustration arises, and is inextricable from the quenched disorder that induces the FM bond. The former type of frustration is called geometrical frustration and the latter spin glass frustration. These two types of frustration can be compared with other traditional magnetic systems as shown in Fig. 1c.

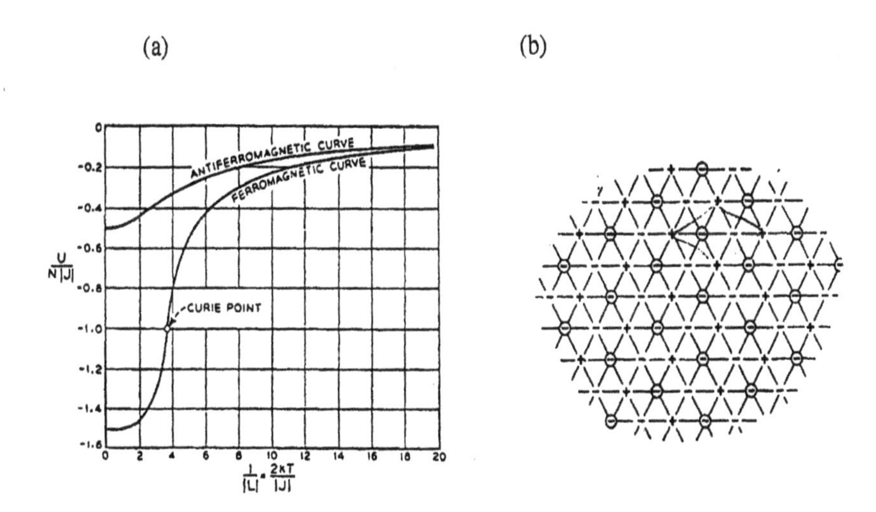


Figure 16.2: (a) Internal energy calculated for the triangular Ising systems for antiferromagnetic (top) and ferromagnetic (bottom) nearest neighbor interactions,. (b) Example of a lowest-energy ground state, showing the spins (circled) which can be flipped at no energy cost (from Wannier [3]).

The above illustration emphasized the effect of frustration on the internal energy, but for experiments, we are mainly interested in the free energy. The difference between the two cases is also manifested here. For the triangle of Ising spins, there are three degenerate lowest-energy states (and their time-reversed pairs), as opposed to one for the FM case. Wannier showed how this local degeneracy grows exponentially with the number of spins in the system, thus leading to macroscopic entropy. For the TIA, 47 % of the total Rln 2 entropy is accounted for in the lowest energy states. An example of such a state is shown in Fig. 2b where it is seen that the circled spins can be reversed with no energy cost. Zero modes are also predicted for Heisenberg systems in cases where a ground state can be defined. In spin glasses the degeneracy caused by frustration also leads to low energy states and associated low-temperature removal of entropy. The low-energy states here are strongly influenced by quenched disorder and yield a $T$-linear specific heat $C(T)$, as shown by Anderson, Halperin, and Varma [6]. Nevertheless, the number of low energy states varies in proportion to the impurity density, which can also be quite small, e.g. $< 1\%$ for CuMn spin glasses. Thus for spin glass frustration, the spectral weight associated with the frustration can be very small and results from structural disorder.

In geometrically frustrated magnets (GFMs), the symmetry incompatibility is between the local antiferromagnetic (AF) interaction, and the global symmetry imposed by a triangular crystal structure. The heurisitic device illustrating this is AF Ising spins on a triangle, but frustration effects are also observed for isotropic spins. There is now a class of materials which exhibit frustration effects and which have common microscopic properties [5]. Microscopically, they have triangle-based magnetic lattices, and isotropic spins (except for the special case of spin ice which is Ising-like, as described below). The macrosopic effects of GFMs are for example i) spin-liquid effects [7, 8], ii) spin-glass states for immeasurably small amounts of quenched disorder [9], and iii) novel phase behavior in applied fields [10, 11, 12].

The other condition besides symmetry-incompatibility necessary for GFMs is under- or marginal-constraint. Underconstraint can arise from rarefaction of interactions. In percolative systems, reduction of $T_c$ is achieved by randomly reducing the mean-field coupling strength - experimentally one does this by preparing a dilute magnet, e.g. randomly placed magnetic spins in a nonmagnetic salt. There is also directed rarefaction, e.g. low-D systems. Random percolation is a trivial way to reduce $T_c$ since this reduces the mean field energy simply by reducing the number of nearest neighbors. In GFMs, on the other hand, $T_c$ is reduced while maintaining a substantial mean-field energy. For nearest-neighbor interactions, a 2D Ising antiferromagnet on a square lattice will order. However, on a kagome lattice, ordering doesn't occur, despite there being the same mean field energy, which is defined by i) the exchange interaction, ii) the spin value, and iii) the number of nearest neighbors (nn), four, in both cases. The difference lies in the connectivity of the next nearest neighbors (nnn). For the kagome case the nnn's are unique for each nn, while for the square lattice they are not and hence the system is overconstrained. So even though the spins are strongly interacting on the kagome lattice, the triangles can be thought of as isolated units to first approximation. Moessner and Chalker have quantified this argument using Maxwellian counting arguments [13].

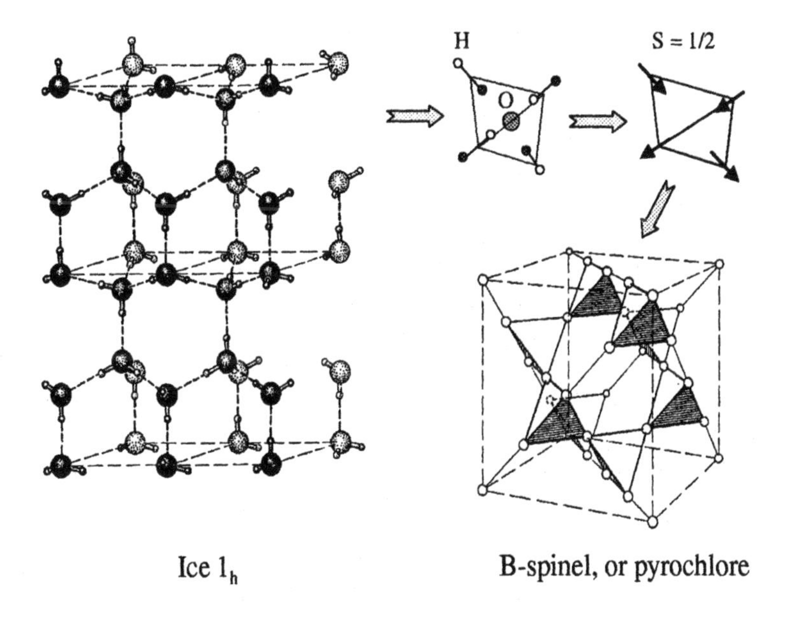


Figure 16.3: Left - crystal structure of ice showing the oxygen atoms (large spheres) and the hydrogen atoms (small spheres) (from Pauling [16]). Also shown is a schematic of the distorted tetrahedron formed by the hydrogens around each oxygen and the analogous spin system. Below is shown the pyochlore lattice (*A* or *B* sites only) (from Anderson [14]).

## 16.3 ORDINARY WATER ICE

The ideas of symmetry-incompatibility and under-constraint come together in ordinary water ice, which is probably the first realization of geometrical frustration. In the 1930's a discrepancy existed between value of entropy of water measured at 298 K measured spectroscopically, 45.10 cal/moleK, and that determined by integration of the specific heat divided by temperature of ice and water from T $\sim$10 K up to 289K, 44.28 cal/moleK [15]. Pauling explained this discrepancy by appealing to the unusual structure of ice as shown in Fig. 16.3 [16]. In ice (Ih), the oxygen atoms occupy a wurzite-type structure with four hydrogen ions placed in an approximate tetrahedral coordination around each O ion. Because the O-O distance is 2.76 Å, significantly greater than twice the O-H distance, 0.96 Å it was deduced that two H ions on each tetrahedra are closer than the other. There are six different arrangements by which this can happen, similar to the three degenerate states of the TIA - Pauling suggested this leads to a residual entropy of Rln $3/2$ = 3.4 J/moleK = 0.81 cal/moleK, which is in good agreement with the above difference of 0.82 cal/moleK. This entropy difference represents frozen-in orientational disorder among the $OH_2$ molecules, which results from the intrinsic interactions of dynamical variables the symmetry of which is incompatible with the extended symmetry of the O atoms.

## 16.4 SPIN ICE IN PYROCHLORES

Harris *et al.* have used the similarity of water ice and the pyrochlore structure, each of which contains corner-sharing tetrahedra (see Fig. 16.3), to suggest that there might exist distinct dynamics associated in a magnetic analog [17]. They showed that a system of Ising doublets with FM interactions occupying the magnetic sites of a pyrochlore lattice maps onto the water ice problem. For such a system, the lowest energy state is two-spins-in and two-spins-out. This is compared to the AF configurations of lowest energy which are all-in, or all-out. We see that there are six possible two-in two-out states, and, in addition, the forcing of a particular configuration on one tetrahedron isn't sufficient to specify the states on neighboring tetrahedra in the pyrochlore structure, which is a result of marginal constraint [13]. By comparison, if the all-in or all-out case has a lower energy, as might occur if superexchange exceeds the dipole-dipole interaction, then specifying one tetrahedron configuration fully determines a state of long-range order. Because the FM state has a large ground state degeneracy, it is geometrically frustrated, and the AF case is not. If the spin is due to a large-moment rare-earth ion, then dipole-dipole energies can easily exceed the superexchange interaction as we will see below.

The pyrochlore compound $RE_2Ti_2O_7$, where $RE$ is a rare earth ion, is an ideal system to study these effects. Here the $Ti^{4+}$ ion is nonmagnetic and all the rare-earths can be substituted on the $A$-site. Special attention is paid to $RE$ = Dy and Ho, since these are both commonly observed to have Ising-type spins, and have FM Weiss constants, and are therefore possible realizations of Harris et al's. spin ice. The susceptibility of $Dy_2Ti_2O_7$, was measured from 1.2 to 4.2 K [18] and shows a broad peak at 1.5 K. For $Dy_2Ti_2O_7$, $\chi(T)$ was extended to lower temperatures and a peak was found at 0.7 K with an inflection point at 0.6 K and a reduction to a small percentage of the peak value below 0.5 K [19]. For $Dy_2Ti_2O_7$, Blöte et al noted the absence of long-range order in this compound and the small size of the specific heat $C(T)$, peak in measurements which extend only up to 1.5 K [19].

The $\chi(T)$ data of Harris *et al.* [17] show that $RE$ = Ho has $\theta_W$ = +1.9 K, i.e. FM. Ramirez *et al.* report $\theta_W$ = +0.5 K for $RE$ = Dy. Thus both compounds are spin-ice candidates. Neutron scattering measurements of the crystal field levels for $Ho^{3+}$ in = $Ho_2Ti_2O_7$ show the first excited state above an Ising-like ground state doublet ($^5I_8, m_J = | \pm 8\rangle$) at 21 meV [20]. This is well-above the energy scale for spin-spin interactions, as inferred from $\theta_W$. These authors point out that since the crystal structure varies little for different $RE$-substitutions, the crystal field parameters obtained from the $RE$ = Ho measurement allows an accurate determination of the splittings for the other compounds. In particular, a similar magnitude of splitting is found for $RE$ = $Dy^{3+}$ ($^6H_{15/2}, m_J = | \pm 15/2\rangle$); for both Ho and Dy, the easy axis is along the line joining the spin with the tetrahedron center. Harris *et al.* performed elastic neutron scattering and $\mu$SR on $Ho_2Ti_2O_7$ [17]. They found no evidence of a magnetic transition in $\mu$SR down to 0.05 K. The neutron measurements showed no evidence for FM order in the temperature-dependence of the nuclear Bragg peak on cooling to 0.35 K but did show field-induced hysteretic behavior in scattering at $\mathbf{Q} = (\frac{1}{2}, \frac{1}{2}, \frac{1}{2})$, suggestive of nontrivial dynamics.

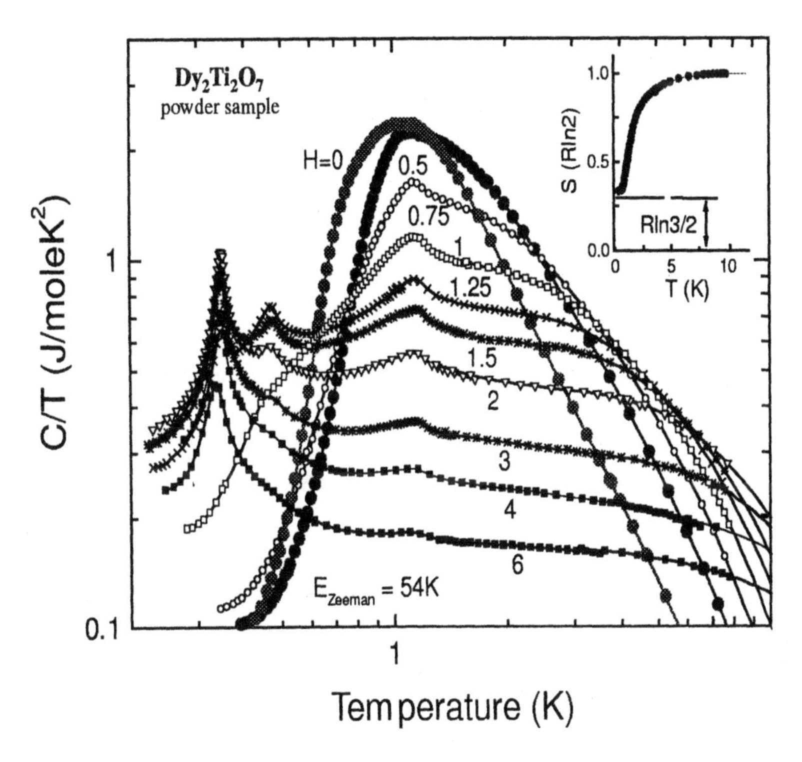


Figure 16.4: Inset: Entropy versus temperature for the spin ice compound $Dy_2Ti_2O_7$. The dashed line shows the prediction of Pauling for water ice. The agreement between this value and the entropy shortfall suggest a zero-temperature disorder similar in nature to the hydrogen disorder in ice. Specific heat versus temperature for the spin ice compound $Dy_2Ti_2O_7$ at several different field values. The sharp field-induced peaks indicate transitions to long range order. The lack of field dependence of the peaks is unusual and suggests ordering among spins transverse to the field.

Thermodynamic evidence for an ice-like state was found in $Dy_2Ti_2O_7$. Ramirez *et al.* extended the $C(T)$ measurements of Blöte *et al.* to 12 K [21]. They found that the total integrated entropy amounted to only (67 + 0.04) % of Rln 2, as shown in Fig. 16.4. Since Pauling predicted for ice a ground state entropy of ln 3/2, normalized to O, this means the recovered entropy when normalized to H, is 70.7 % of Rln 2, and we see good agreement between the recovered entropy in both systems, which is strong evidence that $Dy_2Ti_2O_7$ has a finite ground state entropy, similar to water ice. On application of a large field, transitions are observed in $C(T)/T$ in the form of three sharp peaks at $T$ = 0.34, 0.47 and 1.12 K as shown in Fig. 16.4. These peaks are unusual in two respects. First, they represent only a few percent of the total spin entropy - this is most likely due to the polycrystalline nature of the sample. Second, and most surprisingly, the peak positions do not depend on the magnitude of the applied field. Therefore, independent of the details of possible spin-ordering models, these peaks can not be due to longitudinal spin fluctuations - they are still observable at $H = 6$ T where the Zeeman energy for longitudinal spins is 55 K, i.e. a Boltzmann factor $g\mu_B\tilde{S}H/k_BT$ of 157 for the lowest-temperature peak.

The occurrence of sharp peaks in thermodynamic quantities in finite field when there were none in zero field is rare enough [5], but $Dy_2Ti_2O_7$ represents a special case. A possible explanation rests on the observation that when $\mathbf{H}$ is applied along certain directions, e.g. (110), a large fraction of the spins have their Ising axis transverse to $\mathbf{H}$ [21]. The only effect of $H$ on these spins is through their coupling to the nontransverse spins. Ordering among transverse spins can then occur for fields large enough to remove the frozen randomness of the nontransverse spins.

## 16.5 KAGOME-LIKE SYSTEMS

The above example of spin ice showed the effect of geometrical frustration in an Ising spin system. However, Ising systems are rare and, as was mentioned, there exists a class of GFMs which possess Heisenberg spins. One of the most intensively studied compounds in this class is the quasi-2D kagome-like system, $SrCr_{9p}Ga_{12-9p}O_{19}$. This compound was originally found by Obradors *et al.* to possess a Weiss constant of ~-500 K but no long range order down to 4 K [22]. Ramirez *et al.* showed that a spin glass transition occurred at ~3.5 K, and that below 6 K, the specific heat $C(T) \propto T^2$ [7]. It is now known that while the spin glass response of the susceptibility, $\chi(T)$, is suppressed with a modest magnetic field, $H \simeq 1$ T, $C(T)$ is insensitive to magnetic field up to field values ~6T as shown in Fig. 16.5 [23]. A $T^2$ form is expected for antiferromagnetic magnons in two dimensions, so it is tempting to attribute this behavior to long wavelength excitations in an ordered system. However, neutron scattering measurements by Broholm, Aeppli *et al.* [24] showed no evidence for long range order. Instead, they observed the correlation length to grow only weakly on cooling, reaching its maximum of ~7 Å at 1.5 K. There is presently no explanation of the origin of the $T^2$ behavior.

Although we are unable at the present to understand the nature of the low energy excitations in $SrCr_{9p}Ga_{12-9p}O_{19}$, the behavior of $C(T)$ in Fig. 16.5

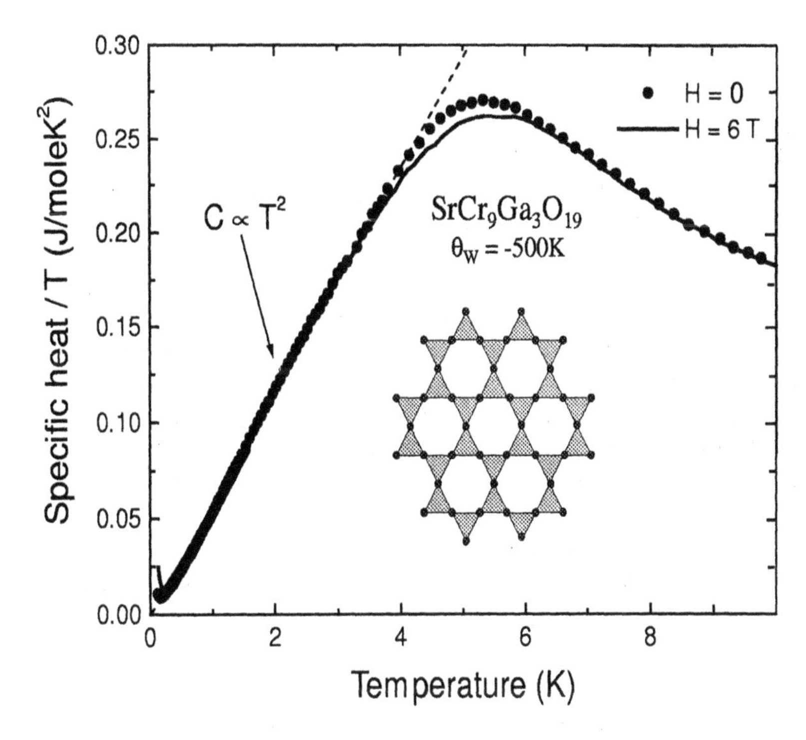


Figure 16.5: Specific heat of the kagome-like compound $SrCr_9Ga_3O_{19}$ versus temperature. The total amount of entropy under the curve is large ($\sim$ 15%) considering a Weiss constant of $\sim$500 K. The lack of field dependence is shown by the data taken in a field of 6 T. The origin of the $T^2$ behavior is not understood at present.

illustrates the large spectral weight downshift in this compound. Despite a Weiss temperature $\theta_W$ $\sim$-500 K, roughly 15 % of the entropy is removed below 10 K, i.e., almost two orders of magnitude lower than $\theta_W$. The anomalous $T^2$ behavior can therefore be seen as one example of the unexpected many-body effects that can arise from spectral weight downshift in a GFM. A clue to the origin of the SWD in kagome systems is found in the exact diagonalization studies of spin-1/2 Heisenberg lattices [25]. In this work, it is found that the kagome system possesses a larger number of excited singlet levels, than does the triangular system. While the $T^2$ behavior seen in experiments on $SrCr_{9p}Ga_{12-9p}O_{19}$ could not be reproduced in these studies, the field independence has been observed [26] and it is felt that correlated singlet modes hold the key to understanding the low-energy excitations in $SrCr_{9p}Ga_{12-9p}O_{19}$.

## 16.6 GEOMETRICAL FRUSTRATION IN NON-MAGNETIC SYSTEMS

As was already emphasized, symmetry incompatibility, marginal constraint, and the resulting SWD are the key signatures of geometrical frustration in magnetism. It is instructive to ask what form these signatures might adopt in a nonmagnetic context, and we take as one example, lattice dynamical systems. In GFMs, the correlated behavior that is suppressed by frustration is an order-disorder transition. In a lattice context we are interested mainly in displacive, or soft mode, transitions, so we ask, how can a soft mode be frustrated and what might the experimental consequences be? This line of thought is motivated by recent experiments on $ZrW_2O_8$ which has been observed to display unusally large, negative thermal expansion (NTE) ($\alpha \equiv (\partial \ln L/\partial T)_P \simeq 10^{-5}/T$), which is also isotropic and constant between 50 and 420 [27]. Since the disappearance of the soft mode results in large NTE in the critical region, we will be asking whether the large temperature region of NTE is the result of frustrating the softening, i.e. frustrating the transition, in order to make formal contact to the magnetic examples already discussed.

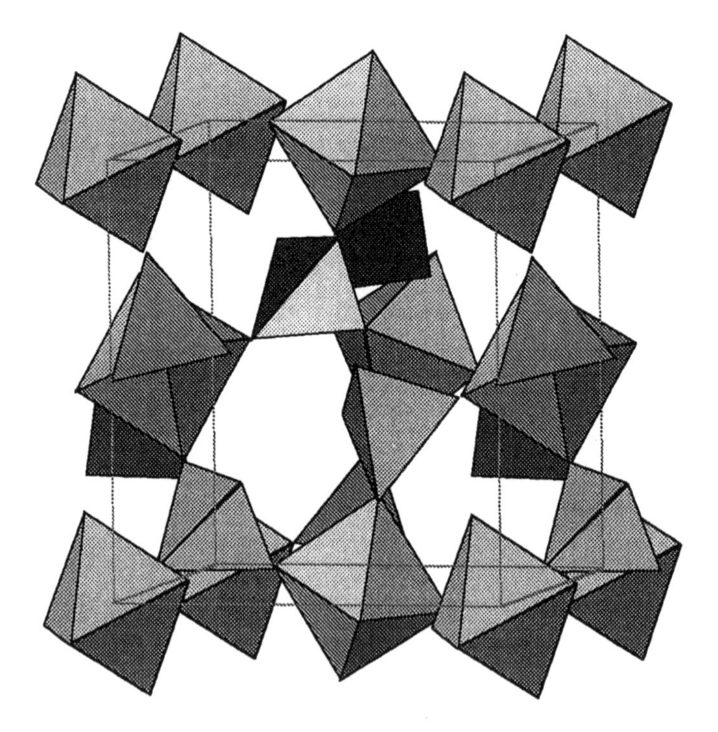

Figure 16.6: Inset: Unit cell of $ZrW_2O_8$ showing the $ZrO^6$ octahedra and the $WO_4$ tetrahedra. The unusual structural feature is the unshared vertices of the $WO_4$ tetrahedra.

The structure of $ZrW_2O_8$, is comprised of $ZrO_6$ octahedra and $WO_4$ tetrahedra (Fig. 16.6). The unusual feature in this cubic material is that every $WO_4$ tetrahedron has one unshared vertex. It is expected that the Debye-Waller factors for the corresponding oxygens, $O_3$ and $O_4$ sites, will be larger that that of the other atoms in the unit cell [28], indicating that modes incorporating these atoms will be highly anharmonic. Materials such as $ZrW_2O_8$ where the atoms are represented as rigid units such as octahedra and tetrahedra frequently exhibit soft mode transitions [29].

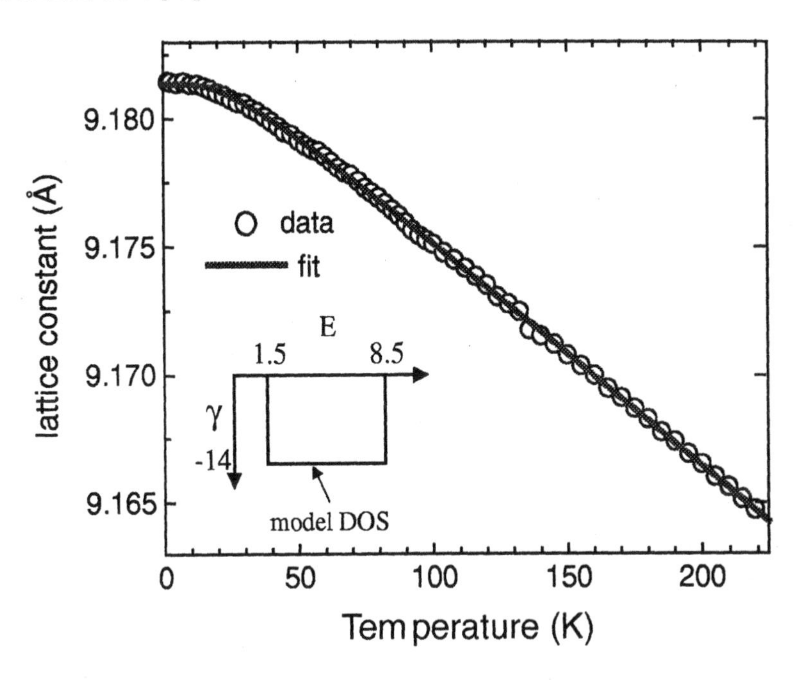


Figure 16.7: The lattice constant versus temperature for $ZrW_2O_8$ (circles). The solid line is a fit to the data using a square density of states, as shown in the inset, and the measured specific heat $C(T)$, and bulk modulus. The only adjustable parameter is the Grüneisen constant (-14) of the model density of states.

Such a transition is a common response to anharmonic motion, which suggests that $ZrW_2O_8$ might exhibit a soft-mode transition. However, using a dynamical matrix method, Pryde *et al.* show that the low-energy modes of $ZrW_2O_8$ do not possess high symmetry features in reciprocal space [30]. Such features have been used in other rigid unit materials to predict the symmetry of soft modes. The absence of soft modes is also an indication of the inability of the system to realize a particular lower symmetry state into which it can transform to remove the anharmonicity. In fact, $ZrW_2O_8$ does not exhibit any structural transformations below 420 K, despite the large anharmonicity.

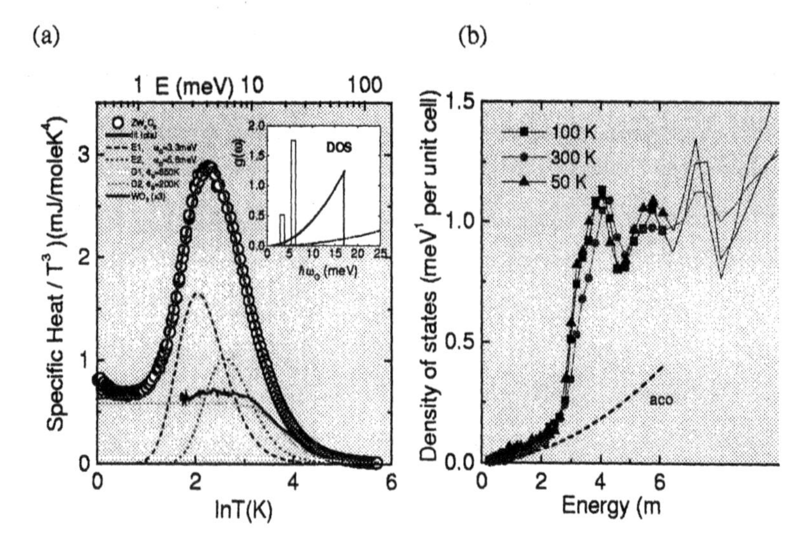


Figure 16.8: (a) Specific heat divided by temperature of $ZrW_2O_8$ from 2 K up to 300 K, versus $\ln T$. The large peak structure is indicative of several Einstein modes with energies in the range of 5 meV. The data were fit by a combination of Debye and Einstein forms, as shown by the smooth lines. (b) Phonon density of states determined by neutron scattering. The peaks correspond in energy to the Einstein peaks seen in specific heat.

The inability of $ZrW_2O_8$ to undergo a soft mode transformation is similar in spirit to the inability of a GFM to order into an AF state. Following the above discussion on frustration and underconstraint, it is useful to ask whether there is a significant shift of spectral weight to low energies as a result of frustration. The dynamical matrix simulations that indicated low-symmetry modes also predicted a large density of states at low frequency (Pryde, Hammonds *et al.* 1996). Measurements of the specific heat of $ZrW_2O_8$ provided the first evidence of such low energy modes [31] (Fig. 16.8). These measurements revealed a contribution to the low-temperature $C(T)$ which is not only large, comprising 6 oscillators/unit cell, but originates from a set of optical modes centered around 3-6 meV, an energy scale much lower than that observed in typical transition metal oxides. Direct measurements of the phonon density of states (Fig. 16.8) confirmed the $C(T)$ measurement, and showed how these low-energy states could lead to NTE [32]. It was found that using the measured density of states, approximated by a square distribution, the measured $C(T)$ and bulk modulus, NTE in $ZrW_2O_8$ could be fit very well by assuming an average Grüneisen coefficient of -14 as shown in Fig. 16.7. Studies of related compounds will help determine the microscopic origin of low energy modes in this material. Further measurements must be done to establish the origin of this low energy scale and in particular the relevance of a GFM-like picture of frustrated soft modes.

ACKNOWLEGMENTS

I would like to thank my collaborators on the experimental work reviewed here, R. J. Cava, G. R. Kowach, C. Broholm, G. Ernst, B. S. Shastry, A. Hayashi, and R. D. Siddharthan.

BIBLIOGRAPHY

[1] K. Binder, and A. P. Young, Rev. Mod. Phys. **58(4)** 801-976 (1986).

[2] J. A. Mydosh, *Spin glasses: an experimental introduction.* (London, Taylor and Francis, 1993).

[3] G.H. Wannier, Phys. Rev. **79** 357 (1950).

[4] R. M. F. Houtappel, Physica **16** 425 (1950).

[5] Ramirez, A. P. (1994). "Strongly Geometrically Frustrated Magnets." Annu. Rev. Mater. Sci. 24: 453-480.

[6] P.W. Anderson, B. Halperin and C. M. Varma, Philos. Mag. **25** 1 (1972).

[7] Ramirez, A. P., G. P. Espinosa and A. S. Cooper, Phys. Rev. Lett. **64** 2070 (1990).

[8] Schiffer, P., A. P. Ramirez, D. A. Huse, P. L. Gammel, U. Yaron, D. J. Bishop and A. J. Valentino, Phys. Rev. Lett. **74** 2379-2382 (1995).

[9] Greedan, J. E., M. Sato, X. Yan and F. S. Razavi (1986). "Spin-glass-like behavior in $Y_2Mo_2O_7$, a concentrated, crystalline system with negligible apparent disorder." Solid State Commun. 59: 895-897.

[10] Hov, S., H. Bratsberg and A. T. Skjeltorp (1980). "Magnetic phase diagram of gadolinium gallium garnet." J. Magn. Magn. Mater. 15-18: 455.

[11] Ramirez, A. P. and R. N. Kleiman (1991). "Low Temperature Specific Heat and Thermal Expansion in the Frustrated Garnet $Gd_3Ga_5O_{12}$." J. Appl. Phys. 69: 5252-5254.

[12] Schiffer, P., A. P. Ramirez, D. A. Huse and A. J. Valentino (1994). "Investigation of the field induced antiferromagnetic phase transition in the frustrated magnet: gadolinium gallium garnet." Phys. Rev. Lett. 73: 25002503.

[13] Moessner, R. and J. T. Chalker (1998). "Low-temperature properties of classical geometrically frustrated antiferromagnets." Phys. Rev. B 58: 12049-12062.

[14] P.W. Anderson, Phys. Rev. **102** 1008-1013 (1956).

[15] Giaque, W. F. and J. W. Stout (1936). "The Entropy of water and the third law of thermodynamics. the heat capacity of ice from 15 to 273 K." J. Amer. Chem. Soc. 58: 1144-1150.

[16] Pauling, L. C. (1945). The Nature of the Chemical Bond. Ithaca, Cornell University Press.

[17] Harris, M. J., S. T. Bramwell, D. F. McMorrow, T. Zeiske and K. W. Godfrey (1997). "Geometrical frustration in the ferromagnetic pyrochlore $Ho_2Ti_2O_7$." Phys. Rev. Lett. 79: 2554-2557.

[18] Cashion, J. D., A. H. Cooke, M. J. M. Leask, T. L. Thorp and M. R. Wells (1968). "Crystal growth and magnetic susceptibility measurements on a number of rare-earth compounds." J. Mater. Sci. 3: 402407.

[19] H. W. J. Blöte, R. F. Wielinga and W. J. Huiskamp, Physica **43** 549-568 (1969).

[20] Siddharthan, R., B. S. Shastry, A. P. Ramirez, A. Hayashi, R. J. Cava and S. Rosenkranz (1999). "Ising pyrochlore magnets: low temperature properties, ice rules and beyond." Phys. Rev. Lett. 83: 1854-1857.

[21] Ramirez, A. P., A. Hayashi, R. J. Cava, R. Siddharthan and B. S. Shastry (1999). "Zero Point Entropy in Spin Ice." Nature 399: 333-335.

[22] Obradors, X., A. Labarta, A. Isalgue, J. Tejada, J. Rodriguez and M. Pernet (1988). "Magnetic frustration and dimensionality in $SrCr_8Ga_4O_{19}$." Solid State Commun. 65: 189.

[23] Ramirez, A. P. (2000). "Entropy Balance and Evidence for Local Spin singlets in a Kagome-like Magnet." Phys. Rev. Lett. 84.

[24] C. Broholm, G. Aeppli, G. P. Espinosa and A. S. Cooper, Phys. Rev. Lett. **65** 3173-3176 (1990).

[25] Lecheminant, P., B. Bernu, C. Lhuillier, L. Pierre and P. Sindzingre (1997). "Order versus disorder in the quantum Heisenberg antiferromagnet on the kagome lattice using exact spectra analysis." Phys. Rev. B 56: 2521-2529.

[26] P. Sindzingre, G. M., C. Lhuillier, B. Bernu, L. Pierre, Ch. Waldtmann and H. U. Everts (2000). "Magnetothermodynamics of the spin-1/2 Kagome Antiferromagnet." Phys. Rev. Lett. 84.

[27] Mary, T. A., J. S. O. Evans, T. Vogt and A. W. Sleight (1996). "Negative Thermal Expansion from 0.3 to 1050 Kelvin in $ZrW_2O_8$." Science 272: 90-92.

[28] Evans, J. S. O., T. A. Mary, T. Vogt, M. A. Subramanian and A. W. Sleight (1996). "Negative thermal expansion in $ZrW_2O_8$ and $HfW_2O_8$." Chem. Mater. 8: 2809-2823.

[29] Dove, M. T., V. Heine and K. D. Hammonds (1995). "Rigid unit modes in framework silicates." Mineral. Mag. 59: 629-639.

[30] Pryde, A. K. A., K. D. Hammonds, M. T. Dove, V. Heine, J. D. Gale and M. C. Warren (1996). "Origin of the negative thermal expansion in $ZrW_2O_8$ and $ZrV_2O_7$." J. Phys. Condens. Matter. 8: 10973-10982.

[31] Ramirez, A. P. and G. R. Kowach (1998). "Specific heat of $ZrW_2O_8$." Phys. Rev. Lett. 80: 4903-4906.

[32] Ernst, G., C. Broholm, G. R. Kowach and A. P. Ramirez (1998). "Phonon density of states and negative thermal expansion in ZWO." Nature 396: 147-149.

# CHAPTER 17

## OLFACTION AND COLOR VISION: MORE IS SIMPLER

J. J. HOPFIELD
Department of Molecular Biology, Princeton University
Princeton, New Jersey 08544, U.S.A.

### ABSTRACT

The human eye has three kinds of receptor cells for color vision. Our entire experience in the subtleties of color in the natural world comes from the different wavelength responses of the light receptors of these three kinds of cells. We can distinguish hundreds to thousands of hues based on these three channels. Since color is experienced as independent of light intensity over a wide range of intensity, our rich experience of color is founded on the ability to measure two ratios at each point in visual space. We assign colors on the basis of those ratios.

Modern molecular biology has shown that the mammalian nose has about 2000 kinds of receptor cells [1, 2]. An odor is characterized by the excitation pattern across the receptor cells [3, 4, 5]. Suppose that it was essential for an olfactory animal to distinguish a million different odors. If we can distinguish 100 colors with three channels and two intensity ratios, we should be able to similarly distinguish 100 odors with 2 intensity ratios, and to distinguish $100^{(n-1)/2}$ odors with $n$ channels. As few as 9 odor channels should enable us to distinguish many million odors.

Under these circumstances, why do we have so many odor receptors? Why has biology chosen a 'large-$n$' route to the olfactory system? We begin this paper with a description of the computational problems faced by the color vision and olfactory systems. The zero-order problems are isomorphic, but the complexities attendant to the real-world situations are so disparate that rather different algorithms are relevant for these two senses. We will show that one way of dealing with this difference is to immensely increase the number of information channels in the olfactory case, leading to an algorithm that utilizes the statistical aspects of the situation [6]. This algorithm can be *simply* implemented by neurobiology [7].

While other algorithms requiring a small number of channels can be constructed, I see no corresponding simple way to implement them in a brain. Thus, 'more' provides a simplicity and effectiveness of neurobiological implementation, in addition to being different [8].

## 17.1 COLOR VISION

Vision is a remote sense. We use our eyes to identify the objects around us and their relationship to us. To remotely identify an object, we need first to identify *properties of an object* from the signals reaching us, and then to use those properties to identify or classify the object. If, for simplicity, we think of an object as having uniform surface properties, then the spectrum of the light reaching our eyes from that object is a product of the spectrum of the illuminating source and the normalized reflection spectrum of the object. In the world in which the mammalian eye evolved, the spectrum of the source was a modification of the spectrum of sunlight (direct sunlight, or sun filtered through a cloud, reflected from the moon, etc.). To begin, think of the spectral shape (but not the intensity) of the illuminating source as fixed. Under these circumstances, the total intensity of light reflected from an object is not an intrinsic property of the object. However, the reflection spectrum is an intrinsic property. When the illuminating light source spectrum is known and fixed, the ratio of the reflected light at two different wavelengths is a measure of the ratio of the reflection coefficients of the object at these two wavelengths, an *intrinsic* property of the object. Thus *an intrinsic property of an object can be deduced* from measuring the light coming to the eye from the same point but at two different wavelengths. By contrast, no intrinsic property of the object can be deduced from the total intensity of light coming from a single point.

Color vision is thus a way to measure the intrinsic normalized reflection spectra of objects. Typical objects in a natural scene have reflection spectra containing a couple broad peaks and valleys. Each of the three kinds of visual receptors (the so-called red, green, and blue cones) has a broad wavelength response. Each has an output that is a convolution of the illuminating spectrum, the reflection spectrum of the object, and the broad spectral response of the photoreceptor. Three channels, two ratios, is far less information than would be necessary to reconstruct the reflection spectrum of the object. However, one must remember that the purpose of our visual system is not to reconstruct the reflection spectrum, but is merely to characterize it well enough to be able to help in the construction and recognition of objects.

What complexities does the color vision system face? Spatial resolution is an issue, for red, green, and blue cone-cells cannot occupy the same point in the focal plane on the retina. As a result, our color vision has worse resolution than our gray-scale vision. (That fact allowed color to be added to the original black-and-white television broadcast signals without increasing bandwidth of a

TV channel. Television images have only the illusion of high resolution color.) A second problem arises from the fact that the illuminating spectrum is not invariant. Sunlight near sunset is very different from the sunlight of noon. The light in the shade of a woods comes from reflections from the trees, and is not that of sunlight itself. The light in the sharp shade of a building near midday comes from atomospheric scattering, which introduces a bias to the solar spectrum. These illumination problems are approximately solved by the nervous system by a comparison of the light reflected from an object with the light reflected from a surrounding region [9]. If that region is regarded as a statistical sample of general objects, then it can be used to construct a normalization of the incident spectrum. The perceived 'color of an object' is then not determined simply by 'the spectrum of the light coming from the object' but it is rather determined by 'the spectrum of light coming from the object relative to the spectrum coming from nearby objects.' An illusion can easily be set up in which two different locations in the visual field are sending exactly the same spectrum to the eye, but one appears red while the other appears green. The existence of such an illusion is a paradox until one remembers that *the purpose of the visual system is to correctly deduce properties of objects, not local properties of images.* A third problem arises from differential adaptation of the red, green, and blue cones (and higher processing areas) by a prolonged exposure to a biased environment. These effects are short-lived and appear to have little functional significance.

## 17.2 OLFACTION

We tend to think of olfaction in primitive terms with the major task one of identifying an object held close to the nose - I will refer to this as the basic olfactory task. However, animals use odor for remote sensing as well as for proximal sensing. For remote objects, wind patterns which bring the scent identify the direction and approximate distance of an odor source. In addition to the basic task, an animal can perform the following:

1) Background elimination. When a weak known odor is thoroughly mixed with an unknown background, the known odor can be identified and its intensity measured.

2) Component separation. When a few known odors are thoroughly mixed (eliminating relative fluctuations), an animal can identify the component odors and their intensities.

3) Odor separation. When odors from different objects are mixed by air turbulence, the fluctuations of the relative contributions to the mixture can be used to separate the odors of multiple unknown objects in the environment.

To complicate matters, all the tasks must be done in the presence of residual adaptation or residual odorant binding to many of the sensory cell types [6].

## 17.3 COMPARATIVE PROBLEMS OF VISION AND OLFACTION

The complications of vision and of olfaction are entirely different. Ordinary visual objects are opaque, so that the spectrum of light arriving at a given point of the retina is that which comes from a single object. The color of the leaves which are occluded by a cardinal do not confuse the identification of the color of the bird. However, a given packet of air has a long history and contains odorants from a variety of sources that have been mixed by turbulence and diffusion. The analogous visual situation would be that of trying to deduce the color of a semitransparent bird that is in front of other colored objects, a problem we cannot visually deal with. By contrast, the visual system can deal with the varied spectral source by using the surrounding areas for normalization.

A competent cook can perform the olfactory component separation task, as in smelling a sauce and listing the four spices that are in it. The eye has no such ability. When light from two known natural objects is mixed, one merely sees a new natural color and has no immediate feeling that this is a mixture of two things one already knows.

The olfactory task 3) above describes the way in which the time-structure of fluctuations of odors can be used to decompose an olfactory scene into multiple objects. Time-structure involving physical motion is used in the visual system to group features into objects ('what moves together, groups together' or 'features that move more or less rigidly with respect to the rest of the visual field should be grouped as an object'). There are no comparable significant uses of time-structure analysis in color computations.

## 17.4 THE LOGARITHMIC DISTRIBUTION OF ODORANT BINDING CONSTANTS

Each of the 2000 receptor cell types expresses a single kind of odorant binding receptor protein on its surface. The binding of a ligand to a receptor is a physical measure on a molecular piece of an object. Information about odorant-receptor binding constants comes chiefly from studies of the threshold for perceptual detection of pure compounds, which vary over an immense range [10]. Humans can detect methanol at a concentration (in air) of $2 \times 10^{-5}$ Moles /liter (M/l), isoamyl acetate at $3 \times 10^{-10}$ M/l, and methoxy-isobutylpyrazine at $2 \times 10^{-14}$ M/l. In a study of 529 odorants [11], the distribution of the logarithms of threshold concentrations was approximately Gaussian (i.e., the distribution of free energies of binding is Gaussian.) The 2 standard deviation ($\pm 2\sigma$) range within which most olfactants lay was 6.8 $\log_{10}$ units, with the extremes separated by 10 $\log_{10}$ units. The threshold for detection of a particular pure odorant will correspond to some minimum coverage $cov_{min}$ of the odorant receptors of a given type (or more probably, a few types) by odorant ligands. Assuming that $cov_{min}$ is the same for all odorants, then the range of perceptual threshold concentrations reflects the range of binding constants. Few binding constants have been measured directly.

For 0.01 % coverage the product of the binding constant (in liters/mole) and the odorant concentration (in moles/liter) is 0.0001. For isoamyl acetate, an odorant with an average threshold, 0.0001 coverage of a receptor type at threshold would imply a binding constant $K = 3 \times 10^5$ l/M. This number can be equivalently stated as the volume available $V_p$ per free odorant molecule when that odorant concentration in air is in equilibrium with half site coverage, namely $V = K \times 1{,}000/6 \times 10^{23}$. For isoamyl acetate $V = 5 \times 10^{-16}$ cc/molecule. $V$ is related to $E_b$ (the binding free energy of contact between the two molecules), $V_b$ (the translational volume to which a bound molecule is restricted) and $\gamma$ (the reduction in angular orientation on binding) by the Boltzmann factor

$$V = V_b \gamma \exp E_b / k_B T. \tag{17.1}$$

$V_b$ is roughly $10^{-25}$ cc, and $\gamma$ about $10^{-3}$. From these expressions the contact binding free energy $E_b$ is on the scale of 29 $k_B T$ or 17 Kcal/M for an average odorant.

The observed enormous range of binding constants is an elementary consequence of the exponential. A 35 % range of binding energies, or 6 Kcal/M, which can easily result from changing the shape of an odorant molecule or of a binding pocket, will result in binding constants ranging from $10^2$ to $10^{11}$ l/M. (It should be noted that these energies are not what a chemist would term binding free energies, for the chemist's definition involves the use of a standard concentration of 1 mole/liter. A binding constant of 1.000 liters/mole is *defined* by chemists *as having a binding free energy of zero* while in fact it requires a binding contact free energy $E_b$ of about 8 Kcal.)

## 17.5 ODORANT MODELING

The pattern of excitation across the receptor cell types describes an odor. To understand such patterns, it is necessary to describe the binding constants of a particular ligand for receptors that do *not* bind it maximally. The best-fit ligand-receptor pairs have a $\pm 2\sigma$ range of binding constants of more than $10^6$ and thus a range of binding free energies of 10 Kcal. The same kind of stereochemistry should produce a similar range of binding constants and binding free energies below the maximal ones. While odors differ markedly in behavior, there is a typical psychophysical range of a factor of 1,000 over which the odor seems to have the same quality, and over which the relationship between the perceived intensity and the actual concentration is approximately a logarithmic or low power law. An odor presented at the concentration where non-linearities begin is driving the cells having the highest affinity receptors into response saturation. For this driving strength we would expect, from the above numbers, that about 1/2 of the cell types would show responses to that odorant and about 1/2 would show no response (i.e., no response visible above noise levels).

Assays are available from studying responses of many cells for an array of odorants both by conventional electrophysiology [12] and by cell biology techniques [13]. For the concentrations of odorants used, 1/4 to 1/2 the cells showed some level of response above noise levels to a particular odorant qualitatively similar to the expectations described above. There is no quantitative summary of experimental information available at present.

The following specific model of binding constant distributions is used for analysis. We presume that the receptor cell sends a signal related to its ligand coverage, and that the ligand coverages due to different ligands are additive [1]. We assume that a sensory cell has a dynamic range of 1,000 in odorant concentration and that the probability distribution of binding constants of the less-than maximal cell classes for a given odor is uniformly spread (on a logarithmic scale) over a range $10^6$ below the binding constant of the most sensitive cell type. While the model is specific, most of the qualitative conclusions of the present analysis rely only on the model being consistent with the experimental fact seen directly in the data of Sicard and of Malnic, namely that *for unrelated odorant pairs a, b there will be hundreds of cell types that respond chiefly to a, hundreds that respond chiefly to b, and hundreds more that respond to both.*

At threshold concentration $c^{\lambda}_{thresh}$ of ligand $\lambda$, only one cell type responds above noise levels in a minimally detectable fashion. This type is associated with receptors having a coverage of $cov_{min}$, so these receptors have a binding constant of $cov_{min}/c^{\lambda}_{thresh}$. The coverages of all other receptor cell types will lie in a range of $10^6$ below $cov_{min}$. When the concentration is raised to 10 $c^{\lambda}_{thresh}$, then 2,000/6 = 333 cell types should respond above noise. These are the cell types that have their binding constants in the range of 0.1-1 times the binding constant of the most strongly binding cell types. As the concentration is raised further more cell types respond, until at concentration 1,000 (in units of the threshold concentration) about half of cell types will be responding above noise levels.

Unrelated molecules have no particular relationship of their binding contacts and binding energies for a given receptor type. Each receptor cell type is independently assigned binding constants for each odorant by the prescription above. Real odors can be related, either because they have dominant chemicals of similar structure (e.g., ethanol and isopropranol), or because they are mixtures in different ratios of the same chemicals (orange and tangerine). While related odors can be

[1] The real situation is somewhat more complicated. The responses may not be additive. The latency of sensory cell response can be odorant specific. The mucous layer and the proteins it contains may play some role in kinetics. The overall situation is reminiscent of the depth perception computation in vision. There are several different algorithms involved, including binocular stereopsis, shape from shading, linear perspective, occlusion, and structure from motion. Visual studies have gained much by analyzing each separately. Similarly, for olfaction it is useful to analyze how concentrations and binding can be used alone.

described within the present context, doing so requires knowing yet unknown details of the mixtures and their binding patterns.

## 17.6 OLFACTORY TASKS IN A LEAST-SQUARED ERROR ALGORITHM

For simplicity, let a known odor $t$ be a single molecular species. This molecule has binding constants $K_i^t$ for receptors on cells of type $i$. When odorant $t$ is present at concentration $c^t$, the fractional coverage $cov_i^t$ of binding sites of type $i$ will be

$$cov_i^t = c^t K_i^t \tag{17.2}$$

Let the fraction of coverage for some unknown odorant be given by $cov_i^u$. This unknown odorant should be identified as $t$ if it is approximately true that there exists a value of $c$ such that

$$cov_i^u \simeq cK_i^t \tag{17.3}$$

for all values of $i$. (This is clearly the case if $u$ happens to be the same molecule as $t$.) In what sense is this approximate inequality to be viewed, given that there will be noise in the system? The simplest assumption (which we will shortly question) is to use a mean square error criterion, and thus to recognize the substance $u$ as being in fact $t$ if the mean square error in Eq. 17.2 is less than some error measure $\epsilon$ (which might itself depend on the value of $c$ thus determined). If such a system knew $N$ odors, and $u$ was picked from among those, it would be necessary on average to search through $N/2$ of them to find which odor was present. To be assured of finding the best fit odorant, all $N$ comparisons must be tried. Since the amount of hardware necessary to memorize $N$ random odors of size $n$ is of order $nN$, this is the minimal size of neural hardware which will be necessary to solve the problem. There are a variety of ways that this simple computation can be implemented in neurobiology with 'wetware' of scale $nN$ and requiring a time of order 1.

The problem of identifying the components of a mixture of two odors can be similarly cast. For two known odors $s$ and $t$, the coverage due to a mixture will be

$$cov_i^{st} = c^s K_i^s + c^t K_i^t. \tag{17.4}$$

Let the fraction of coverage for some unknown binary mixture be $cov_i^u$. This unknown odorant should be identified as the mixture of $s$ and $t$ if it is approximately true that there exist values of $c$ and $c'$ such that

$$cov_i^u \simeq c' K_i^s + cK_i^t. \tag{17.5}$$

For any given $s$ and $t$, simple analysis will yield the best $c$ and $c'$ (mean square error criterion) as long as $n > 2$. However, it is on average necessary to search through $N(N-1)/4$ pairs of possible $s$ and $t$ before finding the correct one, requiring a time of order $N$ on the same hardware.

To compound the computational difficulty, adaptation of responses and continued binding of odorant molecules lead to circumstances in which the response of a subset of receptor cells is badly wrong due to an odor which was recently present. In such a case, a mean-square error measure is inappropriate. If $n = 10$ but two of the channels are badly corrupted by adaptation or continued binding, the correct question for a mean-square error criterion is: 'Is there a $c$ such that some subset of 8 channels meets the mean square error criterion?' To answer this question, all 45 possibilities of selecting the 8 channels must be examined, and the computation requires then 45 times the computing resources.

The combinatorial explosion of separating multiple sources and the difficulty of dealing with realistic sources of error in odor channels suggest that animals, with their limited and slow neural 'wetware', *must not solve the problem in this fashion.* In addition, approaching the problem in this fashion, does not solve the essential problem of trying to recognize a known odor in the presence of an unfamiliar background. We note, however, that no use has yet been made of the large number $n$ of independent channels: Indeed, so far, large-$n$ would appear only to add to the computational difficulties and the necessary size of the resources required.

## 17.7 AN APPROACH THROUGH LARGE-$n$

We develop a representation that describes an odor-induced excitation pattern across receptor cell types *from the viewpoint of a target odor.* For simplicity, let the target $t$ be a single odorant species, and the background be a different single species $b$. Following the odor model above, the coverage of receptors of type $i$ in the presence of concentrations $c_t$ and $c_b$ is given by

$$cov_i = cov_{min}\left(\frac{c^t}{c^t_{thresh}}\right)[1 \text{ or } f^t_i] + cov_{min}\left(\frac{c^b}{c^b_{thresh}}\right)[1 \text{ or } f^b_i]. \quad (17.6)$$

In the argument within $[\cdots]$ in the first term, a '1' is used if it happens that $i$ is the cell type that maximally binds $t$, and otherwise an $f^t_i$ is chosen at random in the range between 1-$10^{-6}$ with a uniform probability distribution in the logarithmic domain. The same is true in the second term for odorant $b$. Equation 17.6 is exactly Eq. 17.2 in a slightly different notation if $c_b = 0$, or like Eq. 17.4 if $s$ is thought of as an unknown background.

If only odorant $t$ is present, then from each cell type $i$ which is appreciably driven the system can 'calculate' the concentration $c_t$ using Eq. 17.3. When an *unknown* odor $u$ is presented that generates a pattern of coverage $cov_i^u$, then for each $i$ which would be driven by the *target* odor we can calculate the concentration $c_i$ of $t$ which would yield that level coverage for cells of type $i$, namely

$$c_i = \frac{c_{thresh}}{f_i^t} \frac{cov_i^u}{cov_{min}}. \tag{17.7}$$

The quantity $c_i$ is the *apparent* concentration of $t$ deduced from the level of receptor occupancy caused by the unknown odorant in a single channel $i$. The observation of $cov_i^u$ is a 'vote' for the presence of $t$ at concentration $c_i$.

Figures 17.1 (a)-(c) show histograms of all the 'votes' $c_i$ when odor $u$ is, in fact, $t$ presented at a concentration 10, 100, and 1,000 (in units of the threshold concentration). A logarithmic scale is used because of the large range of concentrations involved. This scale is also particularly apt for systems that show an approximately logarithmic response as a function of intensity, as many sensory systems do. A Gaussian noise level of 26 % ($\sigma = 0.1$ $\log_{10}$ units) introduces width to the histograms. Because the unknown odor $u$ is in fact $t$, each channel deduces (within noise) the correct concentration from its value of $cov_i^u$. Of the 2000 receptor cell types, about 333, 667, and 1,000 are driven to observable levels by odorant $t$ at these concentrations respectively, corresponding to the total number of 'votes'. The concentration of $t$ can be read from the peak positions, located as expected. As the concentration is decreased another decade, the number of responding cell types approaches one.

Figures 17.1 (d)-(f) show histograms of the same sort, using the same 1,000 channels which are potentially activated by $t$ but calculated for three different odors $b$ unrelated to $t$, each at its saturation level of 1,000 $c_{thresh}^b$. Only those having more than the minimal detectable coverage respond. For each channel, there is a probability of 0.5 of this occurring. Since there is no relationship between $f_i^t$ and $f_i^b$, there is a wide spread of events within these histograms and no sharp peak. A histogram for a weaker presentation of $b$ would be similar, but with fewer events.

Target and non-target odors are highly distinguishable in this representation. Even when the target odor is at a strength of only $3c_{thresh}^t$, the peak in such a histogram for odor $t$ alone has about 150 votes spread over a total concentration range of 0.4 on a logarithmic scale. The probability that an *unrelated odor $u$ at a concentration* 1,000 $c_{thresh}^u$ produces a total number of events greater than 100 in such a range is less than $10^{-6}$. When the target odor $t$ is at strength 10 $c_{thresh}^t$, it becomes astronomically unlikely to confuse $t$ with a random, strong odor. *In this representation basic odor recognition can be accomplished by merely thresholding the total number of votes in a bin of appropriate width.*

The problem of identifying an odor in the presence of an unknown background is examined in Fig. 17.2 (a)-(c). These histograms were generated from an odor mixture in which the target odorant at strengths 10, 100, and 1,000 $c_{thresh}^t$ is mixed with an unrelated background odorant b at strength 1,000 $c_{thresh}^b$. The area

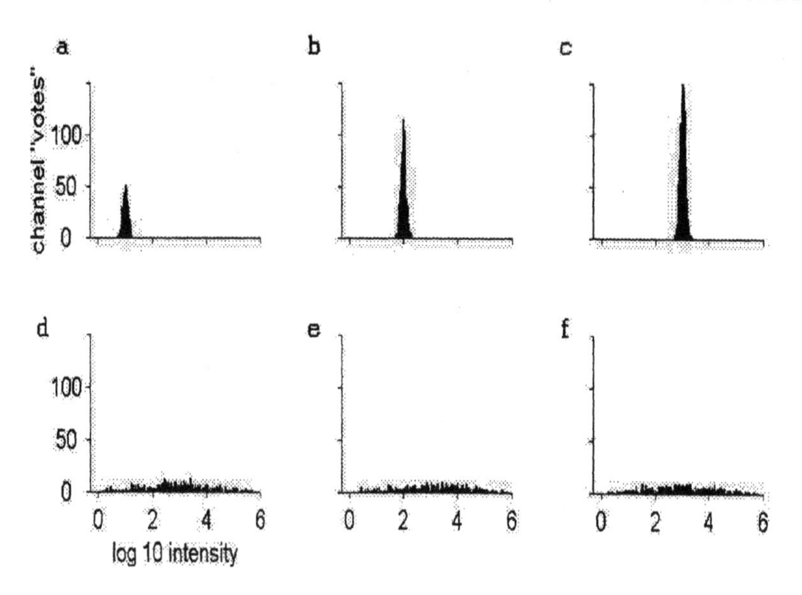


Figure 17.1: Histograms of the number of receptor cell types that 'vote' for various dimensionless intensities (or concentrations) of the target odor. Units of odor strengths are on a logarithmic scale; one unit is factor of 10 in intensity. Threshold intensity for an odorant corresponds to intensity 1, $\log I = 0$. (a)-(c) The target odor itself at concentrations 10, 100, and 1,000. (d)-(f) Three different non-target odors at concentrations 1,000.

of the peak in the histogram is decreased, but a simple threshold on the number of events in a suitable bin width will still distinguish between when the target odor is present and when it is not. The target begins to be detected when its concentration is greater than 3 $c^t_{thresh}$, even in the presence of the saturating concentration of an unknown odor. *There is no way to solve this problem when n is small.*

The statistics of large numbers and the chance occurrence of non-interference make identification in the presence of strong unknown odors possible. No receptor cell type is specific-each has a probability of 0.5 of being at least somewhat activated by a saturating concentration of any odor. This translates to about 333 channels being noticably driven when odor $t$ alone is presented at low concentration 10 $c^t_{thresh}$. However, each channel also has a probability of 0.5 of being negligibly driven by odor $b$ at strength 1,000 $c^b_{thresh}$. Thus, about $167 \pm 13$ channels are still expected to respond to $t$ at concentration 10 $c_t$ *in essentially the same way as they would have in the absence of the unknown background odor*, and many more will do so at higher $c_t$. These channels are responsible for the lowered but appropriately located histogram peaks in Fig. 17.2 (a)-(c) compared to Fig. 17.1 (a)-(c). Channels with responses that are distorted by the presence of the strong background odor produce the broad background.

Figures 17.2 (d)-(f) are like Figs. 17.2 (a)-(c), except that the background

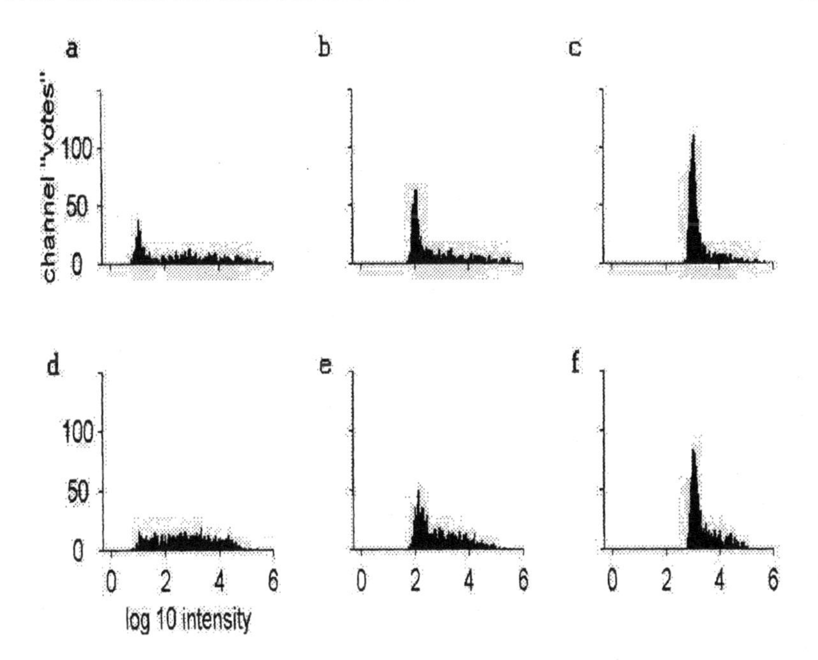


Figure 17.2: Histograms as in Fig. 17.1 (a)-(c), but with the simultaneous presence of a background of strength 1,000 $c^b_{thresh}$. (d)-(f) have a more complex background of total strength 400 in terms of the relevant thresholds, made up of an equal mixture of four non-target odors.

odor is made up of a mixture of four unknown odorants each at a strength 100 (in terms of its threshold). The total strength of the background is now 400. It is much harder to recognize a known single-molecular species in the presence of this complex background, but identification and quantification of a single odorant against a complex background that is several times stronger can still be done by the same method.

Separating a mixture into known components can also be solved in this same fashion. The fact that we could identify a known component of strength 100 in the presence of 4 unknown chemicals each of strength 100 (shown in Fig. 17.2(e)) indicates that a complex mixture of 5 known molecular components could be separated into its five components. When the number of components is raised to 7, this method no longer works. While it is mathematically possible to separate 7 or more known components, it requires more complex processing.

The effects of erroneous data can be understood in terms of these simulations. Sensory channels which are to some extent stuck 'on' are equivalent to channels which have additional input and whose effects are exactly like those of a background odor, contributing a wing to the right in Figs 17.2. Sensory channels that are strongly adapted contribute a wing to the left. Such effects on 50 % of the channels produce little impact on performance other than a small rise in the detection threshold and a small decrease in background rejection capability.

## 17.8 ON THE LARGE NUMBER $n$ OF CELL TYPES: MORE IS SIMPLER

The size of $n$ is qualitatively important. We have intuitive feelings for the case $n = 3$ from color vision. If light from different sources is mixed together, all channels will usually be mixed, and there is no reliable way to tell if a 'target' hue is present or not. For example, light of 5,100 Å and of 4,900 Å wavelengths are quite distinguishable shades of green but drive chiefly green and blue cones. Because of this, a particular mixture of 5,100 Å and 4,500 Å light produces an exact visual color match to 4,900 Å light. $n = 3$ is a small number.

In the case of olfaction, if $n = 5$ and random binding constants picked in the prescribed range, 6 % of odors will have significant binding to only one channel. Odors can then be divided into a set which drive more than one channel, and five sets each containing about 1 % of all molecules. Within each 1 % set all members are utterly indistinguishable. Appropriate levels of five primary odorant chemicals could duplicate any odor. For any primary odorant, unknown backgrounds which drive that unique odorant receptor (and about 50 % of backgrounds will do so) completely prevent the reliable detection of the target. About 25 % of odors will drive two receptor types. For these the ratio of the bindings can be used, generating some reliability, but there is only one ratio, and about 4 % of odors will drive those two receptors in the same ratio. Benefits of large numbers are suggestive, but unreliable. For $n = 100$, many ratios are always available and the statistics already reliable. (For example, the probability at saturation of driving fewer than 25 channels is less than $10^{-6}$.) Were it to turn out that the 2000 receptor types were composed of 100 different families with strong interfamily similarity, the large-$n$ algorithms would still work quite well.

Large $n$ is especially useful when the binding constants are broadly distributed, *producing a naturally sparse representation and appreciable probability of non-interference of two compounds, and thus leaving a statistically significant number of channels that are receiving unambiguous information about each compound* [2]

An elementary algorithm that merely looks for a high peak in an appropriate histogram calculated from receptor cell responses can adequately address the major olfactory tasks. A system was previously described for solving the olfactory 'analog match' problem [4] by using action potential timing relative to an underlying rhythm as the encoding of channel intensity, and using a time-delay network to organize the recognition of a particular odor. The *time-dependent input*

[2] Surprisingly, an ability to recognize words in speech in the presence competing background speech (cocktail party) can be achieved using this same general approach. The two problems are related by a desired invariance (odor intensity and the speed of speaking) and by the fact that there are uncorrupted measures available of properties of individual words when two people are simultaneously speaking [14,15].

to a recognition neuron during one period of the underlying rhythm *is exactly the histogram of* Figs. 17.1 and ??, with the intensity axis converted to time. The time at which each action potential arrives is its 'vote' in this histogram. If the near-simultaneous arrival of perhaps 50 action potentials is necessary to drive a receiving cell to fire, then that cell is performing a thresholding operation on the histogram. Thus *a neural network using time-delays and phase coding of intensity implements all the algorithms in the presence of noise in a time of order unity with order of $Nn$ 'hardware'*. If $n$ were only of order 10 instead of order 1,000, 100-fold less hardware would be necessary, but separating three known odors with this hardware would take of order $N^2$ time, could not deal reliably with strong unknown backgrounds, and would need even more time resources when some channels are corrupted. The large-$n$ algorithm permits the effective and rapid use of the minimal amount of parallel neural hardware.

Undoubtedly, there are other neural network schemes that also produce an effective implementation of this same algorithm. And while neurobiology is not based on 'grandmother cells', this representation does show how easy it would be to solve the problem using a few basic mechanisms available to neurobiology. The central point is that simple schemes can use non-interfering, uncorrupted channels to evaluate the situation, utilizing the fact that in a large system there will be many such channels.

## 17.9 SEPARATION OF TWO UNKNOWN ODORS

The large number of independent channels can be used in conjunction with adapting neurons to solve the problem of separating independent but unknown odors. Adaptation is one mechanism for comparing values of signals at different times. The basis for the algorithm is the fact that all the components of a single odor will fluctuate with the same time course, while two odors that come from different positions in space will be mixed by turbulence, and the two sets of components will have different fluctuations.

The representation essential to a 'two sniff for separation' paradigm can be generated by neurons that adapt. For adapting integrate-and-fire neurons, the cell potentials $u_i$ obey

$$\frac{du_i}{dt} = -\frac{u_i}{\tau} + I_{bias} - Ca_i + I_{sensory,i}, \tag{17.8}$$

where

$$I_{sensory,i} = \begin{cases} \ln(\frac{cov_i}{cov_{crit}}) & : \text{ if } cov_i/cov_{crit} > 1 \\ 0 & : \text{ if } cov_i/cov_{crit} < 1, \end{cases}$$

and

$$\frac{dCa_i}{dt} = -D + \quad \text{action potential term (but } D \text{ shuts off if } Ca_i = 0).$$

The sensory input current of a driven channel is proportional to the logarithm of the coverage of the odorant receptors driving that channel. When $u_i$ reaches a threshold, an action potential of negligible duration is generated, and $u_i$ is then reset to $u_{thresh} - \delta$. The $Ca_i$ term represents an inward $K^+$ current proportional to the internal $Ca^{++}$ concentration. The internal $Ca^{++}$ is depleted at a fixed rate $D$ by a $Ca^{++}$ pump, and is re-supplied by an inward fixed aliquot of $Ca^{++}$ (due to high potential $Ca^{++}$ channels or to the $Na^+$ channels themselves) that enters cell $i$ each time it generates an action potential. Figure 17.3a shows the spike rasters of 80 model neurons responding to two sniffs of a mixed odor.

Because reasonable plotting requires representing only a small fraction of the 2,000 channels, Fig. 17.3 has been presented without measurement noise so that what is happening is visible without such massive statistics being necessary. At $t$ = 100 msec., a constant odor mixture $ax + by$ was introduced, with $a = 50$ and $b = 1000$ in terms of saturation strengths. This mixture might represent a background $y$ that is 20 times as strong as an object $x$. About 65 of the neurons are visibly driven by this odor mixture, fire briefly at a high rate, and then adapt back to their baseline level in spite of the continued presence of the odor. At $t = 500$ milliseconds, a second sniff $\alpha x + \beta y$ is taken, with the object now 50 % stronger in intensity, and the background 20 % stronger (i.e., with $\alpha = 75$, $\beta = 1200$). The suggestible reader may see a subtle change in the raster patterns around that time. This odor remains fixed for the rest of the time, and the neurons again adapt to their basal firing rate.

Figure 17.3b shows the same data in a different representation. With each spike is associated an interspike interval (the interval to the previous spike of that neuron) and an instantaneous frequency (the reciprocal of that interspike interval). Each spike of Fig 17.3a is plotted in Fig. 17.3b. The data after $t = 500$ chiefly fall on three lines. The neurons that are negligibly driven by both odors lie on the bottom straight line. The neurons that are chiefly driven by the stronger odor $y$ generate the low arc. The smaller numbers of neurons that are chiefly driven by the weak 'object' $x$ generate the top arc. *In this representation of the firing pattern, the fact that there are two objects is completely obvious*, in spite of the fact that the raster plot looks devoid of interest in the relevant time interval. An arc in Fig. 17.3b goes up (or down) because the intensity of its corresponding odor went up (or down). If the odors had fluctuated in opposite directions, separation would have been easier.

The procedure succeeds because of the large value of $n$. If there were only 10 different channels involved, no obvious separation of trajectories would be seen. For small $n$, a totally different algorithm suitable for neural networks has been found [14], but it requires many statistically different samples for its input rather than just two, so it would not be as effective for making rapid decisions in a realistic olfactory environment. In addition, it works only when odors have comparable strengths, based on using the input signals in a linear fashion rather than based on using logarithms of inputs.

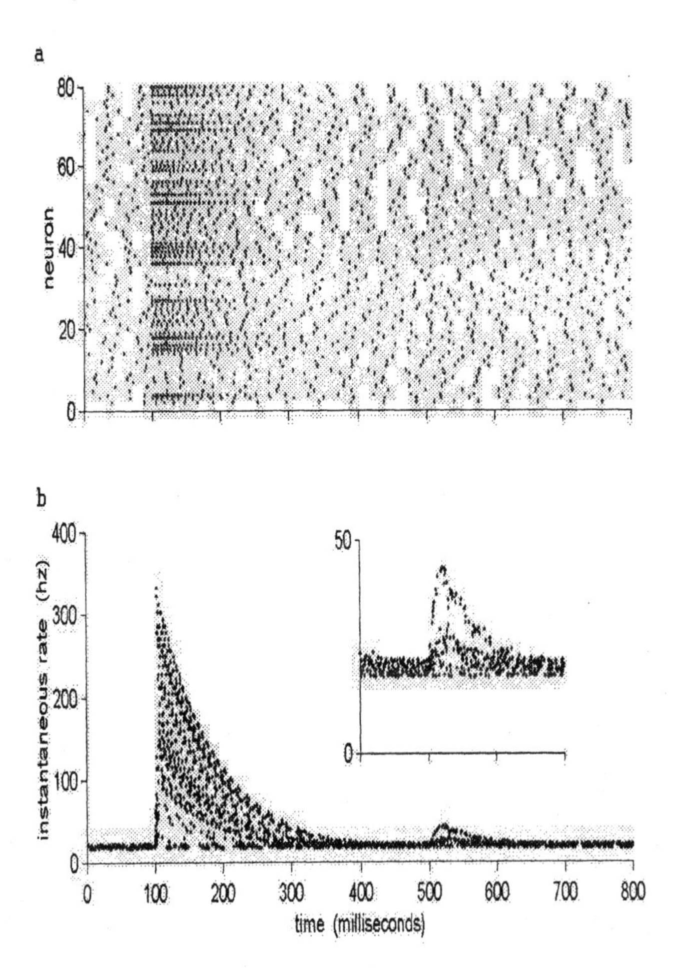


Figure 17.3: 80 adapting, integrate-and-fire neurons are exposed to two sniffs of an odor mixture in which the relative components have changed between the two sniffs. No odor was present before 100 msec. The time period 100-500 has a mixed odor present of the form $50x + 1000y$. At 500 milliseconds, the odor shifts to $75x + 1100y$ (i.e., the weak odor goes up by 50 %, the strong component goes up by 20 %, and their ratio changes by a factor of 1.25). (a) Raster plots of the spikes of the 80 neurons. The sniff at 100 milliseconds strongly activates more than half the neurons, after which they adapt. The changed sniff at 500 milliseconds is almost invisible. (b) The same data as in (a), but the y-axis is the instantaneous firing rate at the time of each action potential. The second sniff is now clearly visible, and most spikes appear to belong to one of three patterns. A 20 % spread in $D$ was included to produce parameter-spread noise.

## 17.10 MORE IS SIMPLER

Neurbiology creates massive amounts of slow, highly parallel hardware. Large-$n$ algorithms (algorithms that will not function when $n$ is small) may be the preferred solution to some of the computational problems of neurobiology.

This research was supported in part by National Science Foundation Grant ECS98-73463.

## BIBLIOGRAPHY

[1] Buck, L. B. and Axel, R. Cell **65**, 175-187 (1991).

[2] Buck, L. B.. Ann. Rev. Neurosci **19**, 517-544 (1996).

[3] Shepherd, B. M. In *Taste, Olfaction, and the Central Nervous System*, ed. Pfaff, D. W., Rockefeller Univ. Press 307-321 (1985).

[4] Holley, A. *In Smell and Taste in Health and Disease*, eds. Getchell, T. C., Doty, R. L., Bartoshuk. L. M. and Snow, J. B. Raven, New York, 329-344 (1991).

[5] Kauer, J. S. Trends Neurosci. **14**, 79-85 (1991).

[6] Hopfield, J. J. Proc. Nat Acad. Sci. (USA) **96**, 12506-12511 (1999).

[7] Hopfield, J. J. Nature **376**, 33-36 (1995).

[8] Anderson, P. W. Science **177**, 393- 396 (1972).

[9] Land, E. H. Proc. Nat. Acad. Sci (USA) **45**, 115-129 (1959).

[10] Cain, W. S., Cometto-Muniz, J. E., and Wijk, R. A. Serby, M. J., and Chobor, K. L., eds. Springer-Verlag, New York, 279-308 (1992).

[11] Devos, M. F., Patte, F., Rouault, J. and Laffort, P., IRL Press, Oxford (1990).

[12] Sicard, G. and Holley, A. Brain Res. **292**, 283-296 (1984).

[13] Malnic, B., Hirono, J., Sato, T. and Buck, L. B. Cell **96**, 713-723 (1999). **10**, 166-172 (1998).

[14] Hopfield, J. J. Proc. Nat. Acad. Sci. (USA) **88**, 6462-6466 (1991).

# CHAPTER 18

# SCREENING AND GIANT CHARGE INVERSION IN ELECTROLYTES

T. T. NGUYEN, A. YU. GROSBERG, AND B. I. SHKLOVSKII

Department of Physics, University of Minnesota
116 Church St. Southeast, Minneapolis, Minnesota 55455

## ABSTRACT

Screening of a macroion such as a charged solid particle, a charged membrane, double helix DNA or actin by multivalent counterions is considered. Small colloidal particles, charged micelles, short or long polyelectrolytes can play the role of multivalent counterions. Due to the strong lateral repulsion at the surface of a macroion, such multivalent counterions form a strongly correlated liquid with a short-range order resembling that of a Wigner crystal. These correlations create additional binding of multivalent counterions to the macroion surface with binding energy larger than $k_BT$. As a result, even for a moderate concentration of multivalent counterions in the solution, their total charge at the surface of macroion exceeds the bare macroion charge in absolute value. Therefore, the net charge of the macroion inverts its sign. In the presence of a high concentration of monovalent salt, the absolute value of the inverted charge can be larger than the bare one. This giant inversion of charge can be observed by electrophoresis or by direct counting of multivalent counterions.

## 18.1 INTRODUCTION

Although the subtitle of this book is "Fifty Years of Condensed Matter Physics", the broad concept of Condensed Matter, indeed the very words, became popular only recently, when the ideas initially spawned in low-temperature solid-state physics started making their way into other fields that deal with systems that are neither 'solid' nor at low-temperatures. The breadth and power of condensed matter theory ideas made it tempting to apply them also to biological problems. One of the most influential efforts in that direction was made by P. W. Anderson [1, 2].

In this work, we discuss a problem in which the ideas from low-temperature physics are applied in a straightforward manner. Specifically, many components of intracellular machinery, including DNA itself, are strongly charged. Screening of a biological macroion by highly charged counterions exhibits an effectively low-temperature behavior, because energies of Coulomb interactions of such counterions on the surface of the macroion exceed $k_BT$ by an order of magnitude or more. One of most interesting manifestations of these lateral interactions is the so-called charge inversion.

Charge inversion is a counterintuitive phenomenon in which a charged particle (a macroion) strongly binds so many counterions in a water solution that its net charge changes sign. As shown below, the binding energy of a counterion with large charge $Z$ is larger than $k_BT$, so that this net charge is easily observable; for instance, it is the net charge that determines linear transport properties, such as particle drift in weak-field electrophoresis. Charge inversion is possible for a variety of macroions, ranging from the charged surface of mica or other solids to charged lipid membranes, DNA or actin. Multivalent metallic ions, small colloidal particles, charged micelles, short or long polyelectrolytes can play the role of multivalent counterions. Recently, charge inversion has attracted significant attention [3, 4, 5, 6, 7, 8, 9, 10, 11].

Charge inversion is of special interest for the delivery of genes to the living cell for the purpose of gene therapy. The problem is that both bare DNA and a cell surface are negatively charged and repel each other, so that DNA does not approach the cell surface. The goal is to screen DNA in such a way that the resulting complex is positive [12]. Multivalent counterions can be used for this purpose. The charge inversion depends on the surface charge density, so the cell surface charge can still be negative when DNA charge is inverted.

Theoretically, charge inversion can be also thought of as an over-screening. Indeed, the simplest screening atmosphere, familiar from linear Debye-Hückel theory, compensates at any finite distance only a part of the macroion charge. It can be proven that this property holds also in non-linear Poisson-Boltzmann (PB) theory. The statement that the net charge preserves the sign of the bare charge agrees with common sense. One might infer that the validity of this statement extends beyond the PB-equation regime. However, it was shown [3, 4, 5] that this common-sense notion fails for screening by $Z$-valent counterions ($Z$-ions) with large $Z$, such as charged colloidal particles, micelles or rigid polyelectrolytes, because there are strong repulsive correlations between them when they are bound to the surface of a macroion. As a result, $Z$-ions form a strongly correlated liquid at the macroion surface, with properties resembling those of a Wigner crystal (WC). The negative chemical potential of this liquid leads to an additional "correlation " attraction of $Z$-ions to the surface. This effect is beyond the mean-field PB theory, and charge inversion is its most spectacular manifestation.

Let us demonstrate the fundamental role of lateral correlations between $Z$-ions in a simple model. Imagine a hard-core sphere with radius $b$ and with negative charge $-Q$ screened by two spherical positive $Z$-ions of radius $a$. One can see that if Coulomb repulsion between the $Z$-ions is much larger than $k_BT$, they are situated on opposite sides of the negative sphere (Fig. 18.1a).

If $Q > Ze/2$, each $Z$-ion is bound, because the energy required to remove it to infinity $QZe/(a+b) - Z^2e^2/2(a+b)$ is positive. Thus, the charge of the whole complex $Q^* = -Q + 2Ze$ can be positive. For example, $Q^* = 3Ze/2 = 3Q$ at $Q = Ze/2$. This example demonstrates the possibility of an almost 300% charge inversion. It is obviously a result of the correlation between $Z$-ions which avoid each other and reside on opposite sides of the negative charge. On the other hand, the description of screening of the central sphere in the PB approximation smears the positive charge, as shown on Fig. 18.1b and does not lead to charge inversion. Indeed, in this case, charge accumulates in a spherically symmetric screening atmosphere only until the point of neutrality, at which point the electric field reverses its sign and attraction is replaced by repulsion.

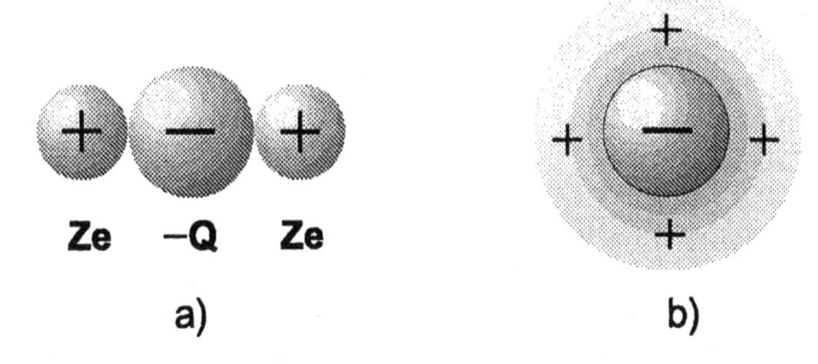


Figure 18.1: a) A toy model of charge inversion. b) PB approximation does not lead to charge inversion.

Weak charge inversion can be also obtained as a trivial result of $Z$-ion discreteness without correlations. Indeed, discrete $Z$-ions can over-screen by a fraction of the "charge quantum" $Ze$. For example, if the central charge $-Q = -Ze/2$ binds one $Z$-ion, the net charge of the complex $Q^* = Ze/2$. This charge is, however, three times smaller than the charge $3Ze/2$ obtained above for screening of the same charge $-Ze/2$ by two correlated $Z$-ions, so that for the same $Q$ and $Ze$, correlations lead to stronger charge inversion.

The difference between charge inversion, obtained with and without correlations, becomes dramatic for a large sphere with a macroscopic charge $Q \gg Ze$. In this case, discreteness by itself can lead to an inverted charge limited by $Ze$. On the other hand, it was predicted [5] and confirmed by numerical simulations [13] that, due to a WC-like short-range order of $Z$-ions on the surface of the sphere, the net inverted charge can reach

$$Q^* = 0.84\sqrt{QZe}, \tag{18.1}$$

i. e. can be much larger than the charge quantum $Ze$. This charge is still smaller than $Q$ because of limitations imposed by the very large charging energy of the macroscopic net charge.

In this paper, we consider systems in which the inverted charge can be even larger than the value predicted by Eq. (18.1). Specifically, we consider the problem of screening by $Z$-ions in the presence of monovalent salt, such as NaCl, in solution. This is a more practical situation than the salt-free one considered in Ref. [4, 5]. Monovalent salt screens long-range Coulomb interactions more strongly than short-range lateral correlations between adsorbed $Z$-ions. Therefore, screening by monovalent salt diminishes the charging energy of the macroion to a greater degree than the correlation energy of $Z$-ions. As a result, the inverted charge $Q^*$ becomes larger than that predicted by Eq. (18.1) and scales *linearly* with $Q$. The amount of charge inversion at strong screening is limited only by the fact that the binding energy of $Z$-ions eventually becomes smaller than $k_BT$, in which case it is no longer meaningful to speak about binding or adsorption. Nevertheless, remaining within the strong binding regime, we demonstrate that the magnitude of the inverted charge can exceed that of the original bare charge, by as much as a factor of 3. We call this phenomenon *giant charge inversion.* Its prediction and theory are the main results of our paper. A more detailed theory of giant charge inversion will be published elswhere [14]).

Since, in the presence of a sufficient concentration of salt, the macroion is screened on a distance scale smaller than its size, the macroion can be thought of as an overscreened surface, with inverted charge $Q^*$ proportional to the surface area. In this sense, the overall shape of the macroion and its surface are irrelevant, at least to a first approximation. Therefore, we consider here the simplest case: screening of a planar macroion surface carrying a negative surface charge density $-\sigma$ by a finite concentration $N$ of a positive $Z$-ion, a concentration $ZN$ of neutralizing monovalent coions, and a large concentration $N_1$ of a monovalent salt. Correspondingly, we assume that all interactions are screened with the Debye-Hückel screening length

$$r_s = (8\pi l_B N_1)^{-1/2},$$

where $l_B = e^2/(Dk_BT)$ is the Bjerrum length, $e$ is the charge of a proton, and $D \simeq 80$ is the dielectric constant of water.

Our goal is to calculate the two-dimensional concentration $n$ of $Z$-ions at the plane as a function of $r_s$ and $N$. In other words, we want to find the net charge density of the plane

$$\sigma^* = -\sigma + Zen \tag{18.2}$$

We are particularly interested in the maximum value of the 'inversion ratio' $\sigma^*/\sigma$ that can be reached at large enough $N$. There is a subtlety regarding the definition of $\sigma^*$. As the entire system, macroion plus overcharging $Z$-ions, is neutralized by the monovalent ions, it is fair to ask, what is the meaning of charge inversion? What is the justification for Eq. (18.2), which disregards monovalent ions?

Under realistic conditions, each $Z$-ion is bound to the macroion surface with a binding energy well in excess of $k_BT$. By contrast, the monovalent ions, which maintain electroneutrality over distances of order $r_s$, interact with the macroion with energies less than $k_BT$. This important distinction led us to define the net charge of the macroion to include the adsorbed $Z$-ions, but not the monovalent ions. This physically justifiable definition is amenable to experimental test. It is conceivable that the strongly adsorbed $Z$-ions are able to resist perturbations from the tip of an atomic force microscope, whereas the neutralizing atmosphere of monovalent ions cannot. One may, in principle, count the adsorbed $Z$-ions, thus directly measuring $\sigma^*$. To give an example, when $Z$-ions are the DNA chains, one can realistically measure the distance between neighboring DNAs adsorbed on the surface. Similar reasoning applies to an electrophoresis experiment in which the electric field is weak enough that the system remains in the linear-response regime. A macroion coated with bound $Z$-ions drifts in the field as a single unit because the field is too weak to affect the strongly bound $Z$-ions. By contrast, the surrounding atmosphere of monovalent ions, smeared over the distance $\sim r_s$, drifts with respect to the macroion. Defining the linear electrophoretic mobility of a macroion as the ratio of effective charge to effective friction, we find that only $Z$-ions contribute to the former, while monovalent ions contribute only to the latter. Most importantly, the *sign* of the effect – whether the macroion moves with or against the field – is determined by the net charge $\sigma^*$. Again, this includes the $Z$-ions, but not the monovalent ones. Furthermore, for a macroion with simple (e.g., spherical) shape, the absolute value of the net macroion charge may be obtained using mobility measurements and the standard theory of friction in electrolytes [15].

In Sec. II we consider screening of a charged surface by compact $Z$-ions such as charged colloidal particles, micelles or short polyelectrolytes, which can be modeled as spheres of radius $a$. We call such $Z$-ions 'spherical'. When adsorbed on the charge surface, spherical ions form a correlated liquid with properties similar to those of the two-dimensional WC (Fig. 18.2).

We begin with the simplest macroion, which is a thin, charged sheet immersed in water solution (Fig. 18.3a). This allows us to postpone the complication related to the image potential that appears in a more realistic macroion, which is a thick insulator charged at the surface (Fig. 18.3b).

We calculate analytically the dependence of the inversion ratio, $\sigma^*/\sigma$, on $r_s$ in the two limiting cases $r_s \gg R_0$ and $r_s \ll R_0$. Here,

$$R_0 = (\pi\sigma/Ze)^{-1/2}$$

is the radius a Wigner-Seitz cell at the neutral point $n = \sigma/Ze$ (as shown at Fig.

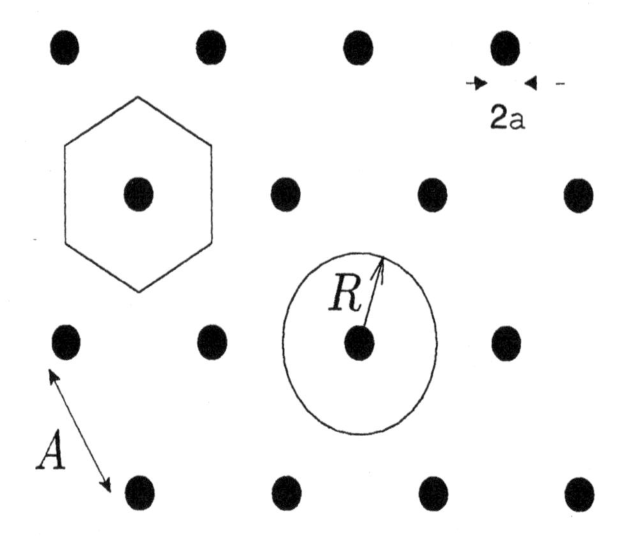


Figure 18.2: Wigner crystal of $Z$-ions (of radius $a$) on the background of surface charge. A hexagonal Wigner-Seitz cell and its simplified version as a disk with radius $R$ are shown.

18.2, we approximate the hexagon by a disk). We find that at $r_s \gg R_0$

$$\sigma^*/\sigma = 0.83(R_0/r_s). \tag{18.3}$$

and in the opposite limit, $r_s \ll R_0$,

$$\frac{\sigma^*}{\sigma} = \frac{\pi (R_0/r_s)^2}{2\sqrt{3}\,\ln^2(R_0/r_s)}. \tag{18.4}$$

Thus $\sigma^*/\sigma$ grows with decreasing $r_s$ and can become larger than 100%. We also present below a numerical calculation of the full dependence of the inversion ratio on $R_0/r_s$.

Let us return to more realistic macroions which have a thick insulating body with dielectric constant much smaller than that of water. In this case, each $Z$-ion has an image charge of the same sign and magnitude. The image charges repel the $Z$-ions and push the WC away from the surface. In this case, charge inversion is studied numerically throughout the entire range of $r_s$. The results turn out to be remarkably simple: the inversion ratio is a factor of 2 smaller than that for the charged sheet immersed in water. Thus, for a thick macroion, using spherical $Z$-ions, one can achieve about 100% charge inversion.

In Sec. III we study adsorption of long rod-like $Z$-ions with negative linear charge bare density $-\eta_0$ on a surface with a positive charge density $\sigma$. (We flipped

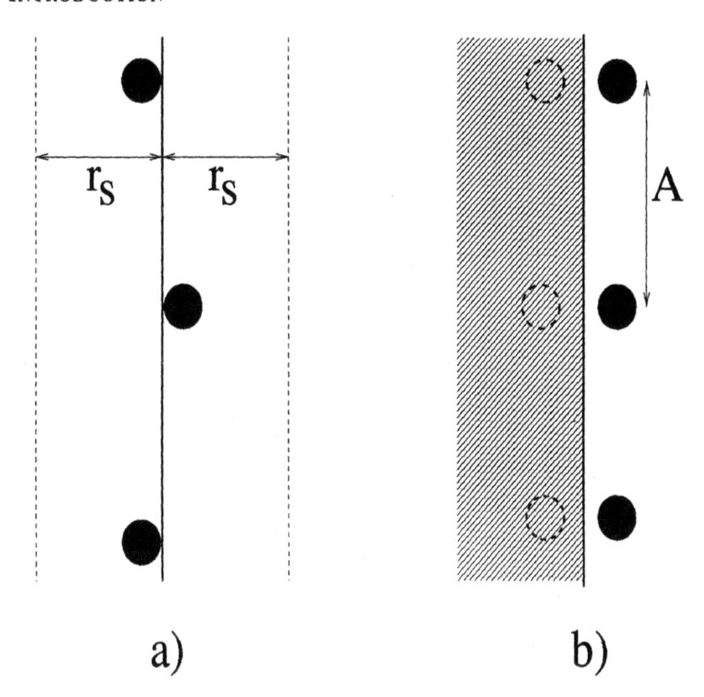


Figure 18.3: Two models of a macroion studied in this paper. $Z$-ions are shown by full circles. a) Thin charged plane immersed in water. The dashed lines show the position of effective capacitor plates related to the screening charges. b) The surface of a large macroion. Image charges are shown by broken circles.

the signs of both surface and $Z$-ion charges to be closer to the practical case when DNA double helices are adsorbed on a positive surface.) Due to the strong lateral repulsion, charged rods tend to be parallel to each other and have a short range order of an one-dimensional WC (Fig. 18.4). In Ref.[16] one can find beautiful atomic force microscopy pictures of almost perfect one-dimensional WC of DNA double helices on a positive membrane.

It is well known that for DNA, the bare charge density $-\eta_0$ is four times larger than the critical density $-\eta_c = -Dk_BT/e$ for Onsager-Manning condensation [17]. According to the solution of the nonlinear PB equation, most of the bare charge of an isolated DNA is compensated by positive monovalent ions residing at its surface so that the net charge of DNA is equal to $-\eta_c$. Generally speaking, the net charge of DNA adsorbed on a charged surface may differ from $-\eta_c$ due to the repulsion of positive monovalent ions condensed on DNA from the charged surface.

It is, however, shown [14] that in the case of strong screening by monovalent salt, $r_s \ll A_0$ ($A_0 = \eta_c/\sigma$), the potential of the surface is so weak that no monovalent ions are released from the DNA surface. In other words, the net

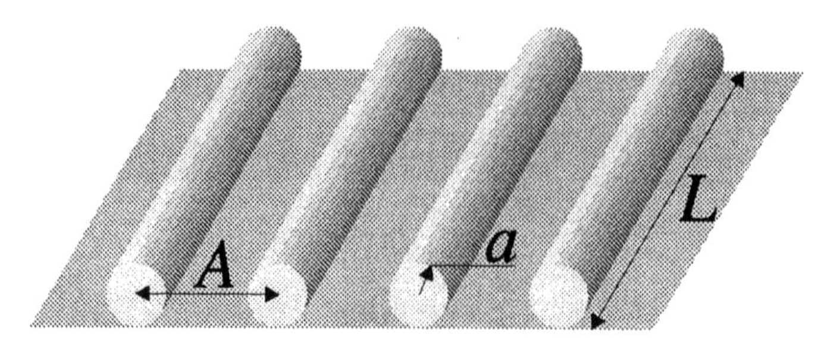


Figure 18.4: Rod-like negative $Z$-ions such as double helix DNA are adsorbed on a positive uniformly charged plane. Strong Coulomb repulsion of rods leads to one-dimensional crystallization with lattice constant $A$.

charge of each adsorbed DNA helix is still equal to $-\eta_c$. Simultaneously, at $r_s \ll A_0$ the Debye-Hückel approximation can be used to describe screening of the charged surface by monovalent salt. Using these simplifications we show that the competition between the attraction of DNA helices to the surface and the repulsion of the neighboring helices results in a negative net surface charge density $-\sigma^*$. The charge inversion ratio reads:

$$\frac{\sigma^*}{\sigma} = \frac{\eta_c/\sigma r_s}{\ln(\eta_c/\sigma r_s)}, \quad (\eta_c \sigma / r_s \gg 1) \tag{18.5}$$

which is equivalent to Eq. (18.4) for the case of 'spherical' $Z$-ions. Thus the inversion ratio grows with decreasing $r_s$ as in the spherical $Z$-ion case. At small enough $r_s$ and $\sigma$, the inversion ratio can reach 400%. This is larger than for spherical ions because, due to the large length of DNA helix, the correlation energy remains large and the WC-like short-range order is preserved at smaller $\sigma r_s$. This phenomenon can be called *giant charge inversion.*

## 18.2 SCREENING OF CHARGED SURFACE BY SPHERICAL $Z$-IONS

Assume that a plane with the charge density $-\sigma$ is immersed in water (Fig. 18.3a) and is covered by $Z$-ions with two-dimensional concentration $n$. Integrating out all the monovalent ion degrees of freedom, or equivalently, considering all interactions screened at the distance $r_s$, we can write down the free energy per unit area in the form

$$\mathcal{F} = \pi\sigma^2 r_s/D - 2\pi\sigma r_s Zen/D + F_{ZZ} + F_{id}, \tag{18.6}$$

where the four terms represent , respectively, the self-interaction of the charged plane, the interaction between the $Z$-ions and the plane, pair-interactions between the $Z$-ions, and the entropy of an ideal gas of two-dimensional $Z$-ions. Using $\sigma^* = -\sigma + Zen$, one may rewrite Eq. (18.6) as

$$\mathcal{F} = \pi(\sigma^*)^2 r_s/D + F_{OCP}, \tag{18.7}$$

where $F_{OCP} = F_c + F_{id}$ is the free energy of the same system of $Z$-ions residing on a neutralizing background with surface charge density $-Zen$, which is conventionally referred to as a 'one-component plasma' (OCP), and

$$F_c = -\pi(Zen)^2 r_s/D + F_{ZZ} \tag{18.8}$$

is the correlation part of $F_{OCP}$. The transformation from Eq. (18.6) to Eq. (18.7) can be simply interpreted as the addition of uniform charge densities $-\sigma^*$ and $\sigma^*$ to the plane. The first addition makes a neutral OCP on the plane. The second addition creates two planar capacitors with negative charges on both sides of the plane which screen the inverted charge of the plane at the distance $r_s$ (Fig. 18.3a). The first term of Eq. (18.7) is just the energy of these two capacitors. There is no cross-term corresponding to the interactions between the OCP and the capacitors because each planar capacitor creates a constant potential, $\psi(0) = 2\pi\sigma^* r_s/D$, at the neutral OCP.

Using Eq. (18.7), the electrochemical potential of $Z$-ions at the plane can be written as

$$\mu = Ze\psi(0) + \mu_{id} + \mu_c,$$

where $\mu_{id}$ and $\mu_c = \partial F_c/\partial n$ are the ideal and the correlation parts of the chemical potential of the OCP. At equilibrium, $\mu$ is equal to the chemical potential $\mu_b$ of the ideal bulk solution because, in the bulk, the electrostatic potential $\psi$ is zero. Therefore, we have:

$$2\pi\sigma^* r_s Ze/D = -\mu_c + (\mu_b - \mu_{id}). \tag{18.9}$$

As we show below, in most practical cases, the correlation effect is rather strong, so that $\mu_c$ is negative and $|\mu_c| \gg k_B T$. Furthermore, strong correlations imply that the short-range order of the $Z$-ions on the surface should be similar to that of a triangular Wigner crystal (WC) since the latter gives the lowest energy for the OCP. Thus, one can substitute the chemical potential $\mu_{WC}$ of the Wigner crystal for $\mu_c$. One can also write the difference of ideal parts of the bulk and the surface chemical potentials of $Z$-ions as

$$\mu_b - \mu_{id} = k_B T \ln(N_s/N), \tag{18.10}$$

where $N_s \sim n/a$ is the bulk concentration of $Z$-ions at the plane. Then, Eq. (18.9) can be rewritten as

$$2\pi\sigma^* r_s Ze/D = k_B T \ln(N/N_0), \tag{18.11}$$

where $N_0 = N_s \exp(-|\mu_{WC}|/k_B T)$ is the concentration of $Z$-ions in the solution next to the charged plane. It is clear that when $N > N_0$, the net charge density $\sigma^*$ is positive, i.e. opposite in sign to the bare charge density $-\sigma$. The concentration $N_0$ is very small because $|\mu_{WC}|/k_B T \gg 1$. Therefore, it is easy to achieve charge inversion. According to Eq. (18.10) at large enough $N$ one can neglect second term on the right-hand side of Eq. (18.9). This gives for the maximal inverted charge density

$$\sigma^* = \frac{D}{2\pi r_s} \frac{|\mu_{WC}|}{Ze}. \tag{18.12}$$

Equation (18.12) has a very simple meaning: $|\mu_{WC}|/Ze$ is the "correlation" voltage across the two parallel capacitors in Fig. 18.3a of plate spacing $r_s$ and total capacitance per unit area $D/(2\pi r_s)$.

To calculate the correlation voltage $|\mu_{WC}|/Ze$, we start from the case of weak screening when $r_s$ is larger than the average distance between $Z$-ions. In this case, screening does not affect the thermodynamic properties of the WC. The energy per $Z$-ion $\varepsilon(n)$ of such a Coulomb WC at $T = 0$ can be estimated as the energy of a Wigner-Seitz cell, because the quadrupole-quadrupole interaction between neigboring neutral Wigner-Seitz cells is very small. This gives

$$\varepsilon(n) = -(2 - 8/3\pi)Z^2e^2/RD \simeq -1.15Z^2e^2/RD,$$

where $R = (\pi n)^{-1/2}$ is the radius of a Wigner-Seitz cell. A more accurate calculation [18] gives the slightly higher energy

$$\varepsilon(n) \simeq -1.11Z^2e^2/RD = -1.96n^{1/2}Z^2e^2/D. \tag{18.13}$$

One can discuss the role of finite temperature on the WC in terms of the inverse dimensionless temperature

$$\Gamma = Z^2e^2/(RDk_BT).$$

We are interested in the case of large $\Gamma$. For example, at a typical $Zen = 1.0\ e/\text{nm}^2$ and at room temperature, $\Gamma = 10$ for $Z = 4$. The Wigner crystal melts [19] at $\Gamma = 130$, so that for $\Gamma < 130$ we are dealing with a strongly correlated liquid. Numerical calculations, however, confirm that at $\Gamma \gg 1$, the thermodynamic properties of strongly correlated liquid are close to those of the WC [20]. Therefore, for an estimate of $\mu_c$ we can still write $F_c = n\varepsilon(n)$ and use

$$\mu_{WC} = \frac{\partial\,[n\varepsilon(n)]}{\partial n} = -1.65\Gamma k_BT = -1.65\frac{Z^2e^2}{RD}. \tag{18.14}$$

We see now that $\mu_{WC}$ is negative and $|\mu_{WC}| \gg k_BT$, so that Eq. (18.12) is justified. Substituting Eq. (18.14) into Eq. (18.12), we get $\sigma^* = 0.83Ze/(\pi r_s R)$. At $r_s \gg R$, charge density $\sigma^* \ll \sigma$, and $Zen \simeq \sigma$, one can replace $R$ by $R_0 = (\sigma\pi/Ze)^{-1/2}$. This yields the result in Eq. (18.3). Thus, for $r_s \gg R_0$, the inverted charge density grows with decreasing $r_s$. Extrapolating to $r_s = 2R_0$, where screening starts to modify the interaction between $Z$-ions substantially, we obtain $\sigma^* = 0.4\sigma$.

Now we switch to the case of strong screening, $r_s \ll R$. It may seem that, in this case, $\sigma^*$ should decrease with decreasing $r_s$ because screening reduces the energy of WC and leads to its melting. In fact, this is what eventually happens. However, there is a range of $r_s \ll R$ where the energy of the WC is still large. In this range, as $r_s$ decreases, the repulsion between $Z$-ions becomes weaker and makes it easier to pack more $Z$-ions on the plane. Therefore, $\sigma^*$ continues to grow with decreasing $r_s$.

Although we can continue to use the capacitor model to deal with the problem, this model loses its physical transparency when $r_s \ll R$, because there is no obvious spatial separation between the inverted charge $\sigma^*$ and its screening atmosphere. Therefore, at $r_s \ll R$, we deal directly with the original free energy (18.6). The requirement that the chemical potential of a $Z$-ion in the bulk solution equals that of $Z$-ions at the surface now reads

$$\frac{\partial F}{\partial n} = \mu_{id} - \mu_b, \tag{18.15}$$

where

$$F = -\frac{2\pi\sigma r_s Zen}{D} + F_{ZZ} \tag{18.16}$$

is the interaction part of the total free energy (18.6) apart from the constant self-energy term $\pi\sigma^2 r_s/D$. According to Eq. (18.10), at large $N$ when

$$\mu_b - \mu_{id} = k_B T \ln(N_s/N) \ll 2\pi\sigma r_s Ze/D, \tag{18.17}$$

we can neglect the difference in the ideal part of the free energy of $Z$-ion at the surface and in the bulk. Therefore, the condition of equilibrium (18.15) can be reduced to the problem of minimization of the free energy (18.16) with respect to $n$. This direct minimization has a very simple meaning: new $Z$-ions are attracted to the surface, but $n$ saturates when the increase in the repulsion energy between $Z$-ions compensates this gain. Since this minimization balances the attraction to the surface with the repulsion between $Z$-ions, the inequality (18.17) also guarantees that thermal fluctuations of $Z$-ions around their WC positions are small. Therefore, $F_{ZZ}$ can be written as

$$F_{ZZ} = \sum_{\mathbf{r}_i \neq 0} \frac{(Ze)^2}{Dr_i} e^{-r_i/r_s}, \tag{18.18}$$

where the sum is taken over all vectors of WC lattice. In the limit $r_s \ll R$, one needs to keep only interactions with the 6 nearest neighbours in Eq. (18.18). This gives

$$F = -\frac{2\pi\sigma r_s Zen}{D} + 3n\frac{(Ze)^2}{DA}\exp(-A/r_s), \tag{18.19}$$

where $A = (2/\sqrt{3})^{1/2} n^{-1/2}$ is the lattice constant of this WC (Fig. 18.2). Minimizing the free energy with respect to $n$, one arrives at Eq. (18.4).

It is clear from Eq. (18.4) that at $r_s \ll R$ the inverted charge continues to grow with decreasing $r_s$. This result could be anticipated for the toy model of Fig. 18.1a if the Coulomb interaction between the spheres is replaced by a strongly screened one. Screening obviously affects repulsion between positive spheres more strongly than their attraction to the negative one and, therefore, makes it possible to keep two $Z$-ions even at $Q \ll Ze$. Above, we studied analytically

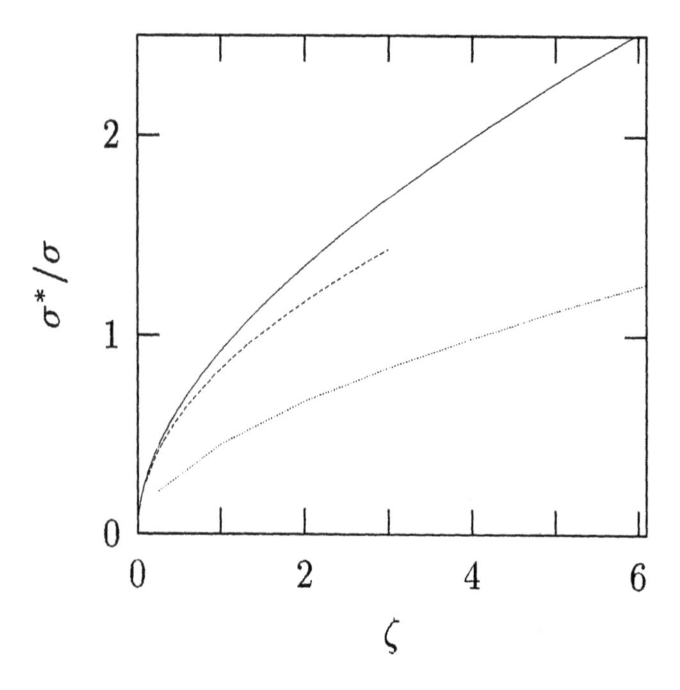


Figure 18.5: The ratio $\sigma^*/\sigma$ as a function of the dimensionless charge $\zeta = Ze/\pi\sigma r_s^2$. The solid curve is calculated numerically for a thin charged plane, the dashed curve is the large $r_s$ limit, Eq. (18.3). The dotted curve is calculated for the screening of the surface of the semispace with dielectric constant much smaller than 80. In this case, image charges (Fig. 18.3b) are taken into account [14].

the two extremes, $r_s \gg R$ and $r_s \ll R$. In the case of arbitrary $r_s$, we can find $\sigma^*$ numerically. Results from minimization of the free energy (18.16) with the help of Eq. (18.18) are plotted as the solid curve in Fig. 18.5. The dotted curve represents numerical results for a thick macroion[14]. In Ref. [14] we showed that the condition of validity of this theory, Eq. (18.17) is met only at $\zeta = Ze/\pi\sigma r_s^2 < 5$. Thus, for a thick macroion, the inversion ratio can reach 100%.

## 18.3 LONG CHARGED RODS AS $Z$-IONS

Consider the problem of screening of a positive plane, with a surface charge density $\sigma$, by negative DNA double helices with a net linear charge density $-\eta_c$. Their length $L$ is smaller than the DNA persistence length $L_p$, so that they may be regarded as straight rods. For simplicity, the charged plane is assumed to be thin and immersed in water, so that we can neglect image charges. Modification of the results due to image charges is given later. We assume that the concentration of monovalent salt is so large that $r_s \ll A$, where $A$ is the lattice constant of the one-dimensional WC (See Fig. 18.4).

At $r_s \ll A$ we can write the free energy per DNA as

$$f = -\frac{2\pi\sigma r_s L\eta_c}{D} + \frac{1}{2}\sum_{\substack{i=-\infty \\ i\neq 0}}^{\infty} \frac{2L\eta_c^2}{D} K_0\left(\frac{iA}{r_s}\right), \tag{18.20}$$

where $K_0(x)$ is the modified Bessel function of 0-th order. The first term of Eq. (18.20) describes the interaction energy of DNA rods with the charged plane, the second term describes the interaction between DNA rods arranged in the one-dimensional WC, where the factor $1/2$ accounts for double counting of the interactions in the sum.

Since the function $K_0(x)$ decays exponentially at large $x$, at $r_s \ll A$ one may keep only the nearest-neighbour interactions in Eq. (18.20). This gives

$$f \simeq -\frac{2\pi\sigma r_s L\eta_c}{D} + \frac{2L\eta_c^2}{D}\sqrt{\frac{\pi r_s}{2A}}\exp(-A/r_s), \tag{18.21}$$

which is similar to Eq. (18.19). To find $A$, we minimize the free energy per unit area, $F = nf$, with respect to $n$, where $n = 1/LA$ is the concentration of DNA helices at the charged plane. This gives:

$$\sqrt{2\pi}\sigma r_s/\eta_c = \sqrt{A/r_s}\exp(-A/r_s). \tag{18.22}$$

As one sees from Eq. (18.22), the lattice constant $A$ of the WC decreases with decreasing $r_s$ and charge inversion becomes stronger. Calculating the net negative surface charge density, $-\sigma^* = -\eta_c/A + \sigma$, we arrive at Eq. (18.5). One can also numerically minimize the free energy (18.20) at all $r_s \leq A$ to find $\sigma^*/\sigma$. The result is displayed as a solid curve in Fig. 18.6.

In the more realistic case of a thick macroion, repulsion from image charges must be taken into consideration. This has been done elsewhere [14]. The result for $\sigma^*/\sigma$ is represented by the dotted curve in Fig. 18.6. It is clear that, in the case of DNA, at a given value of $\eta_c/\sigma r_s$, image charges play an even smaller role than in the case of spherical $Z$-ions. However, image charges reduce the maximal possible value of $\eta_c/\sigma r_s$ significantly. It is limited by the condition that DNA

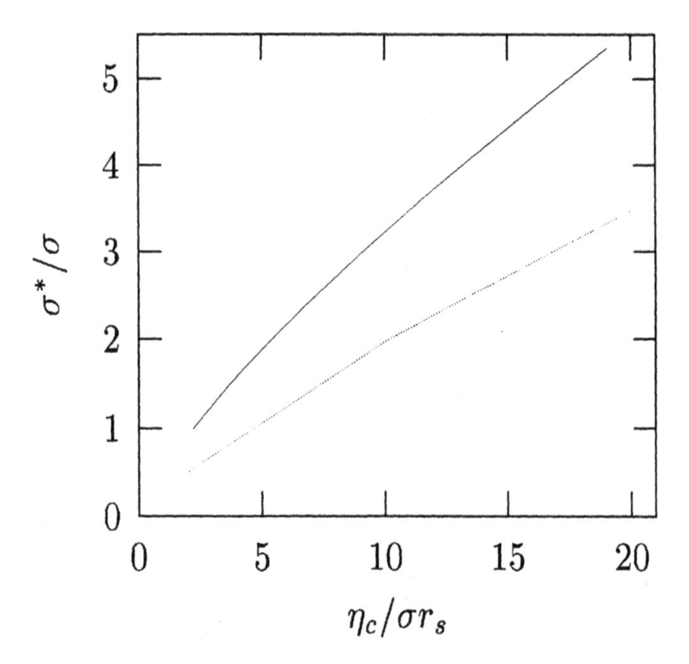


Figure 18.6: The ratio $\sigma^*/\sigma$ as a function of $\eta_c/\sigma r_s$. The solid curve is calculated for a charged plane by numerical solution of Eq. (18.20). The dotted curve is calculated for the screening of the surface of the semi-space with dielectric constant much smaller than 80. In this case image charges are taken into account [14].

is strongly bound to the surface, similar to that of Eq. (18.17) for the case of spherical ions. We have shown in Ref. [14] that this condition is

$$\eta_c/\sigma r_s \ll \sqrt{8\pi L/s l_B} \quad (L \leq L_p), \tag{18.23}$$

where $s = \ln(N_{s,DNA}/N_{DNA})$ is the entropy loss (in units of $k_B$) per DNA due to its adsorption to the surface. $N_{s,DNA}$ and $N_{DNA}$ are the three-dimensional concentration of DNA at the charged surface and in the bulk, respectively. It is clear that the maximal $\eta_c/\sigma r_s$ and maximal inversion ratio grow with $L$. For $L = L_p = 50$ nm and $s = 3$, the maximal $\eta_c/\sigma r_s = 25$. Therefore, according to the dotted curve in Fig. 18.6, the inversion ratio for a thick macroion $\sigma^*/\sigma$ can reach 4. Such inversion can indeed be considered as giant.

We are grateful to R. Podgornik and I. Rouzina for useful discussions. This work was supported by NSF DMR-9985985.

## BIBLIOGRAPHY


[1] P. W. Anderson, Science, **177**, 393 (1972); Proc. Natl. Acad. Sci. (USA), **80**, 3386 (1983).

[2] P. W. Anderson, "Basic notions of condensed matter physics" Addison-Wesley, (1984), p. 262; P. W. Anderson, D. L. Stein, Proc. Natl. Acad. Sci. (USA) **81**, 1751 (1984).

[3] J. Ennis, S. Marcelja and R. Kjellander, Electrochim. Acta, **41**, 2115 (1996).

[4] V. I. Perel and B. I. Shklovskii, Physica A 274, 446 (1999).

[5] B. I. Shklovskii, Phys. Rev. E**60**, 5802 (1999).

[6] E. M. Mateescu, C. Jeppersen and P. Pincus, Europhys. Lett. **46**, 454 (1999); S. Y. Park, R. F. Bruinsma, and W. M. Gelbart, Europhys. Lett. **46**, 493 (1999).

[7] J. F. Joanny, Europ. J. Phys. B **9** 117 (1999); P. Sens, E. Gurovitch, Phys. Rev. Lett. **82**, 339 (1999).

[8] R. R. Netz, J. F. Joanny, Macromolecules, **32**, 9013 (1999).

[9] R. R. Netz, J. F. Joanny, Macromolecules, **32**, 9026 (1999).

[10] Y. Wang, K. Kimura, Q. Huang, P. L. Dubin, W. Jaeger, Macromolecules, **32** (21), 7128 (1999).

[11] T. Wallin, P. Linse, J. Phys. Chem. **100**, 17873 (1996); **101**, 5506 (1997).

[12] P. L. Felgner, Sci. American, **276**, (6) 102 (1997).

[13] R. Messina, C. Holm, K. Kramer, *private communication.*

[14] T. T. Nguyen, A. Yu. Grosberg, B. I. Shklovskii, cond-mat/0002305, J. Chem. Phys., *to appear, 15 July 2000.*

[15] R. J. Hunter, *Foundations of colloid science*, Vol. 1, Oxford University Press (1986).

[16] Ye Fang, Jie Yang, J. Phys. Chem. B **101**, 441 (1997).

[17] G. S. Manning, J. Chem. Phys. **51**, 924 (1969).

[18] L. Bonsall, A. A. Maradudin, Phys. Rev. B**15**, 1959 (1977).

[19] R. C. Gann, S. Chakravarty, and G. V. Chester, Phys. Rev. B **20**, 326 (1979).

[20] H. Totsuji, Phys. Rev. A **17**, 399 (1978).

# CHAPTER 19

# FOREST FIRES AND LUMINOUS MATTER IN THE UNIVERSE

PER BAK[1,2] AND KAN CHEN[1,2]

[1]Niels Bohr Institute, Blegdamsvej 17, Copenhagen, Denmark
[2]Department of Computational Science, Faculty of Science
National University of Singapore, Singapore 117543

## ABSTRACT

The forest fire model is a reaction-diffusion model in which energy, in the form of trees, is injected uniformly, and burns (is dissipated) locally. It was introduced as a toy model of turbulence in the hope of catching the essential phenomenology [1, 2]. More generally, it could serve to illustrate how complex structures emerge in systems with many components that are driven out of equilibrium by an energy flux.

It was originally thought that the dissipative field would form a fractal structure, but recently it was shown that the spatial distribution of the fires forms a novel geometrical structure in which the fractal dimension varies continuously from zero to three as the length scale increases from the smallest scale $l = l_0$ to a correlation length $l = \xi$ [3]. We suggest that this picture applies to the "intermediate range" of turbulence where it provides a natural interpretation of extended scaling that has been observed at small length scales [4].

Unexpectedly, the scaling form may also account for the distribution of luminous matter in the universe [5]. Comparison with data from the SARS red-shift catalogue, and the LEDA database provides a good fit with a correlation length $\xi \sim 300$ Mpc. The geometrical interpretation is clear: At small distances, the universe is zero-dimensional and point-like. At distances of the order of 1 Mpc the dimension is unity, indicating a filamentary, string-like structure; when viewed at larger scales it gradually becomes 2-dimensional wall-like, and finally, at and beyond the correlation length, it becomes uniform.

The forest fire model has a cousin, the Drossel-Schwabl model [6], where new fires are ignited at a small rate. It exhibits traditional SOC behaviour, mimicking the power-law distribution of real forest fires in the US and Australia [7], and also reproduces the power-law distribution of individuals affected by measles epidemics [8].

## 19.1 INTRODUCTION

The second half of the previous century was certainly defined by the discoveries in solid state physics more than anything else. However, Stephen Hawking recently expressed the view that Complexity is the science of the 21st century. While solid state physics deals with uniform arrangements of very many atoms, mostly in equilibrium, complexity deals with the emergence of novel and surprising phenomena in systems with many degrees of freedom that are driven out of equilibrium. "More is different" as Anderson eloquently defined the phenomenon of emergence [1]. How can complex phenomena like galaxies, planetary systems, and our earth with its geophysics, life, and human social activities emerge from the big bang following the simple laws of physics? Why did not everything explode into a simple gas, or collapse into a solid?

Complex behaviour is generally believed to emerge in open dissipative systems, where energy is continuously injected, and eventually is dissipated locally. Turbulence in liquids is the canonical example, but unfortunately the Navier Stokes equations is quite difficult to study analytically and numerically. It has been suggested that complex behaviour may be associated with a self-organized critical state, where the dynamics takes place in terms of intermittent avalanches, rather than progressing in a smooth and gradual way [10, 2].

The forest fire model is a simple toy model of a turbulent system. It is a discrete model defined on a lattice, in the best traditions from Ising-like models in condensed matter physics. In our vision, the forest fires would represent the intermittency observed in turbulence, and power-law spatial and temporal correlations would naturally occur if the model would operate at the self-organised state. It turns out that, while the model is indeed critical in the sense that it has a diverging correlation length, there is no fractal self-similarity below the correlation length, as is usually the case for critical systems. The criticality can therefore not be described as a fixed point in the Wilson sense. On the contrary, the apparent dimension changes as one steps further and further backwards, and considers things at a larger and larger length scale. At some point, the correlation length is reached, and the distribution of fires remains uniform beyond that length. Thus, as one change the scale of observation $l$, different classes of fractal objects are observed, ranging from points to lines to walls, and finally to a homogeneous set.

While this was not what we were looking for, it may nevertheless turn out to represent the actual scaling in turbulence. In the scaling regime, beyond the correlation length, the energy dissipation field in turbulence is known to be homogeneous with great accuracy. However, there appears to be an "intermediate range", rather than a single lower cut-off length, where the interesting intermittent behaviour that we usually associate with turbulence takes place. The correlations do not follow power laws, but are characterised by smoothly varying effective exponents. Only the relative moments follow scale-independent ratios. This can readily be interpreted as a scale-dependent exponent for the dimension of the dissipative field, as observed in our model.

On a quite different front, there has been much controversy about the spatial distribution of luminous matter in the universe. The apparent hierarchical structure has led many researchers to believe that the distribution could be self-similar, or fractal [11]. On the other hand, the uniformity of the background radiation requires that the universe be homogeneous at the largest scale, contradicting the simple self-similar scaling picture. Perhaps, the luminous matter in the universe obeys this new type of geometry, which unifies both observations. While we hesitate to claim that the universe should be viewed as one giant forest fire, we do suggest that the novel scaling picture may represent a quite general geometrical form for non-equilibrium dissipative systems.

## 19.2 THE FOREST FIRE MODEL

The forest fire model is defined as follows. On a 3-dimensional cubic lattice, trees representing energy are grown randomly at empty sites at a rate $p$. During one time unit, trees burn down (leaving room for new trees), and ignite neighbouring trees. These processes represent dissipation and diffusion of energy, respectively. After a transient period, the system enters a statistically stationary state with a complex distribution of fires.

We simulated both the two- and three-dimensional versions with sizes up to $2048^2$ and $1024^3$, respectively, although here we will be concerned only with the three-dimensional model. Up to half a million time steps were used to collect the statistics in each simulation. Special care is needed for the simulation of a sustained forest fire due to the fact that, for a given system size $L$, the fire dies out if the rate is not sufficiently high. We shall see that this feature is important for applications to the spread of diseases, and to the existence of light matter in the universe. If artificially restarted (say, by just adding a fire), the system often goes into a state with global oscillations. Typically we can only study systems sizes that are much larger than the correlation length. This contrasts with conventional critical phenomena, where information for length scales up to the size of the system can be obtained for systems that are smaller than the correlation length: they exhibit finite-size scaling. Here, the critical behaviour throughout the scaling region collapses as soon as the correlation length reaches the system size. Thus, the range of scales $l$ that we can study is squeezed by the inequalities $1 < l < \xi \ll L$, making the calculation numerically demanding despite the simplicity of the model. Our simulations involved more than a billion sites!

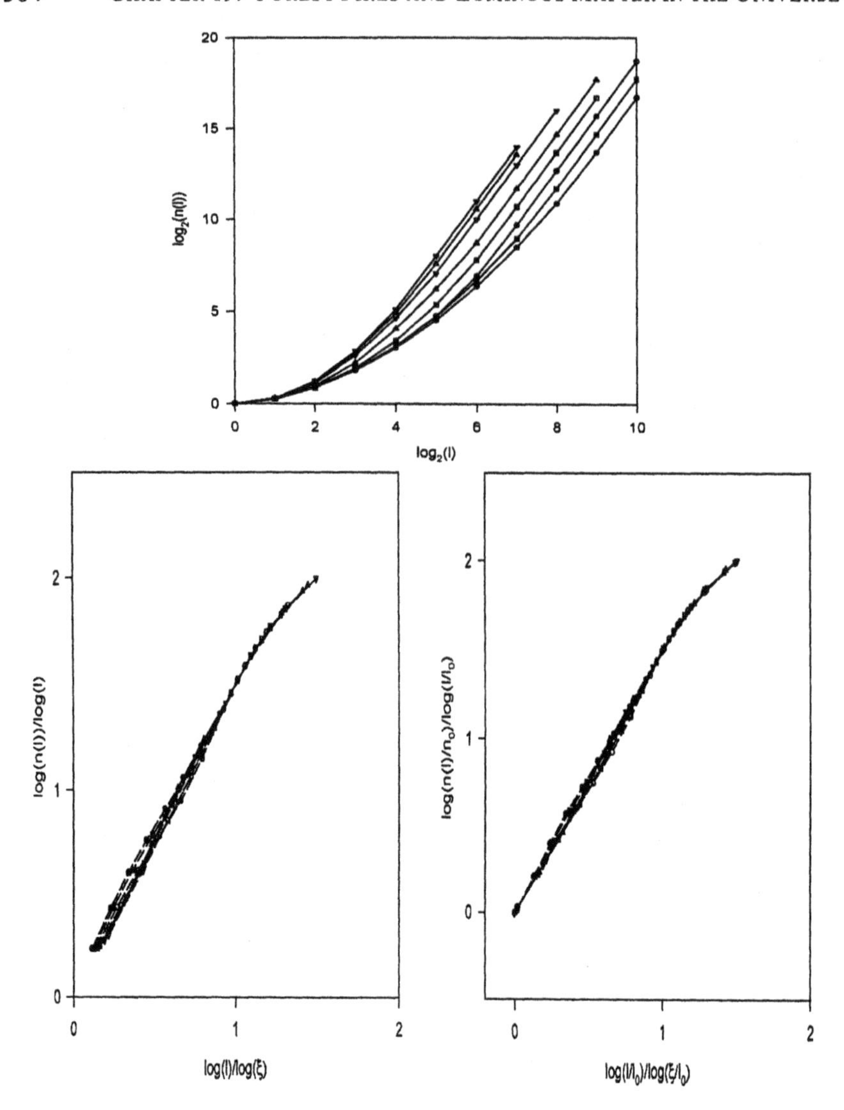


Figure 19.1: (a) (Upper panel) $\log(n(l))$ vs $\log(l)$ for various system sizes and tree growth rates $p$ for the 3D forest fire model: filled-circle $(1024^3, p = 1.25 \times 10^{-4})$, filled-square $(1024^3, p = 2.5 \times 10^{-4})$, open-circle $(1024^3, p = 5 \times 10^{-4})$, open-square $(512^3, p = 0.001)$, filled-triangle $(512^3, p = 0.002)$, inverted filled-triangle $(256^3, p = 0.005)$, open-triangle $(128^3, p = 0.0075)$, inverted open-triangle $(128^3, p = 0.01)$. (b) [lower left panel] $\log(n(l))$ vs $\log(l)$ for various $\log(n(l))/\log(l)$ vs. $\log(l)/\log(\xi)$ with the same set of data. The correlation lengths are given by $\xi = (0.77p)^{-2/3}$. (c) [lower right panel] $\log(n/n_0)/\log(l/l_0)$ vs $\log(l/l_0)/\log(\xi/l_0)$, with the lower cut-off $l_0 \propto p^{0.03}$.

The average amount of fires $n(l)$ within boxes of size $l$ which contain fires was measured. Figure 19.1a shows $\log(n)$ vs $\log(l)$ for a wide range of $p$ and $L = 128, 256, 516, 1024$ for $d = 3$. There is no linear regime, indicating that there is no well-defined fractal dimension, in contrast to our original claim. The slopes of curves generally increase with $l$, and saturate at a value of 3 at larger $l$ value, indicating that the distribution becomes uniform beyond a length scale that we identify as the correlation length. A unified description can be obtained by plotting $\log(n)/\log(l)$ vs $\log(l)/\log(\xi)$ for the same data (See Fig. 19.1b, where good data collapse, involving length scales extending over three orders of magnitude, is obtained for $\xi = (0.77p)^{-2/3}$ ($\nu = 2/3$). Actually, a slightly better fit is obtained if the lower cut-off is allowed to depend on $p$, so that in principle there are two adjustable exponents. Figure 19.1c shows $\log(n/n_0)/\log(l/l_0)$ vs $\log(l/l_0)/\log(\xi/l_0)$ with $l_0 \propto p^{0.03}$.

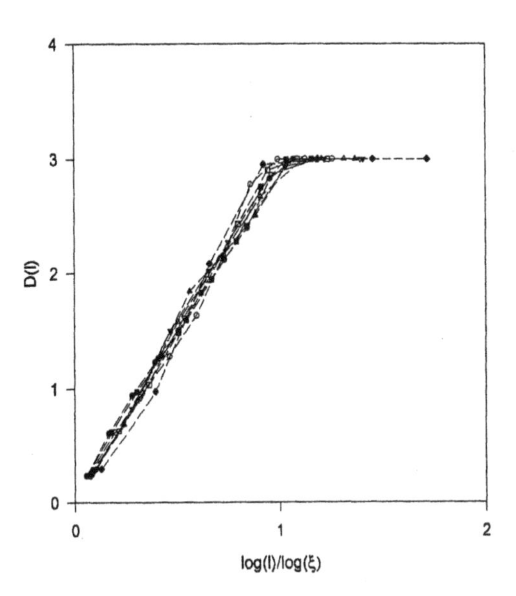


Figure 19.2: The length-dependent fractal dimension $D(l) = d\log(n(l))/d\log(l)$ (calculated from differences between the data points in Fig. 19.1) vs $\log(l)/\log(\xi)$ for the data sets used in Fig. 19.1, with $\xi = (0.77p)^{-2/3}$.

The apparent fractal dimension $D(l)$ at the length scale $l$ is given by the derivative $d\log(n)/d\log(l)$, which we plot in Fig. 19.2. It is clear that the fractal dimension increases approximately linearly with $\log(l)$,

$$D(l) \sim 3\log(l/l_0)/\log(\xi/l_0), \tag{19.1}$$

and saturates sharply at a value of 3 at the correlation length $\xi$, and remains 3 beyond the correlation length. This simple formula is our main result! The amount of fires within a box of size $l$ becomes

$$\log(n) \sim \left(\frac{3}{2}\frac{\log(l/l_0)}{\log(\xi/l_0)}\right)\log(l/l_0) \quad . \tag{19.2}$$

The exponent $\nu$ can be derived by the argument of energy conservation, which requires that the number of fires $n(\xi)$ in a box of size $\xi$ times the number of boxes of that size, $(L/\xi)^d$, scales as $pL^d$. Since the fractal dimension $D(l)$ is linearly dependent on $\log(l)$, we have $n(\xi) = \xi^{d/2}$; this leads to $\nu = 2/d$. The data collapse shown in Figs. 19.1b and 19.2 indicates bona fide scaling, although of a quite novel and unique nature, without self-similarity under a change of scale. Whether one would call this novel state critical is a matter of taste.

The behaviour throughout the scaling region $l \ll \xi$ is influenced by both the correlation length and the smallest length scale of dissipation. For example, one can estimate the correlation length by measuring the increase of dimensionality from one small length scale to another. In contrast, for conventional critical phenomena, the properties up to the correlation length are those of the critical state, and for $l \ll \xi$ there is no way to detect $\xi$. As we shall see, this might allow us to estimate the correlation length of the universe from observations at smaller accessible length scales

The forest fire model was originally thought of as a "toy" model of turbulence. Recently, deviations from fractal, or multi-fractal, scaling has been interpreted as "extended self-similarity" in an intermediate dissipative range between the Kolmogorov length and the inertial range. Perhaps one might understand this phenomenon geometrically in terms of the concept of scale-dependent dimension introduced here. In particular, data presented by Benzi *et al.* [4] seem to indicate a logarithmic dependence of scaling exponents versus length scale. Other experiments showing a dimension depending on the Reynolds number [12, 13] might alternatively be interpreted as a scale dependent dimension. So far, we have not reached an analytical understanding of this simple, and completely novel, type of scaling. Traditional renormalization group analysis based on rescaling of length scale will not apply here, and it is our belief that a quite different framework might be needed.

## 19.3 SCALE-DEPENDENT DIMENSION OF LUMINOUS MATTER

Uniformity of the background radiation requires that the universe must be homogeneous at the largest scale. This is known as the cosmological principle. However, a decade ago, Coleman and Pietronero [14] suggested that the universe at length scales $L$ up to a couple of Mpc is fractal with fractal dimension, $D \sim 1.2$. Subsequent studies seemed to confirm this picture. In an extensive and very thorough work, Sylos-Labini *et al.* [11] interpret existing data as an indication of a fractal dimension of two, extending to 50 Mpc. Even though there is a general agreement about the existence of fractal-like galactic structures at moderate scales, there is still intense debate whether or not the universe is homogeneous at very large scales and, if so, how the transition to homogeneity takes place [15, 16]. The value of the homogeneity scale and the matter distribution within such scale have great cosmological consequences.

We shall see that a careful analysis of galaxy maps indicates that the geometrical structure of luminous matter in the universe is very similar to that of the forest-fire model. Our alternative form provides a better fit to the data than conventional models. The underlying picture is one in which luminous matter is being created and destroyed in an ongoing non-equilibrium, dynamical process. This similarity is appealing: it suggests that the universe shares the basic characteristic features of other dynamical systems, so perhaps the dynamics of the universe is not unique, but belongs to a more general universality class of non-equilibrium turbulent systems.

We suggest that the length-scale dependent behaviour observed in this model may be sufficiently general that it is worthwhile to make a detailed comparison with real astronomical data. The underlying viewpoint is that the galactic dynamics is turbulent, with stellar objects interacting with one another in reaction-diffusion type processes through shock waves, super-novae explosions, galaxy mergers, etc. Apart from an overall amplitude, there are only two fitting parameters in our proposed galaxy distribution, namely the upper length scale $\xi$, where the distribution becomes uniform, and the lower cutoff, $l_0$, where the distribution becomes point-like.

In their seminal work, Sylos-Labini, Pietronero and coworkers [11] analysed several database catalogues of galaxy maps. From the databases, they created volume-limited (VL) samples containing all galaxies exceeding a certain absolute luminosity within a given volume. Then they calculated the conditional density $\Gamma^*(l)$, which is the average density of galaxies within a sphere of size $l$. This quantity corresponds to the density $n(l)$ defined above, divided by the volume $l^3$. Thus, from equation 2 we immediately obtain the following prediction for $\Gamma^*(l)$:

$$\log\left(\Gamma^*(l)\right) \sim \left(\frac{3}{2}\left(\frac{\log(l/l_0)}{\log(\xi/l_0)}\right) - 3\right) \log(l/l_0) \quad . \tag{19.3}$$

Namely, on a log-log plot there is a pure quadratic dependence, rather than the linear dependence found for self-similar fractal structures.

We have fitted the above expression to the conditional densities extracted by Pietronero *et al.* from two widely different data bases with consistent results. The LEDA database is a heterogeneous compilation of data from the literature containing more than 200,000 galaxies. The Stromlo-APM red shift survey (SARS [17]) consists of 1797 galaxies. Figure 19.3 shows results from the fits, with two different cut-offs for the LEDA database. The labelling follows Sylos Labini *et al.*, with the numbers representing the lower luminosity cut-offs. Obviously, there are larger fluctuations for the sparser, but perhaps higher quality, Stromlo-

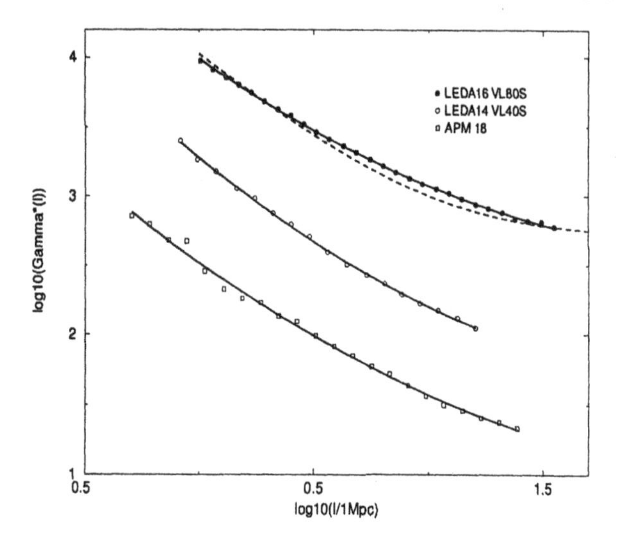


Figure 19.3: Conditional average densities for various galaxy catalogues (arbitrary scale), as derived by Sylos Labini *et al* [11], compared with fits to equation 1, yielding $\xi = 260$ Mpc from the LEDA16 data, $\xi = 275$ Mpc from the LEDA14 data, and $\xi = 380$ Mpc from the APM data. The broken line is a conventional fit to equation 4 with $\gamma = 1.3$, $r_0 = 10$ Mpc.

APM data set than for the LEDA database. The fits are very good in view of the fact that the only fitting parameters are the upper and lower length scales, $\xi$ and $l_0$, respectively. In contrast to conventional critical phenomena, the correlation length enters the expression for length scales below the correlation length. We are therefore able to fit the correlation length to the data, despite the fact that data are unavailable at and beyond the projected correlation length.

The upper length scale is the one at which the curves become flat, $d = 3$. The three fits yield very consistent values of this length scale ($\xi = 260$ Mpc from the LEDA16 data, $\xi = 275$ Mpc from the LEDA14 data, and $\xi = 380$ Mpc from the APM data). The empirical logarithmic scale dependence of the dimension can be seen directly by re-plotting the data in figure 3: Figure 19.4 shows $D(l) = 2 \times (\log(\Gamma^*(l)/\Gamma(l_0))/\log(l/l_0) + 3)$. All data sets yield linear behaviour. The correlation length is found by linear extrapolation to the point where $D(l)$ assumes the value of 3. The dimensions derived from the intense galaxies, LEDA 16 and APM 18, are essentially identical, but the LEDA 14 data yield a somewhat steeper scale dependence. However, they all converge at essentially the same homogeneity length. We predict a sharp crossover to uniformity, i. e. a sharp kink in the curve, at this length, which will be readily observable once data become available.

The lower cut-off $l_0$ is the scale at which the slope of the curves in the figure assumes the value of $-3$. We find $l_0 = 370$ light-years, $l_0 = 3700$ light-years, and

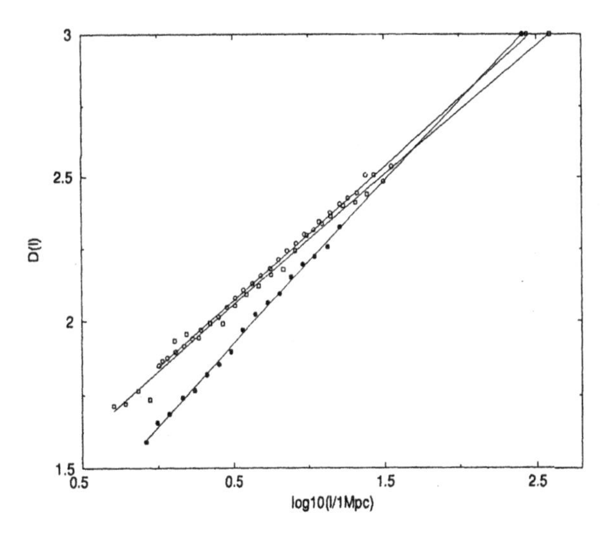


Figure 19.4: Scale dependent dimension $D(l)$ derived from the data points in figure 1 as explained in text. We conjecture that future data points follow the straight lines, and saturate sharply to $D = 3$ at the correlation length.

$l_0 = 330$ light years for the three samples, respectively. This scale is determined with less precision than the correlation length $\xi$. It is not clear how well our scaling form applies to the analysis of the galaxy distribution at small length scales.

The geometry of the luminous set is *not fractal* when viewed over the entire range of scales, since there is no self-similarity for different scales. Nevertheless, the scale-dependent dimension has a clear geometrical interpretation: At small distances, the universe is zero-dimensional and point-like. Indeed, energy dissipation takes place on individual point objects, like stars and galaxies. At distances of the order of 1 Mpc the dimension is unity, indicating a filamentary, string-like structure; when viewed at larger scales it gradually becomes 2-dimensional wall-like, and finally at the correlation length $\xi$, it becomes uniform.

It might be instructive to compare with more conventional interpretations of the large scale structure [18]. The conditional density can be related to a correlation function $g(r)$ through [11]

$$\Gamma^*(l) \sim \langle n \rangle (1 + g(l)), \tag{19.4}$$

where $\langle n \rangle$ is the mean density of galaxies. For instance, the field theory of de Vega *et al.* [19] yields an expression of this form. The correlation function is often assumed to be of the form $g(l) = (r_0/l)^\gamma$. Figure 19.3 also shows a fit to this expression, with $r_0 = 10$ Mpc and $\gamma = 1.3$. The fit is clearly inferior, flattening out at length scales that are too small. This is in accordance with the observations by Sylos Labini *et al.* that the value of fitted parameter $r_0$ depends heavily on the range of length scales used. At larger scales, the difference between the two fits is even more pronounced; when more data become available in the near future, one should be able to discriminate even better between the two pictures. In this traditional view, there is a smooth crossover to homogeneity when the *amplitude*, expressed in terms of $r_0$ reaches unity. In contrast, following our "critical phenomena" viewpoint, there is a sharp, possibly exponential, cutoff of the non-uniform part of the correlation function at the *correlation length*.

This has some important cosmological consequences. In the traditional formulation, one usually visualises that the amplitude $r_0$ of the power-law fluctuations increases with time, starting from the time of the decoupling of radiation from hadronic matter, leading to an increase of the cross-over length to homogeneity. In our phenomenology, the correlation length $\xi$ is the only parameter, so it is this quantity that is increasing with time. The average density of galaxies in the universe is equal to the density within the correlation length, i.e. $\langle n \rangle = \Gamma^*(\xi) \sim 1/\xi^{3/2}$. Thus, once the correlation length has been determined, one knows the density of galaxies. Assuming that the entire density of hadronic matter scales as that of the luminous galaxies studied here, one might get an estimate of the mass of the universe. From the fit to LEDA 14 one gets that the density of galaxies in the entire universe with apparent magnitude greater than 14 is $\langle n \rangle = 2 \times 10^{-3} Mpc^{-3}$. From the fit to the APS 18 data we find that the density of galaxies with apparent luminosity greater than 18 is $\langle n \rangle = 3 \times 10^{-4} Mpc^{-3}$. The traditional fits above yield much larger values for the density of galaxies in the universe, depending on the range of length scales used in the fit [11]. It will be more than exciting to have these novel predictions checked when data extending to several hundred Mpcs become available in the next decade!

## 19.4 FOREST FIRES AND MEASLES

Soon after the inception of the forest fire model, it became clear that it did not show the usual scaling associated with self-organised critical behaviour. A long debate on the criticality or non-criticality of the model took place in the past ten years, culminating in the discovery of the novel type of scaling described above. However, in the meantime, Drossel and Schwabl [6, 20] came up with a different version, in which fires are injected at a small, but finite rate. Their version exhibits standard, non-controversial SOC. All trees connected with the igniting spark burn instantaneously. The model can be exactly solved in one dimension [21, 20] in terms of a cascade process, or actually an inverse cascade process where small forests grow and join other forests, until they become large and burn.

This model is of particular interest since it turns out that the two-dimensional version describes the distribution of thousands of real forest fires in the US and Australia! Malamed, Morein, and Turcotte [7] collected data in three different regions in the USA, and in Australia. They found power-law distributions, with exponents ranging form 1.3 to 1.5 over up to 5 orders of magnitude, compared with the somewhat smaller slope, 1.16 of the two-dimensional Drossel-Schwabl model.

Rhodes and Anderson [8] studied the outbreak of measles epidemics in isolated populations on the Faroe Islands and Bornholm in Denmark, and on Iceland. An exponent 1.28 was found (with much less accuracy than for the forest fires). Perhaps we should view the spread of diseases as a forest fire. The power-law distribution implies that the disease "knows" its environment when it hits a population, since it has been there before. For populations exceeding 50 millions, there will always be some cases of measles, indicating that this represents the correlation length. In order to have the disease become extinct, the correlation length must be increased to become worldwide, by vaccination or limiting the transmission, just as the fire goes extinct if the correlation length approaches the system size. To summarise: diseases become extinct when the correlation length becomes worldwide.

## ACKNOWLEDGMENTS

We are grateful, and deeply indebted, to Luciano Pietronero for patiently explaining the intricacies of the scaling of luminous matter in the universe. Indeed, our way of thinking follows closely the path laid out by Luciano and his co-workers. PB thanks the National University of Singapore for great hospitality. KC is grateful to the support and hospitality of Niels Bohr Institute, where part of this work was done.

## BIBLIOGRAPHY

[1] P. Bak, K. Chen, and C. Tang, Phys. Lett. **147**, 297 (1990).

[2] P. Bak and K. Chen, Scientific American, **264**, 46-53 (1991).

[3] K. Chen, and P. Bak, CONDMAT 9912417.

[4] R. Benzi, S. Ciliberto, C. Baudet, and G. R. Chavarria, Physica D **80**, 385 (1995).

[5] P. Bak and K. Chen, ASTRO-PH 0001443.

[6] B. Drossel and F. Schwabl, Phys. Rev. Lett. **69**, 1629 (1992).

[7] B. D. Malamud, G. Morein, and D. L. Turcotte, Science **281**, 1840 (1998).

[8] C. J. Rhodes and R. M. Anderson, Nature **381**, 600 (1996).

[9] P.W. Anderson, Science **177**, 393 (1972).

[10] P. Bak, C. Tang, and K. Wiesenfeld, Phys. Rev. Lett. **59**, 381 (1987); Phys. Rev. A. **38**, 364 (1988); for a review see P. Bak, *How Nature Works: The Science of Self-Organized Criticality,* (Copernicus, New York, 1996; Oxford, 1997).

[11] F. Sylos Labini, M. Montuori, L. Pietronero, *Phys. Rep.* **293**, 61 (1998), and these proceedings. While these authors do suggest that the universe is fractal up to reachable length scales, their main point is that there is no indication of homogeneity at those scales. Indeed, we agree completely with this view. Our projected correlation length comes from extrapolation of our formula.

[12] R. R. Prasad and K. R. Sreenivasan. Phys. Fluids A**2**, 792 (1990).

[13] P. Tong and W. I. Goldberg, Phys. Fluids **31**, 2841 (1988).

[14] P. H. Coleman, L. Pietronero, R. H. Sanders, *Astron. Astrophys.* **200**, L 32 (1988).

[15] A. Gabrielli, F. Sylos Labini, R. Durrer, *ASTRO-PH/9905183*

[16] For a lucid discussion on these different views, see J. Gaite, A. Dominguez, J. Perez-Mercader, *Astrophys. J.* **522**, L5 (1999).

[17] J. Loveday, B. A. Peterson, G. Efstatiou, S. J. Maddox, *Astrophys. J.* **390**, 338 (1992).

[18] J. P. E. Peebles, *The Large Scale Structure of the Universe,* (Princeton University Press 1980)

[19] An alternative viewpoint is that the scaling is fractal due to a gravitational stability, and can be derived by mapping an appropriate field theory to the Ising model (H. J. de Vega, N. Sanchez, F. Combes, *Nature* **383**, 56 (1996). These authors associate the mass density with energy fluctuations, rather than the order parameter.

[20] B. Drossel, S. Clar, and F. Schwabl, Phys. Rev. Lett. **71**, 3799 (1993).

[21] M. Paczuski and P. Bak, Phys. Rev. E, **48**, R3214 (1993).

# Chapter 20

# Complexity in Cosmology

L. Pietronero [1,2], M. Joyce [1,2], M. Montuori [1,2], F. Sylos Labini [2,3]

[1]Dipartimento di Fisica, Università di Roma "La Sapienza", I-00185 Roma, Italy
[2]INFM Sezione Roma I, I-00185 Roma, Italy
[3]Dépt. de Physique Théorique, Université de Genève
CH-1211 Genève, Switzerland

## Abstract

In this lecture, we discuss the impact of the ideas of modern statistical physics on the characterization of fractal galaxy structures and their theoretical implications in cosmology. First, we give a brief overview of the field of complex structures and mention the main models that can be considered as "Active Principles" for the generation of fractal and scale-invariant structures in physical phenomena. Then we show the impact of these new concepts on the description of the correlation properties of galaxy structures. The observed fractal properties of these structures have deep implications for the overall picture of the large scale universe. In particular, this new perspective implies the use of new theoretical concepts and methods for the understanding of the nature and dynamics of galaxy clustering. Finally, we discuss a new approach to the cosmological problem, in which the fractal nature of galaxy distribution is considered explicitly.

## 20.1 Introduction

"More is different": This epochal paper of 1972 by Phil Anderson [1] has set the paradigm for what has now evolved into the science of Complexity. The idea that "reality has a hierarchical structure in which at each stage entirely new laws, concepts, and generalizations are necessary, requiring inspiration and creativity to just as great a degree as in the previous one" has set a new perspective in our view of natural phenomena. The reductionist view focuses on the elementary bricks of which matter is made, but then these bricks are put together in marvelous structures with highly structured architectures. Complexity is the study of these architectures which depend only in part on the nature of the bricks, but also have their fundamental laws and properties which cannot be deduced from the knowledge of the elementary bricks.

In physical sciences, the geometric complexity of structures often corresponds to fractal or multifractal properties [2]. It is not clear whether this is an intrinsic unique property or it is due to the fact that we can only recognize what we know. Perhaps in the future, we shall see much more, but for the moment one of the elements we can identify in complexity is its fractal structure. Considering then the dynamical processes often associated with complex structures, we have as basic concepts: chaos, fractals, avalanches [3] and $1/f$ noise. Often, complex structures arise from processes that are strongly out of equilibrium and dissipative. There is a broad field, however, that lies in-between equilibrium and non-equilibrium phenomena. This is the field of glasses and spin glasses, which leads to highly complex landscapes and to the concept of frustration [4]. Finally an important field in which these ideas can also be applied is that of adaptation via evolution, which is characterized by a degree of self-organization and a critical balance between periods of smooth evolution and dramatic changes.

## 20.2 FRACTAL STRUCTURES AND SELF ORGANIZATION

In Fig. 20.1 we show examples of regular and irregular structures. In the top left panel, we have a distribution of points characterized by a small-scale granularity that turns, at larger scales, into a well-defined background density with a specific structure corresponding to an excess density around the center. This structure can be characterized by its position, size and intensity. One can also define a density profile along a line as shown in the top right panel. This profile may be well-approximated by a smooth (analytical) function: In this case, it is a constant plus a Gaussian. If we consider the dynamical evolution of our structure including the specific interactions between its constituent points, we can write a differential equation for the smooth function of the density profile. In this perspective the structure is essentially represented by the three elements: position, size and intensity (amplitude). The typical result of this study is to understand whether the structure moves, if it becomes more or less extended or more or less intense. This is the traditional approach to the study of structures based on the implicit assumption of regularity or analyticity which has been the one adopted in Statistical Physics before the advent of Critical Phenomena in the seventies.

In the bottom left of Fig. 20.1 we show instead a strongly irregular structure for which all the concepts used to characterize the previous picture lose their meaning. There is no background density: There are structures in many zones and at various scales but it is not possible to assign to them a specific size or intensity. This situation is also illustrated by the density profile (bottom right panel) which is highly irregular at any scale. All the previous concepts and theoretical

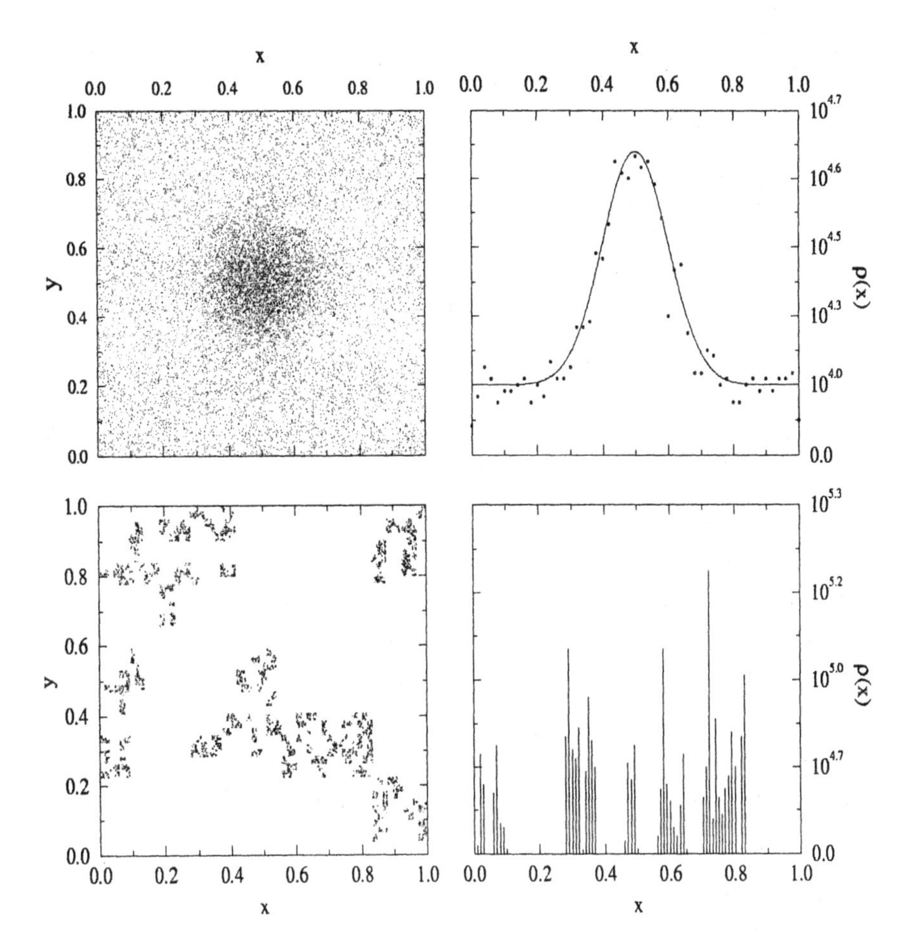


Figure 20.1: Example of analytical and non-analytic structures. *Top panels:* (Left) A cluster in a homogeneous distribution. (Right) Density profile. In this case the fluctuation corresponds to an enhancement of a factor 3 with respect to the average density. *Bottom panels:* (Left) Fractal distribution in the two dimensional Euclidean space. (Right) Density profile. In this case the fluctuations are non-analytical and there is no reference value, i.e. the average density. The *average density scales* as a power law from any occupied point of the structure.

methods lose their meaning. In order to give a meaningful characterization of the properties of this structure, one has to look at it from a fresh perspective. This structure, which consists of a simple stochastic fractal, has its regularity in the scale transformation. This naturally leads to power-law correlations characterized by an exponent, the fractal dimension. Also from a theoretical point of view, the understanding of the origin of the irregular or fractal properties cannot arise from the traditional differential equation approach. It requires, rather, new methods of the type associated with the renormalization group [2, 5, 6].

The physics of scale-invariant and complex systems is a novel field that embraces topics from several disciplines, ranging from condensed matter physics to geology, biology, astrophysics and economics [5]. This broad interdisciplinarity reflects the fact that these new ideas allow us to look at natural phenomena in a radically new and original way, and leads to unifying concepts that are independent of the detailed structure of the systems. The objective is the study of complex, scale-invariant structures, that appear both in space and time in a large variety of natural phenomena. New types of collective behaviors arise and their understanding represents one of the most challenging areas in modern statistical physics.

The activity in this field is a collaborative effort involving numerical simulations, analytical and experimental work. It may be characterized by research at three levels:

(i) Mathematical or geometrical level. This consists in applying the methods of fractal geometry in new areas to get new insights into important unresolved problems, to gain a better overall understanding. Such an approach permits us to investigate scientifically many phenomena characterized by intrinsic irregularities, which have been previously neglected because of the lack of an appropriate framework for their mathematical description. Examples of this type may be found in the geophysical and astrophysical data. In particular, we will discuss in some detail the scaling properties of galaxy correlations and the much-debated question of the fractal-versus-homogeneous Universe at large scales.

(ii) Development of physical models: The Active Principles for the generation of Fractal Structures.

Computer simulations represent an essential method in the physics of complex and scale-invariant systems. A large number of models have been introduced to focus on specific physical mechanisms which can lead spontaneously to fractal structures. Here we list those, which, in our opinion, represent the active principles for processes that generate scale-invariant properties based on physical processes. In Ref.[5] one can find many papers on these models:

- Diffusion Limited Aggregation (DLA, 1981)
- Dielectric Breakdown Model (DBM, 1984) These models are the prototype of the so-called fractals in which an iteration process based on Laplace equation leads spontaneously to very complex structures like that shown in Fig. 20.2.
- Cluster-Cluster Aggregation (Cl-Cl, 1983)
- Invasion Percolation (IP, 1983)
- The Sandpile model (1987)

The concept of self-organization is common to all the models discussed here but it has been especially emphasized in relation to the sandpile model.

- The Kardar-Parisi-Zhang model of surface growth (KPZ, 1986)

- The Bak-Sneppen model. (BS, 1993)

Figure 20.2: Dielectric Breakdown Model in cylindrical geometry. Example of a highly complex structure arising from a simple growth model in which the growth probability is proportional to the local electric field [6]. The zebra stripes around the structure represent equipotential lines.

The properties of these models are summarized in Table 20.1. In addition to these simplified models, we know that fractal structures are naturally generated in fluid turbulence, as described by the Navier-Stokes Equations, as the fractional portion of space in which dissipation actually occurs. Also, studies of gravitational instabilities suggest that gravity with random initial conditions may be enough to generate fractal clustering. However, the connections between the two important problems of turbulence and gravitational clustering, on the one hand, and the simplified models listed above, on the other, are only indirect. Each phenomenon and model mentioned seems to belong to a different universality class.

| | Self-Organiz. | Disorder Q/S | Fractals | Avalanch. | $d_c$ |
|---|---|---|---|---|---|
| DLA('81) | • | S | • | | $\infty$ |
| DBM ('84) | • | S | • | | $\infty$ |
| Sandpile ('87) | • | S | | • | 4 |
| Inv. Perc. ('83) | • | Q | • | • | 6 |
| Bak-Sneppen ('93) | • | Q | ? | • | ? |
| KPZ ('86) | • | S - (Q) | (•) | | $\infty$ (?) |

Table 20.1: Summary of the properties of the main models which represent the "Active Principles" for the generation of fractal and scale invariant structures. All of them are self-organized. The disorder can be quenched (Q) or stochastic (S). Some of them lead to fractals and/or to avalanches with scale invariant distribution. In the last column $d_c$ represents the upper critical dimension.

Actually the space-time distribution of earthquakes and the famous Gutenber-Richter (hereafter GR) power-law relation between number and magnitude suggest that earthquake physics could also represent an interesting playground for complex systems [7]. This is only partially true because the power-law (GR) distribution of earthquakes implies scale-invariant properties. From this, one can argue that the properties of small earthquakes may be related, in some sense, with the probability of a large earthquake. On the other hand, the concept of predictability is related to finite-time and -space correlations between events. These properties are usually non-universal, and not easy to address with the known methods of statistical physics. For example, different models can give similar GR power-law relations, but be completely different with respect to finite-time and -space correlation. It is for these basic reasons that the application of ideas and methods of statistical physics to earthquake physics has not been very successful up to now in improving the concepts related to predictability [7].

(iii) Development of theoretical understanding

At a phenomenological level, scaling theory inspired by usual critical phenomena has been successfully used. This is essential for the rationalization of results derived from computer simulations and experiments. This method allows us to identify the relations between different properties, and to focus on the essential ones. From the point of view of the formulation of microscopic fundamental theories, the situation is still in evolution. With respect to usual equilibrium-statistical mechanics, these systems are far from equilibrium and their dynamics is intrinsically irreversible. This situation does not seem to lead to any sort of ergodic theorem and the temporal dynamics has to be explicitly considered in the theory. This, together with the concept of self-organization as compared to criticality, represents the main new elements for the formulation of microscopic theories.

## 20.3 RECONSTRUCTING THE PUZZLE

In Fig. 20.3 we report two recent observations of galaxy distributions in three dimensional space. Specifications about observation techniques and data representation can be found in [9]. For our discussion here it is enough to say that these distributions are constructed in such a way to be observer independent and to provide a reasonable representation of galaxy clustering. Points represent galaxies and both distributions are far from homogeneous.

The first is a better sample (statistically), but is shallower, extending to $\sim$ 150 $Mpc/h$ compared to $\sim$ 500 $Mpc/h$ in the second [1]. To give an idea of the scales involved, we note that the size of a single galaxy is a fraction of $Mpc$ (typically 0.01 $h^{-1}Mpc$), while the Hubble radius (a sort of total size of the observable universe in the standard picture) is 3,000 to 5,000 $h^{-1}Mpc$. So a length-scale extending to 500 $h^{-1}Mpc$ is quite a large size (only one decade smaller than the entire universe).

[1]The unit of Megaparsec is defined as 1 $Mpc = 3.08 \times 10^{18}$ km and $0.5 \leq h \leq 1$ is the normalized value of the Hubble Constant.

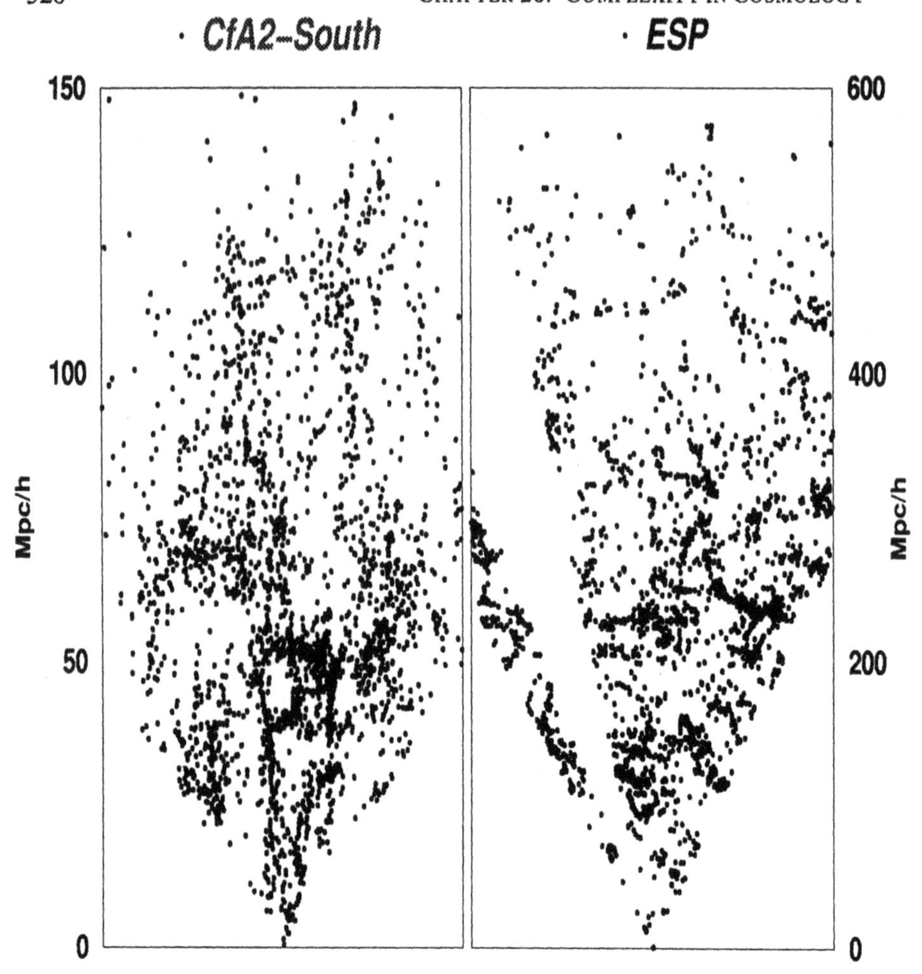


Figure 20.3: *Left Panel.* Redshift space distribution of galaxies in 30°-thick slice in declination from the CfA2-South Redshift Survey. Our galaxy lies at the lower vertex of the plot. The limits in Ascension Right are $0^h < \alpha < 4^h$ and $20^h < \alpha < 24^h$ and in Declination are $-2.5° < \delta < 27.5°$. In this region there are 2920 galaxies. The total solid angle of the survey is $\Omega \sim 1\ sr$ and the apparent magnitude limit is $m_B = 15.5$. *Right Panel.* As the left panel but the ESP survey (Vettolani *et al.*, 1997 Astron.Astrophys. 325, 954). The total solid angle of the survey is $\Omega \sim 0.006\ sr$ and the apparent magnitude limit is $m_B = 19.4$. This strip is $1° \cdot 20°$ thich . In this region there are 3175 galaxies. The conical empty region on the left is an observational selection effect.

It is well established and accepted by everybody that galaxy clustering, as observed for example in Fig. 20.3, is fractal at least up to some scale. The controversy is about the implications of this result and the eventual existence of a crossover to a homogeneous distribution. The question of matter homogeneity on large scale is a very important one, since it is the basic assumption of the standard cosmological model. As we see in the figure, however, galaxy distribution is far from homogeneous on small scale, and the large scale structures (filaments and walls) which we recognize in the figure appear to be limited only by the boundary of the sample in which they are detected.

There is currently an acute debate on the result of the statistical analysis of large scale features. The standard interpretation asserts that large-scale homogeneity can be inferred from isotropy of angular data [16], while the precise determination of the homogeneity scale can be achieved through the analysis of the three dimensional catalogs of galaxies. The standard approach consists of the evaluation of the two-point correlation function $\xi(r)$ [16]. From the results of such an analysis, it is broadly believed that galaxies have a fractal distribution extending up to $\sim$ 10 $h^{-1}Mpc$, with a well-defined crossover to homogeneity beyond that scale. However, considering also other arguments, the value of the homogeneity scale can be very different according to different authors, ranging from 20 $h^{-1}Mpc$ [16] up to more than 300 $h^{-1}Mpc$ (see [17]) [2].

In the past years, we have proposed a solution to this apparent puzzle. In our view, the problem lies in the standard statistical analysis, and more specifically in the interpretation of the corresponding results. The main point in this discussion is that galaxy structures are fractal, irrespective of what the crossover scale is, and the importance of this fact has never been properly appreciated.

On the observational side, the available data in cosmology have been growing exponentially in the last ten years. In the coming decade, we will have an incredible amount of new data, in particular, from the three-dimensional observations (via redshift) of matter distribution. Cosmology therefore is based more on observational and testable grounds than ever in the past. On the other hand, statistical physics is in a position to provide the concepts that are crucial to study and understand these data. Such a project requires close interaction between three different lines, (i) Data Analysis, (ii) N-body simulations, (iii) Formulations of simple physical models to make the first steps towards a real theoretical understanding.

### 20.3.1 SCALING PROPERTIES AND DATA ANALYSIS

Given a distribution of points, the first basic question concerns the possibility of defining a physically meaningful average density. In fractal-like systems such a quantity depends on the size of the sample, and it does not represent a reference value, as in the case of an homogeneous distribution. Basically, a system cannot be homogeneous below the scale of the maximum void present in a given sample. However, the complete statistical characterization of highly irregular structures is the objective of Fractal Geometry [2].

[2] See also the web page *http://pil.phys.uniroma1.it/debate.html* where all these materials are available.

The major problem from the point of view of data analysis is to use statistical methods that are able to properly characterize the scale of self-similar distributions. We briefly discuss the main methods and we refer the interested reader to [8, 9] for a more complete discussion of the matter.

Let

$$n(\vec{r}) = \sum_i \delta(\vec{r} - \vec{r}_i) \tag{20.1}$$

be the number density of points in the system (the index $i$ runs over all the points) and let us suppose that we have an infinite system. If the presence of an object at the point $\vec{r}_1$ influences the probability of finding another object at $\vec{r}_2$, these two points are correlated. Hence, there is a correlation at distance $r$ if

$$G(r) = \langle n(\vec{0})n(\vec{r})\rangle \neq \langle n\rangle^2 \tag{20.2}$$

where we average taking all occupied points of the system as the origin and over the total solid angle, supposing statistical isotropy. On the other hand, there is no correlation if

$$G(r) = \langle n\rangle^2. \tag{20.3}$$

The proper definition of $\lambda_0$, the *homogeneity scale*, is the length scale beyond which the average density becomes well-defined, i.e. at which there is a crossover towards homogeneity with a flattening of $G(r)$. The length-scale $\lambda_0$ is related to the typical dimension of the largest voids in the system.

The traditional way to describe correlations for analytical (non fractal) structures is via the dimensionless function:

$$\xi(r) = \frac{\langle n(0)n(r)\rangle - \langle n\rangle^2}{\langle n\rangle^2}\,. \tag{20.4}$$

In the case of a fractal distribution, the average density $\langle n\rangle$ in the infinite system is zero, so that $G(r) = 0$ and $\lambda_0 = \infty$, and consequently $\xi(r)$ is not defined. In this case, the only well-defined quantity characterizing the two-point correlations is the function $\Gamma(r)$:

$$\Gamma(r) = \lim_{R_s \to \infty} \frac{\langle n(r)n(0)\rangle_{R_s}}{\langle n\rangle_{R_s}} \tag{20.5}$$

where $R_s$ is the size of a generic finite sample of the system, and $\langle ...\rangle_{R_s}$ indicates the average over all the occupied points of the sample (so that $\langle n\rangle_{R_s}$ is the average density of the sample). This function measures the average density of points at a distance $r$ from another occupied point, and this is the reason why it is called the conditional average density. Obviously, in the case of a distribution for which $\lambda_0$ is finite, $\Gamma(r)$ provides the same information as $G(r)$, i.e. it characterizes the correlation properties for $r < \lambda_0$ and the crossover to homogeneity.

Therefore, for a finite sample of points (e.g. galaxy catalogs), the first analysis to be done concerns the determination of $\Gamma(r)$. Such a measurement is necessary to distinguish between the two cases: (1) a crossover towards homogeneity in the sample shown by a flattening of $\Gamma(r)$, and hence an estimate of $\lambda_0 < R_s$ and $\langle n \rangle$; (2) a continuation of the fractal behavior ($R_s \ll \lambda_0$). Obviously, only in case (1) is it physically meaningful to study the correlation function $\xi(r)$ (Eq. 20.4), and extract from it the length scale $r_0$ ($\xi(r_0) = 1$), which is related to the intrinsic homogeneity scale $\lambda_0$. The functional behavior of $\xi(r)$ with distance then gives information on the correlation length of the density fluctuations.

Let us analyze the case $R_s \ll \lambda_0$. This is the so-called "fractal" case, and it is compatible with both the situation of $\lambda_0$ finite, but $R_s \ll \lambda_0$ (a sample size that is smaller than the homogeneity scale), or the situation in which $\lambda_0 \to \infty$, i.e. the case of a fractal distribution at any scale. It is easy to show [8, 9] that in this case (and in a spherical sample), Eq. 20.5 becomes

$$\Gamma(r) = \frac{BD}{4\pi} r^{D-3} \tag{20.6}$$

with $B$ the lower cut-off of the structure, which does not depend on the sample size. This shows that $\Gamma(r)$ is practically independent of the sample size. On the other hand, it is possible to show that $B$ is related approximately to the average distance between nearest-neighbor points in the system.

As already mentioned, in the fractal case ($R_s \ll \lambda_0$), the sample estimate of the homogeneity scale, through the value of $r$ for which the sample-dependent correlation function $\xi(r)$ (given by Eq. 20.7) is equal to 1, is meaningless: This estimate is the so-called "correlation length" $r_0$ [11] in the standard approach, where the assumption of homogeneity is implicitly used. The correlation function $\xi(r)$ is given operatively by

$$\xi(r) = \frac{\langle n(r)n(0)\rangle_{R_s}}{\langle n \rangle_{R_s}^2} - 1 = \frac{\Gamma(r)}{\langle n \rangle_{R_s}} - 1\,. \tag{20.7}$$

The basic point in the present discussion is that the mean density of the sample, $\langle n \rangle_{R_s}$, used in the normalization of $\xi(r)$, is not an intrinsic quantity characterizing the system, but is a function of the finite size $R_s$ of the sample. In fact, from Eq. 20.6, the expression for $\xi(r)$ in a sample from a fractal distribution is

$$\xi(r) = \frac{D}{3}\left(\frac{r}{R_s}\right)^{D-3} - 1\,. \tag{20.8}$$

From Eq. 20.8 it follows that $r_0$ (defined as $\xi(r_0) = 1$) is a linear function of the sample size $R_s$

$$r_0 = \left(\frac{D}{6}\right)^{\frac{1}{3-D}} R_s\,. \tag{20.9}$$

Hence, it is a spurious quantity that is simply related to the sample's finite size and devoid of physical meaning.

Finally, we emphasize that the fractal dimension is estimated by fitting $\xi(r)$ with a power-law, which in fact, as one can see from Eq. 20.8, for a fractal is power-law only for $r \ll r_0$ (or $\xi \gg 1$). For larger distances there is a clear deviation from the power-law behavior due to the definition of $\xi(r)$. Again this deviation is due to the finite size of the observational sample and does not correspond to any real change in the correlation properties. It is easy to see that, if one estimates the exponent at distances $r \approx r_0$, one systematically obtains a higher value of the correlation exponent (and a smaller value of the fractal dimension, as $D \approx 1.2$ in [11]) due to the break of $\xi(r)$ in a log-log plot.

The fact that galactic structures are fractal, no matter what the homogeneity scale $\lambda_0$, has important implications for the interpretation of several phenomena such as the luminosity bias, the mismatch galaxy-cluster, the determination of the average density, the separation of linear and non-linear scales, etc., and on the theoretical concepts used to study such properties. We also note that the properties of dark matter are inferred from the ones of visible matter, and hence they are closely related. If one changes perspective on the statistical properties for galaxies and clusters, this necessarily implies a change of perspective on the properties of dark matter.

### 20.3.2 IMPLICATIONS OF FRACTAL STRUCTURE UP TO $\lambda_0$

This is clearly an important point which is at the basis of the understanding of galaxy structures and, more generally, of the cosmological problem. We distinguish two different approaches: direct tests and indirect tests. By direct tests, we mean the determination of the conditional average density in three-dimensional surveys, while with indirect tests we refer to other possible analyses, such as the interpretation of angular surveys, the number counts as a function of magnitude or of distance or, in general, the study of non-average quantities, i.e. when the fractal dimension is estimated without making an average over different observers (or volumes). While in the first case one is able to have a clear and unambiguous answer from the data, in the second one is only able to make some weaker claims about the compatibility of the data with a fractal or a homogeneous distribution. For example the paper of Wu *et al.* [17] is mainly concerned with compatibility arguments, rather than with direct tests. However, even in this second case, it is possible to understand some important properties of the data, and to clarify the role and the limits of some underlying assumptions which are often uncritically adopted. Here, we do not enter into the details of the discussion of the real data, but consider briefly the main theoretical consequences of the three cases, which we separate as follows: (i) $\lambda_0 \approx 50\ Mpc/h$, (ii) $\lambda_0 \approx 300\ Mpc/h$ and (iii) $\lambda_0 \approx 1000\ Mpc/h$,

- (1) The fractal extends only up to $\sim 50$. It is beginning to be recognized in the literature on galaxy catalogs [17] that this is the minimal distance to which fractal structure extends, but usually still without consideration of the real consequences of this.

The standard approach to galaxy distribution has identified very small "correlation lengths", 5 $Mpc/h$ for galaxies and 25 $Mpc/h$ for clusters. The first of these (which is supposed to be known with high precision) are certainly inconsistent with a fractal extending to scales, at least, five time larger. We have shown that this inconsistency is conceptual and not due to incomplete data or weak statistics. Hence, in this hypothesis one has to abandon all the concepts related to these length scales. Some of these are:

(i) The estimate of the matter density in clusterized objects (visible + dark), which has been claimed to be $\Omega \approx 0.2 - 0.3$, decreases considerably.

(ii) The normalization of $N$-body simulations is often performed to some length-scale or amplitude of fluctuations, which are related to 5 $Mpc/h$ and 25 $Mpc/h$.

(iii) Concepts like the galaxy-cluster mismatch and the related luminosity bias, as well as the understanding of the clustering via the bias parameter $b$ (i.e. linear or non-linear - "stochastic bias" - amplification of $\xi(r)$) lose any physical meaning.

(iv) The interpretation of the velocity field is also based on the linear approximation which certainly cannot hold at scales smaller than 30 - 50 $Mpc/h$.

In summary major modifications are necessary in the current description of the origin and dynamics of large scale structures and for the role of dark matter.

- (2) The fractal extends up to 300 - 500 $Mpc/h$. In this case the standard picture of gravitationally induced structures after electromagnetic decoupling is untenable. There is no time to create such large scale correlated structures via gravitational instability, starting from Gaussian initial conditions.

- (3) The fractal extends up to $> 1000$ $Mpc/h$, and homogeneity does not exist, at least in the distribution of galaxies (i.e. visible matter). In this extreme case a new picture for the global metric [10] (and see next section) is then necessary.

For some questions the fractal structure leads to a radically new perspective and this is hard to accept. But it is based on the best data and analyses available. Up to a certain scale, it is neither a conjecture nor a model, it is a fact. The theoretical problem is that there is no dynamical theory to explain how such a distribution of matter could have arisen from the smooth initial state that is expected to exist in the very early universe. However, this is a different question. The fact that something is hard to explain theoretically has nothing to do with whether it is true or not. Facing a hard problem is far more interesting than hiding it under the rug by an inconsistent procedure.

Indeed, this will be the key point to understand in the future, but first we should agree on how this new 3D data should be analyzed. Further, the eventual crossover to homogeneity, if it exists, has also to be identified with our approach. If, for example, homogeneity is really found say at $\sim 100\ Mpc/h$, then clearly all our criticisms of the standard methods and results still hold fully. In summary, the standard method cannot be used, either to disprove homogeneity, or to prove it. One simply has to change methods.

## 20.4 FRACTAL COSMOLOGY IN AN OPEN UNIVERSE

As we have discussed the clustering of galaxies is well characterized by fractal properties, with the presence of a possible crossover to homogeneity still a matter of considerable debate. We now consider the cosmological implications of a fractal distribution of matter. The usual prejudice is that this cannot be compatible with a Friedman metric and with isotropic radiation. We challenge this misconception which is based on a traditional view of fluctuations, and show that a fractal structure can lead to surprising new ways of describing the metric and the compatibility between isotropic radiation and inhomogeneous matter.

Independently of the data, resistance to the fractal picture is certainly to a considerable degree due to the conviction that it is incompatible with the framework of the standard theories and in particular with the high degree of isotropy of the microwave background radiation (see Ref. [10] for references). In this respect one should note that in standard models the origin of radiation and baryonic matter is completely separate, with the latter being created in a dynamical process ('baryogenesis') completely distinct from the origin of the primordial radiation bath. The isotropy of the latter is therefore not fundamentally tied to the distribution of the matter, and the only real constraint is how much any such distribution actually perturbs the radiation. We consider the grounds for these theoretical biases against the possible continuation of a fractal distribution to arbitrarily large scales. Our central result is that a fractal distribution for matter, *even when there is no upper cut-off to homogeneity*, can in fact be treated in the framework of an expanding universe Friedmann cosmology.

A fractal being a self-similar and an intrinsically fluctuating distribution of points at all scales, appears to preclude the description of its gravitational dynamics in the framework of the Friedmann-Robertson-Walker (FRW) solutions to general relativity [11]. The problem is often stated as being due to the incompatibility of a fractal with the Cosmological Principle, where this principle is identified with the requirement that the matter distribution be isotropic and homogeneous [14]. This identification is in fact very misleading for a non-analytic structure like a fractal, in which all points are equivalent statistically, satisfying what has been called a Conditional Cosmological Principle [2, 8]. The obstacle to applying the

FRW solutions has in fact solely to do with the lack of homogeneity. One of the properties of a fractal of dimension $D$, however, is that the average density of points in a radius $r$ about any occupied point decreases as $r^{D-3}$, so that asymptotically the mass density goes to zero [2]. An approximation which therefore *may* describe the large scale dynamics of the universe in the case that the matter has such a distribution continuing to all scales is given by neglecting the distribution of matter at leading order, relative to the small but homogeneous component coming from the cosmic microwave background. We will now show that is indeed a good perturbation scheme, and calculate the physical scale characterizing its validity.

Consider first the standard FRW model with contributions from matter and radiation, for which the expansion rate is

$$H^2(t) = \left(\frac{\dot{a}}{a}\right)^2 = \frac{8\pi G}{3}\left(\rho_{rad} + \rho_{mat}\right) - k/a^2 \qquad (20.10)$$

where $a(t)$ is the scale factor for the expansion, and $\rho_{rad} \propto 1/a^4$ is the radiation density, and $\rho_{mat} \propto 1/a^3$ the (homogenous) matter density. The constant $k = -H_0^2 a_o^2(1 - \Omega_r - \Omega_m)$, where $H_0$ ($a_o$) is the expansion rate (scale factor) today and $\Omega_r$ ($\Omega_m$) is the ratio of the radiation (matter) energy density today to the 'critical' density $\rho_c = \frac{3}{8\pi G}H_0^2$. The sign of $k$ determines whether the universe is closed ($k > 0$) or open ($k < 0$), with $k = 0$ corresponding to a 'critical' spatially flat universe. Given the temperature of the CBR (see Ref. [10] for references), we have [3] $\Omega_r h^2 \approx 2.3 \times 10^{-5}$ (where $h$ is the Hubble constant in units of 100Mpc/km/s, with a typical measured value of $h \approx 0.65$ ; see Ref. [10, 11] for references). If we make the simple and natural assumption that galaxies trace the mass distribution, the value of $\Omega_m$ depends directly on the determination of the scale of the cross-over to homogeneity. If the observed fractal distribution continues to a scale $R_{homo}$, above which it turns over to homogeneity, one has

$$\Omega_m = \Omega_{10}\left(\frac{10}{R_{homo}}\right)^{3-D} \qquad (20.11)$$

where $\Omega_{10}$ is the average density of matter (relative to critical) in a sphere of radius $10Mpc/h$ about a galaxy, $D$ is the fractal dimension, and $R_{homo}$ is measured in $Mpc/h$. For $R_{homo}$ sufficiently large that $\Omega_m < \sqrt{\Omega_r}$ the $\rho_m$ term in (20.10) is always sub-dominant, and there is no matter-dominated era. For simplicity we now consider the limit in which $R_{homo} \to \infty$. Solving for the scale factor, we then have

$$a(t) = a_o(2H_0\Omega_r^{\frac{1}{2}} t)^{\frac{1}{2}}\left(1 + \frac{1-\Omega_r}{2\Omega_r^{\frac{1}{2}}}H_0 t\right)^{\frac{1}{2}} \qquad (20.12)$$

[3] We will neglect here, for simplicity, the minor modifications due to massless or low mass neutrinos, which can easily be incorporated in our analysis.

which shows how the early-time radiation-dominated behaviour ($a \propto t^{\frac{1}{2}}$) changes to the linear law $a \propto t$ at $t \approx 2H_0^{-1}\sqrt{\Omega_r}$ (red-shift $z \sim 1/\sqrt{\Omega_r}$ where $1 + z = a_o/a$). One may then show [10] how, in each of these two phases (dominated by the radiation and curvature, respectively), the fractal can be treated consistently as a perturbation to this FRW solution, in which matter is completely neglected.

In the curvature-dominated phase, the radiation is negligible and, at scales well within the horizon, one can use Newtonian gravity to describe the solution and perturbation to it when the self-gravity of the matter is included. The leading solution is simply the free expansion of the fractal, with every point moving radially away from its neighbor at a constant velocity proportional to its distance. In Ref. [10] we have shown that when one takes the self energy of the fractal into account, there is always a scale above which the fractal will be well-described by continued Hubble flow for all subsequent times. This scale is ultimately determined by the three-point properties of the distribution, as it basically depends on the local departures from spherical symmetry. A fractal with a very weak three-point correlation is one that has a very isotropic angular projection and it can be compatible with the above criterion. Only at a scale below this (which is a few $Mpc$ observationally) will the local self-gravity of the fractal 'win' over the kinetic energy of the Hubble flow (leading to the formation of very small scale structures - galaxies and clusters).

It is simple also to derive an expression for the peculiar velocity (small compared to the Hubble flow velocity $v_H$) which implies a simple linear relation, just as in standard perturbed homogeneous cosmology [11], between the local force and the velocity perturbation $(\Delta\vec{v}/v_H)(\vec{R}) \propto (\vec{F}(\vec{R})/R)$. This relation, for which there is apparently observational support, is usually used to determine an unknown constant (the 'bias' factor) [13]. In the present framework it can in principle be used to extract information about the total mass density and the three-point correlation properties.

We have thus seen in Ref. [10] that an open FRW universe is always a good approximation beyond some finite scale if matter is distributed as a simple fractal up to an arbitrarily large scale. In particular, such an open model - because it is dominated by the kinetic energy of the Hubble flow - can explain naturally how large structures can co-exist with almost perfect Hubble flow. We further note a few of its other striking features: *(i)* Since the Universe is to a good approximation in completely free expansion at large scales with $a(t) \propto t$, we have the deceleration parameter $q_o \approx 0$. This is a good fit to recent supernovae observations [20]. Rather than being due to the effect of an unknown 'anti-gravitational' component which mysteriously cancels the decelerating effect of the matter on the expansion [12], the effect is due to the decay towards zero of the matter density on such scales. *(ii)* The expansion age of the Universe is $t_0 = H_0^{-1} \approx 10\ h^{-1} \approx 15$ billion years, larger by 50% than in the standard matter dominated case. This value is comfortably consistent with the estimated age of globular clusters (the oldest known astrophysical objects) $11.5 \pm 1.3$ billion years [15]. *(iii)* The size of the horizon today is $R_H(t_0) \approx -\frac{1}{2}cH_0^{-1}\ln\Omega_r \approx 20,000\ Mpc/h$, a factor of about three larger than in the standard case.

In summary, we have seen that the scale-invariant or fractal nature of galaxy clustering represents an extremely interesting example of complex structures. The introduction of the methods of modern statistical physics into this area of astrophysics is having a dramatic impact and leads to a new perspective for the interpretation of observations, computer simulations and for the formation of novel cosmological theories. The new data expected in the next few years will provide a crucial test for the various scenarios.

## BIBLIOGRAPHY


[1] P.W. Anderson, Science **177** 393 (1972).

[2] B.B. Mandelbrot, *The Fractal geometry of Nature"*, (Freeman, San Francisco, 1982).

[3] See the article by Per Bak and Kan Chen in this volume.

[4] See the article by Marc Mézard in this volume.

[5] C.J.G. Evertsz *et al.* Eds, *Fractal geometry and Analysis*, (World Scientific, Singapore 1996).

[6] A. Erzan, L. Pietronero and A. Vespignani, Rev Mod. Phys **67** 545 (1995).

[7] Y.Y. Kagan, "Is Earthquake Seismology a Hard, Quantitative Science ?", Pure and Applied Geophysics **155** 233 (1999).

[8] P.H. Coleman & L. Pietronero, Phys. Rep. **213** 311 (1992).

[9] F. Sylos Labini, M. Montuori & L. Pietronero, Phys. Rep. **293** 66 (1998).

[10] M. Joyce, P.W. Anderson, M. Montuori, L. Pietronero & F. Sylos Labini , Europhys. Lett. in print (2000) (May 1st).

[11] P.J.E. Peebles, *Principles of Physical Cosmology"*, (Princeton Univ. Press, 1993).

[12] P. J. E. Peebles, Nature **398** 25 (1999).

[13] M. Strauss & J. A. Willick, Physics Reports, **261** 5 (1995).

[14] P. Coles, Nature **391** 120 (1998).

[15] B. Chaboyer, Physics Reports, **307** 23 (1998).

[16] *Critical dialogues in Cosmology* Ed. By N. Turok, (World Sci., Singapore, 1997). See the papers by M. Davis (p.13) and L. Pietronero *et al.* (p.24).

[17] K.K. Wu, O. Lahav & M. Rees, Nature **397** 225 (1999).

[18] M. Chown, New Scientist **2200** 23 (1999).

[19] G. Smoot *et al.* , Ap. J. **464** L1 (1992).

[20] S. Perlmutter *et al.*, Ap. J. **517** 565 (1999).

# CHAPTER 21

# STATISTICAL PHYSICS AND COMPUTATIONAL COMPLEXITY

SCOTT KIRKPATRICK[1] AND BART SELMAN[2]

[1]IBM, Thomas J. Watson Research Center, Yorktown Heights, NY 10598
[2]Computer Science Deptartment, Cornell University, Ithaca, NY 14850

ABSTRACT

"NP-complete" problems are at the core of many computational tasks of practical interest, which means that the cost of solution will grow exponentially with problem size in the worst case (provided P $\neq$ NP). Although heuristics can be quite effective on such problems in most cases, there is a growing appreciation that these problems contain phase transitions, and at the phase boundaries exponential complexity becomes the typical outcome, not just the worst case. Using methods from statistical physics, a much better understanding of such phase transition phenomena in computational problems has been obtained in recent years. We will review several key results in this area, thereby illustrating some of the deep connections between computer science and statistical physics. The seminal work by Fu and Anderson [13, 14] provided the initial impetus for much of this work.

## 21.1 INTRODUCTION

Many key computational tasks of practical interest have been shown to be computationally intractable. Such problems, found for example, in planning, scheduling, machine learning, hardware design, and computational biology, generally belong to the class of NP-complete problems. To solve an NP-complete problem, it is widely believed that the computational resource requirements grow exponentially with problem size, at least in the worst case. The typical case behavior of these problems is often much more difficult to characterize, but is more relevant from a practical perspective.

Fu and Anderson [13, 14] first conjectured a deep connection between NP-complete problems and models studied in statistical physics. More recently, we have shown that NP-complete problems can exhibit phase transition phenomena, analogous to those in physical systems, with the hardest problem instances occurring at the phase boundary. However, the exact relationship between phase transition phenomena and computational properties has remained unclear. For example, phase transitions have also been observed in computationally "easy" problems (*i.e.*, ones that are not NP-complete).

In this paper, we will discuss recent results that give a precise characterization of the relationship between phase transition phenomena and typical case computational complexity. More specifically, we show that when the underlying computational task exhibits a continuous phase transition, resource requirements grow only polynomially with problem size, while a special type of discontinuous phase transition corresponds to an exponential growth in typical resource requirements, characteristic of truly hard problem instances.

These results illustrate the potential benefits of exploring the rich connections between computer science and statistical physics.

## 21.2 SATISFIABILITY AND HARD-PROBLEM INSTANCES

As our computational task, we will consider the Boolean satisfiability (SAT) problem, an archetypal NP-complete problem. In the SAT problem, one is given a formula in Boolean logic and the task is to determine whether the formula is satisfiable. We will consider Boolean formulae written in a special form, called conjunctive normal form. Each formula consists of a series of clauses conjoined together (logical "AND"), where each clause is a disjunction (logical "OR") of literals. A literal is a Boolean variable or its negation (logical "NOT"). In the $k$-SAT problem, each clause contains exactly $k$ literals. An example of a formula consisting of three clauses and two Boolean variables, $A$ and $B$, is (($A$ OR (NOT $B$)) AND ((NOT $A$) OR $B$) AND ((NOT $A$) OR (NOT $B$))). The formula is *satisfiable*, because, *e.g.*, the assignment $A$ set to FALSE and $B$ set to FALSE satisfies this formula. One method for checking whether a formula with $N$ Boolean variables is satisfiable is to check for each of the $2^N$ truth assignments, whether there is one that satisfies all the clauses in the formula. If none is found, then the formula is *unsatisfiable*.

One might imagine that there are much more clever ways of determining whether a formula is satisfiable, for example, methods that don't search through the space of all possible truth assignments. However, one of the key results in computational complexity theory shows that it is highly unlikely that such a clever algorithm exists. More specifically, Cook [9] showed the $k$-SAT problem with $k \geq 3$ to be NP-complete. As a consequence, there does not exist any procedure that does significantly better on all Boolean formulae than one that exhaustively checks all truth assignments (assuming P $\neq$ NP, a widely believed, but as of yet unproven, conjecture [7]). In other words, no matter how clever an algorithm one develops for Boolean satisfiability, there will be formulae on which the algorithm takes time exponential in the number of variables. Note that NP-competeness is a worst-case notion. In practice, one is often more interested in what happens in a "typical" case or "average" case scenario. We now turn our attention to the behavior of algorithms on such typical Boolean formulae.

Our initial interest in the satisfiability problem arose from early reports that many satisfiability problems are easily solvable. For example, Goldberg [19] describes a class of random SAT problems that are surprisingly easy for the Davis-Putnam satisfiability procedure, which is one of the most widely used complete algorithms for satisfiability testing [11]. Goldberg's work led to an extensive theoretical exploration of his particular random instance model. In Goldberg's model, each Boolean clause is generated by selecting literals with some fixed probability. This leads to clauses of varying length. A rigorous analysis, reported in a series of papers [15, 16, 33], has shown that in this model, the *average-case* complexity is polynomial for *almost all* choices of parameter settings. In other words, it is difficult to generate computationally hard problem instances. Note that this does not mean that hard instances do not exist; it is simply means that such instances are extremely rare, and, in fact, may never be observed in practice.

In [28, 25, 37], we show how by using a different model for generating random formulae, called the *fixed-clause-length* model, one can easily generate hard problem instances. Consider generating a random 3-SAT problem. Each clause is generated by randomly selecting three variables from among $N$ variables; each of these variables is negated with probability 0.5. We generate a total of $M$ clauses. We found that the key in generating computationally hard instances is the ratio between $M$ and $N$. Fig. 21.1 shows the median cost of solving randomly generated instances at different ratios of variables to clauses. The data was obtained by running the Tableau method, which is a highly efficient implementation of the Davis-Putnam (DP) procedure [10]. We measure the solution cost in number of recursive calls to the DP procedure. This measure is proportional to the actual run time of the algorithm but is machine independent. We see that the cost peaks at around a ratio of 4.3 clauses per variable.[1] Our experimental data shows that at this point the cost of determining satisfiability grows exponentially with size of the formulae.

Figure 21.2 gives the fraction of formulae that are unsatisfiable as a function of the ratio of clauses to variables for randomly generated formulae with 50 variables. At low ratios, few clauses compared to the number of variables, almost all instances are satisfiable (*i.e.*, the unsatisfiable fraction is almost zero). At relatively high ratios of clauses to variables, in a sense too many constraints, almost all

[1] For large $N$, this ratio converges to around 4.25. [10, 25].

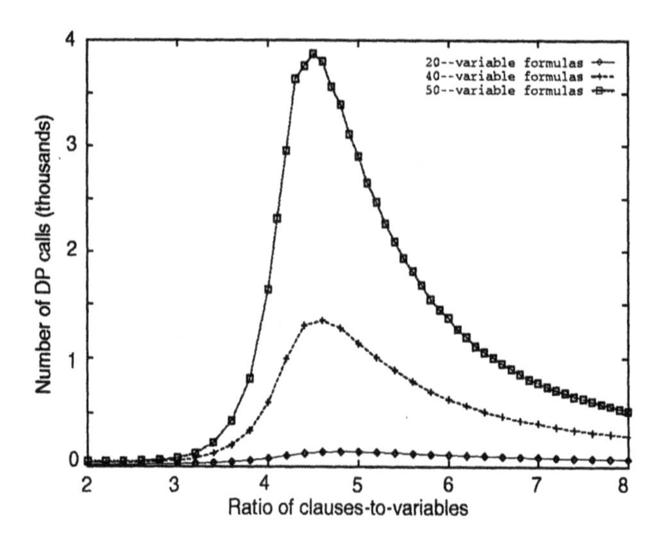


Figure 21.1: Solving 3SAT instances.

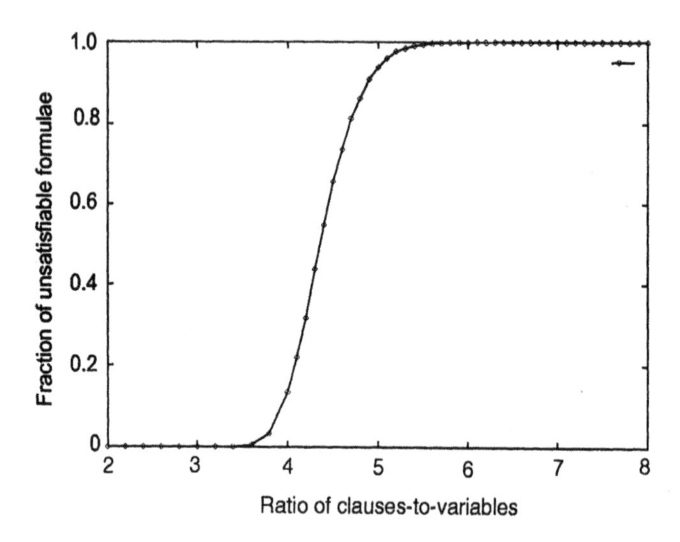


Figure 21.2: Fraction of unsatisfiable 3SAT problems.

randomly generated instances are unsatisfiable. A sudden change occurs around the critical ratio of 4.3. At this ratio, a "phase transition" from the mostly satisfiable phase to the mostly unsatisfiable phase takes place. From Figure 21.1, we see that the phase-transition region coincides with the area with the hardest problem instances. In this region, the instances are again critically constrained. Below, we will take a closer look at what happens inside the phase transition region. A similar relation between exponential scaling and phase transition phenomena has been observed for a range of NP-complete problems. See for example, [6, 21].

The randomly generated critically constrained problem instances have been used extensively in the study and development of algorithms for graph coloring and satisfiability testing [36]. A key question is whether the results obtained for such instances are at all indicative of the behavior of the algorithms on more structured, real-world instances. The results of the DIMACS Challenge on Satisfiability Testing [38] suggest that the behavior of algorithms on hard random problems can indeed be representative of the behavior on more structured problems. The DIMACS Benchmark Problem Set contained several hard random instances and numerous more structured problems. The satisfiability algorithms fell in two categories: complete systematic procedures, and incomplete stochastic methods. These methods complement each other, in that there are problem classes where the stochastic methods are best, whereas on other problem classes the systematic methods are superior. However, within each category, algorithms that were fastest on the hard random instances usually also performed best on the more structured problems. Apparently, the hard random instances do exercise the various time critical parts of the algorithms. Therefore, the performance of algorithms on such hard random instances is a reasonably good indicator of the overall performance on a more diverse set of problem instances.

Aside from being useful as benchmark problems, there is also evidence that critically constrained problems may occur naturally in real-world applications. Nemhauser [32] studied a large airline scheduling problem, involving approximately 500 planes. The original schedule was obtained with a heuristic method and therefore only an approximation of the optimal solution. After a substantial computational effort, Nemhauser's group found a provably optimal solution. It was expected that such an optimal solution would lead to a savings of one or more planes over the heuristically obtained schedule. However, quite surprisingly, the optimal schedule did not save a single plane. The explanation appears to be that the problem had become critically constrained: Because of economic factors, the airline had assigned additional routes to planes that were idle during parts of the day in the original schedule. So, external factors can give rise to critically constrained real-world planning and scheduling problems. The ratio of constraints to variables will probably differ from the critical ratios found in hard random instances, because of the inherent internal structure of real-world problems. The reader is encouraged to consult any of the following additional references [21, 4, 8, 10, 18, 22, 27, 39, 41].

## 21.3 CONNECTIONS TO STATISTICAL PHYSICS

It has long been apparent that NP-complete problems have something in common with the models discussed in the literature of the statistical mechanics of disordered media. Loosely speaking, the optimization problems around which most NP-complete decision problems are formulated can be translated into the problem of finding ground states of appropriately constructed systems with many simple degrees of freedom. And the fact that "spin glasses," known to have very long relaxation times under Metropolis-style evolution at any low temperature, are found in many models of dynamical systems with quenched-in random interactions, has made establishing a connection between "glassiness" and the complexity defined by computational cost in computer science a natural objective. However, the problems defined in the two fields are not at all the same. The most significant difference is that complexity classes such as P and NP are defined in terms of the existence or nonexistence of at least one worst-case instance, while relaxation times and glassy behavior in statistical mechanics are characteristics of the most probable configurations of large systems, computed in practice as averages over partition functions (*i.e.*, over all initial conditions, or over all ways of defining the random system given some control parameters). This extra step of defining a probability measure over instances of a combinatoric problem in order to analyze average case complexity is one which until recently the computer science community was reluctant to take. Or, stated more concretely, relatively few of such average case compelxity results have been obtained in computer science. This is partly due to the difficulty of chosing the "right" underlying probability distribution for the ensemble.

Let's consider the simplest case in which to make the connection between a known NP-complete problem and a spin glass — partitioning a random graph with weighted edges into two equal sets of nodes, while minimizing the cost of the edges which cross the boundary between the two partitions. This is the example treated by Fu and Anderson [13]. Each node $i$ can be in only on of two states, which we might call the "left" partition or the "right" partition. This can be represented by a spin $S_i$ for each node, with $S_i = -1$ meaning that node $i$ is in the "left" partition, and $S_i = 1$ meaning the "right" partition. If $J_{ij}$ is the weight associated with each edge in the graph (many of these will be zero), the cost of a particular partition can be calculated as

$$H_1 = \sum_{i,j} J_{ij}(1 - S_i S_j)/2. \tag{21.1}$$

We need a second interaction term to enforce the constraint that the sizes of the two partitions are equal, or nearly so. Let

$$H_2 = \lambda \sum_{i,j} (S_i S_j), \tag{21.2}$$

and the Hamiltonian for the model equivalent to a particular graph bipartitioning problem is

$$H = H_1 + H_2. \tag{21.3}$$

For sufficiently large $\lambda$, the second term only allows configurations where the number of nodes in the "left" partition is equal to that in the "right" partition. The model which results is an infinite-range Ising antiferromagnet with random ferromagnetic interactions added (to keep edges connected by large weights in the same partitions). From the study of the SK models, the simplest spin glasses, this is known to give rise to a spin glass phase at low temperatures. We can define a similar Hamiltonian for the SAT problem.

We shall follow the notation used in the series of papers by Monasson and Zecchina. Since the variables in $k$-SAT are Boolean, we can use Ising spins to represent them. Let $S_i = +1$ if the variable $x_i$ is set to true, and $S_i = -1$ if it is false. We will capture the set of $M$ clauses in a formula in a single $M \times N$ matrix of random interactions, $\Delta$, with $\Delta_{l,i} = +1$, if the $l^{\text{th}}$ clause contains $x_i$, $\Delta_{l,i} = -1$, if the $l^{\text{th}}$ clause contains $\neg x_i$ (here $\neg$ denotes logical negation), and otherwise $\Delta_{l,i} = 0$.

Then the number of clauses not satisfied provides a natural Hamiltonian for this problem, and is given by

$$E[\Delta, S] = \sum_{l=1,M} \delta \left[ \sum_{i=1,N} \Delta_{l,i} S_i \; ; \; -k \right] \tag{21.4}$$

where $\delta[i; j]$ is the Kronecker symbol.

To fix the number of variables in each clause to be $k$ we add the constraint

$$\sum_{i=1,N} \Delta_{l,i}^2 = k \quad \text{for all} \quad l = 1, ..., M \tag{21.5}$$

This can either be viewed as an Ising Hamiltonian with random external fields, or studied by expanding out the Kronecker delta, in which case we obtain interactions between up to $k$ spins at a time. This gives a sense in which $k = 2$ (only pairwise interactions) is clearly distinct from $k \geq 3$.

## 21.4 A CLOSER LOOK AT THE PHASE TRANSITION

Figure 21.3 shows the phase transition for 3SAT for several different values of $N$ (the number of variables). Note how the threshold function sharpens up for larger values of $N$. In [25], we show that the threshold has characteristics typical of phase transitions in the statistical mechanics of disordered materials.

"Finite-size scaling" analysis is a useful tool in the study of phase transition phenomena. This approach is based on rescaling the horizontal axis by a factor

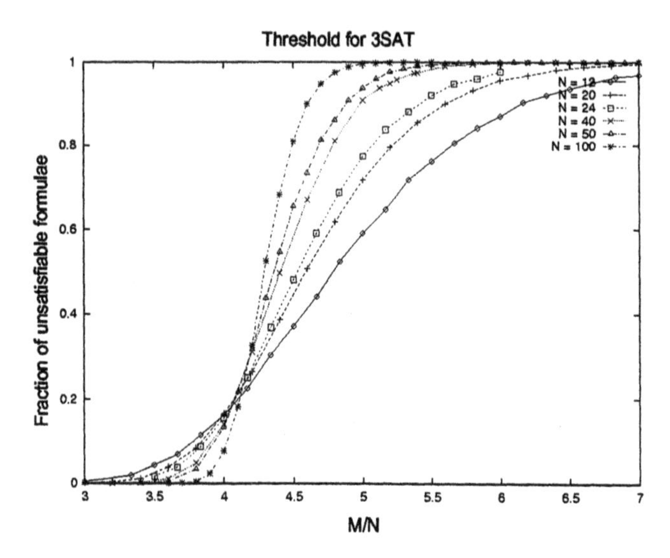


Figure 21.3: The 3SAT phase transition sharpens up.

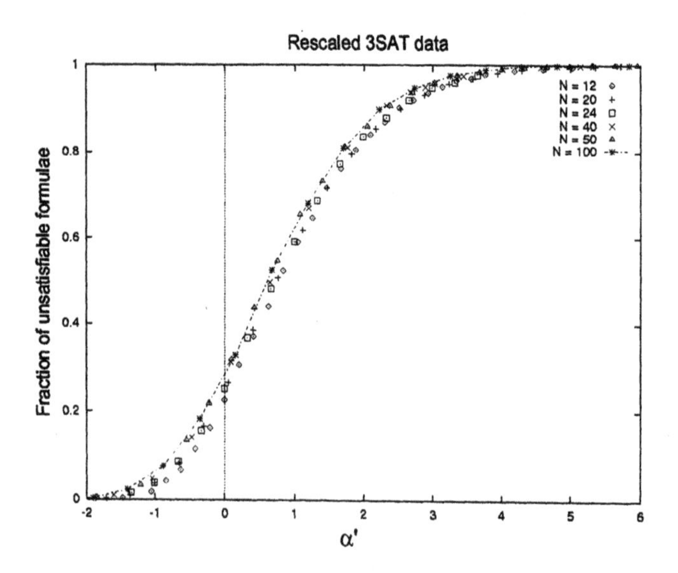


Figure 21.4: Rescaling the 3SAT phase transition.

that is a function of $N$. The function is such that the horizontal axis is stretched out for larger $N$. So, in effect, rescaling "opens up" the phase-transition for higher values of $N$, and thus gives us a better look inside the transition. Figure 21.4 shows the result of rescaling the curves from Figure 21.3. Along the rescaled horizontal axis is $\alpha' = N^{1/\nu}(\alpha - \alpha_c)/\alpha_c$, where $\alpha = M/N, \nu = 1.5, \alpha_c = 4.2$. We see that the original curves are rescaled into a single universal curve. Very recent theoretical work by Wilson [40] claims that the current experimental data has led to an underestimate of the critical exponent $\nu$, since $\nu \rightarrow 1$ as $k$ becomes large is predicted by "annealed" arguments as well as obtained in the replica-symmetric solution of Monasson and Zecchina, while Wilson's construction gives the critical regime a breadth always characterized by $\nu >= 2$. Wilson's results may require values of $N$ that are not accessible experimentally. As a result, some questions remain open about the correct way to define large $k, N$ limits in this problem.

Finite-size scaling can also be used to study properties other than satisfiability in the critical region. See [35] for a rescaling of computational cost curves, and [34] for a rescaling of the prime implicate function.

## 21.5 MIXTURES OF 2-SAT AND 3-SAT PROBLEMS

As noted earlier, the $k$-SAT problem is NP-complete for $k \geq 3$. For $k = 2$, there exists a linear time algorithm for determining the satisfiability of Boolean formulae [3]. This method cleverly avoids checking all possible truth assignments and works effectively (scaling linear in $N$) on all possible 2-SAT problems.

In order to understand what occurs between $k = 2$ and $k = 3$, we have studied [30, 2, 31] formulae containing mixtures of 2- and 3-clauses: consider a random formula with M clauses, of which $(1-p)M$ contain two literals and $pM$ contain 3 literals, with $0 \leq p \leq 1$. This "2+$p$–SAT" model smoothly interpolates between 2–SAT ($p = 0$) and 3–SAT ($p = 1$). The problem is NP–complete, since any instance of the model for $p > 0$ contains a sub-formula of 3-clauses, but our interest here is in the complexity of "typical" problem instances.

We seek $\alpha_c(2+p)$, the threshold ratio $M/N$ of the above model at fixed $p$. We know $\alpha_c(2) = 1$ and $\alpha_c(3) \simeq 4.25$. The formulae cannot be almost always satisfied if the number of 2–clauses (respectively 3–clauses) exceeds $N$ (resp. $\alpha_c(3)N$). As a consequence, the critical ratio must be bounded by $\alpha_c(2+p) \leq \min\left(\frac{1}{1-p}, \frac{\alpha_c(3)}{p}\right)$.

The $2+p$–SAT model can be mapped onto a diluted spin glass model with $N$ spins $S_i$: $S_i = 1$ if the Boolean variable $x_i$ is true, $S_i = -1$ if $x_i$ is false. And, again, with any configuration we associate an energy $E$, or cost-function, equal to the number of clauses violated. Random couplings between the spins are induced by the clauses. *The most important result of the replica approach [29] is the emergence, in the large $N, M$ limit and at fixed p and $\alpha$, of order parameters describing the statistics of optimal assignments, which minimize the number of* *violated clauses.* In this section, we give an overview of the results from statistical mechanics.

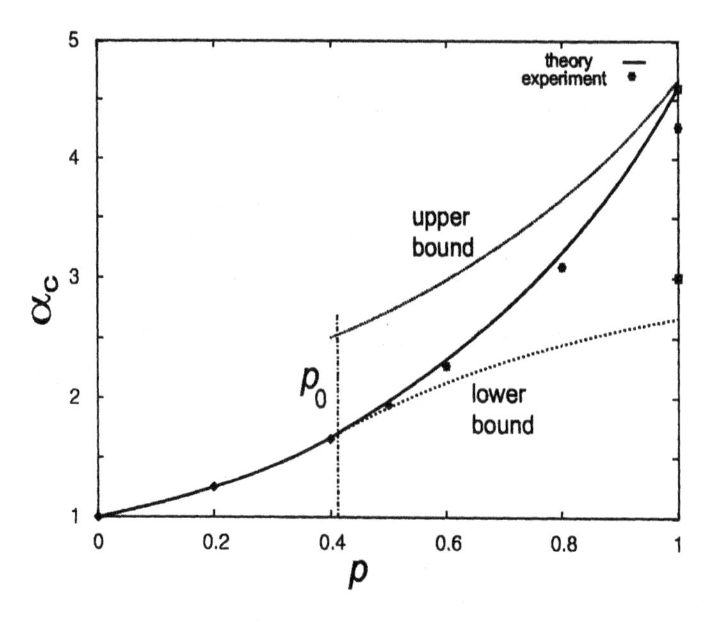


Figure 21.5: Theoretical and experimental results for the SAT/UNSAT transition in the 2+p-SAT model.

Consider an instance of the $2+p$-SAT problem. We use the $\mathcal{N}_{GS}$ ground state configurations to define

$$m_i = \frac{1}{\mathcal{N}_{GS}} \sum_{g=1}^{\mathcal{N}_{GS}} S_i^g \tag{21.6}$$

the average value of spin $S_i$ over all optimal configurations. Clearly, $m_i$ ranges from $-1$ to $+1$ and $m_i = -1$ (respectively $+1$) means that the corresponding Boolean variable $x_i$ is always false (resp. true) in all ground states. The distribution $P(m)$ of all $m_i$ gives the microscopic structure of the ground states. The accumulation of magnetizations $m$ around $\pm 1$ represents a "backbone" of almost completely constrained variables, whose logical values cannot vary from solution to solution, while the center of the distribution $P(m \simeq 0)$ describes weakly constrained variables. The threshold $\alpha_c$ will coincide with the appearance of an extensive backbone density of fully constrained variables $x_i$, with a finite probability weight at $m = \pm 1$. A simple argument shows that the backbone must vanish when $M/N = \alpha < \alpha_c$. Consider adding one clause to a SAT formula found below $\alpha_c$. If there is a finite fraction of backbone spins, there will be a finite probability that the added clause creates an UNSAT formula, which cannot occur.

For $\alpha < \alpha_c$, the solution exhibits a simple symmetry property, usually referred to as Replica Symmetry (RS), which leads to an order parameter which is precisely the magnetization distribution $P(m)$ defined above. An essential qualitative difference between 2-SAT and 3-SAT is the way the order parameter $P(m)$ changes at the threshold. This discrepancy can be seen in the fraction $f(k, \alpha)$ of Boolean variables which become fully constrained, at and above the threshold. As said above, $f(k, \alpha)$ is identically null below the threshold. For 2-SAT, $f(2, \alpha)$ becomes strictly positive above $\alpha_c = 1$ and is continuous at the transition : $f(2, 1^-) = f(2, 1^+) = 0$. On the contrary, $f(3, \alpha)$ displays a discontinuous change at the threshold : $f(3, \alpha_c^-) = 0$ and $f(3, \alpha_c^+) = f_c(3) > 0$.

While for the continuous transitions, the exact value of the threshold can be derived within the RS scheme, for the discontinuous case the RS prediction gives only upper bounds. The exact value of the threshold can be predicted only by a proper choice of the order parameter at the transition point, *i.e.*, by a more general symmetry breaking scheme, a problem which is still open. However, the predictions of the RS equations, such as the number of solutions, remain valid up to $\alpha_c$, and the RS prediction for the nature of the threshold should be qualitatively correct.

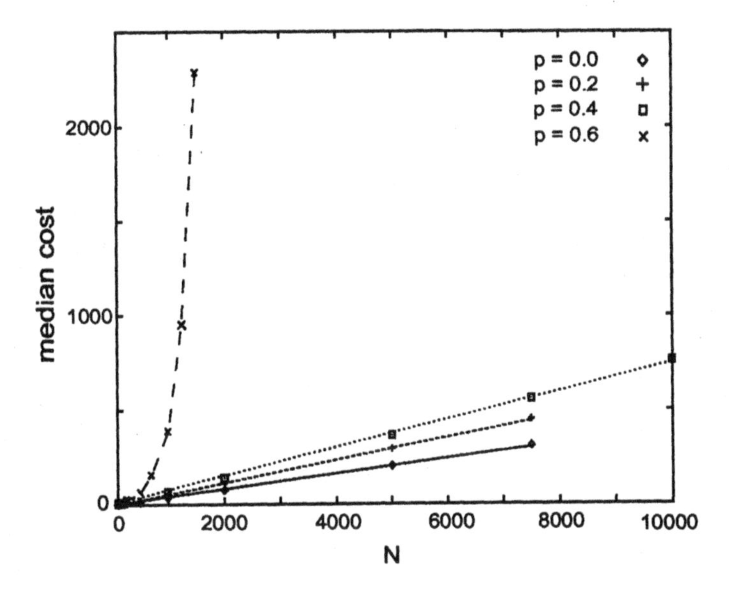


Figure 21.6: Median computational cost of proving a formula SAT or UNSAT using the Tableu procedure [10], for $p$ ranging from 0 to 1.

For the mixed $2 + p$–SAT model, the key issue is therefore to understand how a discontinuous 3–SAT-like transition may appear when increasing $p$ from zero up to one and how it affects the computational cost of finding solutions near threshold. Applying the method of [29], we find for $p < p_0$ (according to various estimates, $p_0$ lies between 0.4 and 0.416), there is a **continuous** SAT/UNSAT transition at $\alpha_c(2+p) = \frac{1}{1-p}$. This has recently been verified by rigorous analysis up to $p = 0.4$ [1]. The RS theory appears to be correct for $\alpha < \alpha_c(2+p)$, and thus gives both the critical ratio and the typical number of solutions, as in the $k = 2$ case. The SAT/UNSAT transition should coincide with a replica symmetry breaking transition, as discussed in [29]. So, for $p < p_0$, the model shares the characteristics of random 2–SAT.

For $p > p_0$, the transition becomes **discontinuous** and the RS transition gives an upper bound for the true $\alpha_c(2+p)$. The RS theory correctly predicts a discontinuous appearance of a finite fraction of fully constrained variables which jumps from 0 to $f_c$ when crossing the threshold $\alpha_c(2+p)$. However, both values of $f_c(2+p)$ and $\alpha_c$ are slightly overestimated, *e.g.*, for $p = 1$, $\alpha_c^{RS}(3) \simeq 4.60$ and $f_c^{RS}(3) \simeq 0.6$ whereas experiments give $\alpha_c(3) \simeq 4.27$ and $f_c(3) \sim 0.5$. A replica symmetry breaking theory will be necessary to predict these quantities. For $p > p_0$, the random 2+p–SAT problem shares the characteristics of random 3–SAT.

Figure 21.5 shows our results for the location of the SAT/UNSAT transition for the 2+p-SAT model. The vertical line at $p_0$ separates the continuous from the discontinuous transition. The full line is the replica-symmetric theory's predicted transition, believed exact for $p < p_0$, and the diamond data points are results of computer experiment and finite-size scaling. The other two lines show upper and lower bounds obtained in [1], while the stronger upper bound due to [26], and the best known lower bound, due to [17], are indicated by square data points (at $p = 1.0$).

Figure 21.6 gives data on the scaling behavior of the Davis-Putnam style satisfiability procedure, Tableau [10], on the 2+P-SAT problem for a range of values of $p$. (For each value of $p$, we give the cost at the phase transition boundary.) The main observation to make is the linear scaling for $p < p_0$ and the exponential scaling for higher values of $p$. This is consistent with the analytical results that show that the behavior of the overall ensemble of clauses changes at a the critical ratio $p_0$, where it the system switches form the polynomial charateristics of 2-SAT to the exponential scaling for 3-SAT.

## BIBLIOGRAPHY

[1] Achlioptas, D., Kirousis, L., Kranakis, E., and Krizanc, D. Rigorous results for random $(2+p)$–SAT. *Proc. RALCOM 97*, 1–10 (1997).

[2] Anderson, P.W. Solving problems in finite time. *Nature*, Vol. 400(8), 1999.

[3] Aspvall, B, Plass, M.F. and Tarjan, R.E. *Inf. Process. Lett.* 8, 121 (1979)

[4] Baker, A. Intelligent Backtracking On Constraint Satisfaction Problems: Experimental and Theoretical Results. Techn. Report. CIS-TR-95-08, Univ. of Oregon, Computer & Information Science, 1995.

[5] Buro, M. and Kleine-Büning, H. Report on a SAT competition. Techn. Rep. # 110, Dept. of Math. and Inform., University of Paderborn, Germany (1992).

[6] Cheeseman, P., Kanefsky, B., Taylor, and William M. Where the really hard problems are. *Proc. IJCAI-91* (1991) 331–336.

[7] Clay Mathematics Institute. Millennium Prize Problems. See www.ams.org/claymath/.

[8] Clearwater, S.H., Huberman, B.A., Hogg, T. Cooperative solution of constraint satisfaction problems. *Science*, 254 (1991).

[9] Cook, S.A. The complexity of theorem-proving procedures. *Proc. STOC-71* (1971) 151–158.

[10] Crawford, J.M. and Auton, L.D. Experimental results on the cross-over point in satisfiability problems. *AAAI-93* (1993) 21–27. (Ext. version in Artif. Intel.)

[11] Davis, M. and Putnam, H. A computing procedure for quantification theory. *J. of the ACM*, 7 (1960) 201–215.

[12] Dubois, O., and Boufkhad, Y. A general upper bound for the satisfiability threshold of random K–SAT formulas. *Journal of Algorithms 24*, 395-420 (1997)

[13] Fu, Y.-T., and Anderson, P. W. Applications of statistical mechanics to NP-complete problems in combinatorial optimization. *J. Phys.*, **A**19, 1605 (1986).

[14] Fu, Y.-T., and Anderson, P. W. In *Lectures in the Sciences of Complexity*, D. Stein (ed.), (Addison-Wesley, 1989), p. 815.

[15] Franco, J. and Paull, M. Probabilistic analysis of the Davis Putnam procedure for solving the satisfiability problem. *Discrete Applied Math.* 5 (1983) 77–87.

[16] Franco, J. Elimination of infrequent variables improves average case performance of satisfiability. *SIAM J. Comput.*, 20(6) (1991) 1119–1127.

[17] Frieze, A., and Suen, S. Analysis of two simple heuristics on a random instance of K–SAT. *Journal of Algorithms 20* (1996), 312-335.

[18] Gent, I.P. and Walsh, T. Computational phase transitions in real problems. Research Report, Dept. of AI, Edinburgh (1995).

[19] Goldberg, A. On the complexity of the satisfiability problem. Courant Comp. Sci. Rep., No. 16, New York University, NY, 1979.

[20] Hayes, B. Can't get no satisfaction. *American Scientist* Vol. 85(2), 108–112 (1996). http://www.amsci.org/amsci/issues/Comsci97/compsci9703.html

[21] Hogg, T., Huberman, B.A. & Williams, C. (eds.) Frontiers in problem solving: phase transitions and complexity. *Artificial Intelligence* Vol. 81 (I & II) (1996).

[22] Hogg, T. and Williams, C.P. Expected gains from parallelizing constraint solving for hard problems. *AAAI-94* (1994) 331–336.

[23] Kamath, A., Motwani, R., Palem, K., and Spirakis, P. Tail bounds for occupancy and the satisfiability threshold. *Random Structures and Algorithms* 7, 59 (1995)

[24] Kirkpatrick, S., Gelatt, C.D., and Vecchi, M.P. Optimization by simulated annealing. *Science*, 220 (1983) 671–680.

[25] Kirkpatrick, S. and Selman, B. Critical Behavior in the Satisfiability of Random Boolean Expressions. *Science*, 264 (May 1994) 1297–1301. Also p. 1249: "Mathematical Logic: Pinning Down a Treacherous Border in Logical Statements" by B. Cipra.

[26] Kirousis, L., Kranakis, E., and Krizanc, D. Approximating the unsatisfiability threshold of random formulas. *Proceedings of the 4th European Symposium on Algorithms*, (1992), 27-38.

[27] Larrabee, T. and Tsuji, Y. Evidence for a satisfiability threshold for random 3CNF formulas. *Proc. AAAI Symp. on AI and NP-hard problems*, Palto Alto, CA (1993).

[28] Mitchell, D., Selman, B., and Levesque, H.J. Hard and easy distributions of SAT problems. *Proc. AAAI-92*, San Jose, CA (1992) 459–465.

[29] Monasson, R., and Zecchina R. *Phys. Rev.* E 56, 1357 (1997).

[30] Monasson, R., and Zecchina R. Kirkpatrick, S., Selman, B., Troyansky, L. Phase Transition and Search Cost in the $2 + p$ SAT Problem. Proc. of *PhysComp96*, Toffoli, T., Biafore, M., Leão, J., eds., Boston (1996).

[31] Monasson, R., and Zecchina R., Kirkpatrick, S., Selman, B., Troyansky, L. Determining computational complexity from characteristic 'phase transitions'. *Nature*, Vol. 400(8), 1999, 133–137.

[32] Nemhauser, G., Barnhart, C., Hane, C., Johnson, E., Marsten, R., and Sigismondi, G. The fleet assignment problem. AI/OR Org. Workshop, Plymounth, VT (1994).

[33] Purdom, Jr. P.W., and Brown, C.A. Polynomial average-time satisfiability problems. *Inform. Sci.*, 41 (1987) 23–42.

[34] Schrag, R. and Crawford, J. Implicates and prime implicates in Random 3-SAT. *Artificial Intelligence*, 81 (1-2) 192–222 (1996).

[35] Selman, B. and Kirkpatrick, S. Critical behavior in the cost of K-satisfiability. *Artificial Intelligence* Vol. 81, 273–295 (1996).

[36] Selman, B., Levesque, H.J., and Mitchell, D. A new method for solving hard satisfiability problems. *Proc. AAAI-92*, San Jose, CA (1992) 440–446.

[37] Selman, B.; Mitchell, D.; and Levesque, H. Generating Hard Satisfiability Problems. *Artificial Intelligence*, Vol. 81, 1996, 17–29.

[38] Trick, M. and Johnson, D. (Eds.) *Proc. DIMACS Challenge on Satisfiability Testing.* Piscataway, NJ, 1993. (DIMACS Series on Discr. Math.)

[39] Williams, C.P. and Hogg, T. Using deep structure to locate hard problems. *Proc. AAAI-92,* San Jose, CA (1992) 472–277.

[40] Wilson, D.B. The empirical values of the critical k-SAT exponents are wrong. xxx.lanl.gov/abs/math.PR/0005136.

[41] Wolfram, S. Universality and complexity in cellular automata. *Physica D,* 10 (1984) 1–35.

www.ingramcontent.com/pod-product-compliance
Ingram Content Group UK Ltd.
Pitfield, Milton Keynes, MK11 3LW, UK
UKHW020239161224
452563UK00007B/238